# Advanced Ceramic Coatings and Interfaces

# Advanced Ceramic Coatings and Interfaces

*A Collection of Papers Presented at the 30th International Conference on Advanced Ceramics and Composites January 22–27, 2006, Cocoa Beach, Florida*

Editors
Dongming Zhu
Uwe Schulz

General Editors
Andrew Wereszczak
Edgar Lara-Curzio

A JOHN WILEY & SONS, INC., PUBLICATION

Published by John Wiley & Sons, Inc., Hoboken, New Jersey
Published simultaneously in Canada.



For general information on our other products and services please contact our Customer Care Department within the U.S. at 877-762-2974, outside the U.S. at 317-572-3993 or fax 317-572-4002.

Wiley also publishes its books in a variety of electronic formats. Some content that appears in print, however, may not be available in electronic format.

***Library of Congress Cataloging-in-Publication Data is available.***

ISBN-13 978-0-470-08053-5
ISBN-10 0-470-08053-1

10 9 8 7 6 5 4 3 2 1

# Contents

### Non-Destructive Evaluation of Thermal and Environmental Barrier Coatings



### Ceramic Coatings for Spacecraft Applications



### Multifunctional Coatings and Interfaces

# Preface

The Symposium on Advanced Ceramic Coatings for Structural, Environmental and Functional Applications was held at the 30th Cocoa Beach International Conference on Advanced Ceramics and Composites in Cocoa Beach, Florida, during January 22–27, 2006. A total of 85 papers, including 10 invited talks, were presented at the symposium, covering broad ceramic coating and interface topic areas and emphasizing the latest advancement in coating processing, characterization and development.

The present volume contains thirty contributed papers from the symposium, with topics including thermal and environmental barrier coating development and testing, modeling and life prediction, non-destructive evaluation, multifunctional coatings and interfaces, as well as functionally graded materials, highlighting the state-of-the-art ceramic coatings technologies for various critical engineering applications.

We are greatly indebt to the members of the symposium organizing committee, including Yutaka Kagawa, Karren L. More, Robert Vaßen, Seiji Kuroda, Yong-Ho Sohn, Irene T. Spitsberg, and Jennifer L. Sample, for their assistance in developing and organizing this vibrant and cutting-edge symposium. We also would like to express our sincere thanks to manuscript authors and reviewers, all the symposium participants and session chairs for their contributions to a successful meeting. Finally, we are also grateful to the staff of The American Ceramic Society for their efforts in ensuring an enjoyable conference and the high-quality publication of the proceeding volume.

DONGMING ZHU
UWE SCHULZ

# Introduction

This book is one of seven issues that comprise Volume 27 of the Ceramic Engineering & Science Proceedings (CESP). This volume contains manuscripts that were presented at the 30th International Conference on Advanced Ceramic and Composites (ICACC) held in Cocoa Beach, Florida January 22–27, 2006. This meeting, which has become the premier international forum for the dissemination of information pertaining to the processing, properties and behavior of structural and multifunctional ceramics and composites, emerging ceramic technologies and applications of engineering ceramics, was organized by the Engineering Ceramics Division (ECD) of The American Ceramic Society (ACerS) in collaboration with ACerS Nuclear and Environmental Technology Division (NETD).

The 30th ICACC attracted more than 900 scientists and engineers from 27 countries and was organized into the following seven symposia:

- Mechanical Properties and Performance of Engineering Ceramics and Composites
- Advanced Ceramic Coatings for Structural, Environmental and Functional Applications
- 3rd International Symposium for Solid Oxide Fuel Cells
- Ceramics in Nuclear and Alternative Energy Applications
- Bioceramics and Biocomposites
- Topics in Ceramic Armor
- Synthesis and Processing of Nanostructured Materials

The organization of the Cocoa Beach meeting and the publication of these proceedings were possible thanks to the tireless dedication of many ECD and NETD volunteers and the professional staff of The American Ceramic Society.

Andrew A. Wereszczak and Edgar Lara-Curzio
General Editors

*Oak Ridge, TN (July 2006)*

# Advanced Thermal Barrier Coating Development and Testing

# RELATION OF THERMAL CONDUCTIVITY WITH PROCESS INDUCED ANISOTROPIC VOID SYSTEMS IN EB-PVD PYSZ THERMAL BARRIER COATINGS

A. Flores Renteria, B. Saruhan
Institute of Materials Research,
German Aerospace Center
Linder Hoehe, Porz-Wahnheide
Cologne, NRW 51147, Germany

J. Ilavsky
X-Ray Operations and Research (XOR), Experimental Facilities Division,
Advanced photon Source (APS)
Argonne National Laboratory
9700 S. Cass Avenue, bldg 438E
Argonne, IL 60439, USA

## ABSTRACT

Thermal barrier coatings (TBCs) deposited by Electron-beam physical deposition (EB-PVD) protect the turbine blades situated at the high pressure sector of the aircraft and stationary turbines. It is an important task to uphold low thermal conductivity in TBCs during long-term service at elevated temperatures. One of the most promising methods to fulfil this task is to optimize the properties of PYSZ-based TBC by tailoring its microstructure. Thermal conductivity of the EB-PVD produced PYSZ TBCs is influenced mainly by the size, shape, orientation and volume of the various types of porosity present in the coatings. These pores can be classified as open (inter-columnar and between feather arms gaps) and closed (intra-columnar pores). Since such pores are located within the three-dimensionally deposited columns and enclose large differences in their sizes, shapes, distribution and anisotropy, the accessibility for their characterization is very complex and requires the use of sophisticated methods. In this work, three different EB-PVD TBC microstructures were manufactured by varying the process parameters, yielding various characteristics of their pores. The corresponding thermal conductivities in as-coated state and after ageing at 1100C/1h and 100h were measured via Laser Flash Analysis Method (LFA). The pore characteristics and their individual effect on the thermal conductivity are analysed by USAXS which is supported by subsequent modelling and LFA methods, respectively. Evident differences in the thermal conductivity values of each microstructure were found in as-coated and aged conditions. In summary, broader columns introduce higher values in thermal conductivity. In general, thermal conductivity increases after ageing for all three investigated microstructures, although those with initial smaller pore surface area show smaller changes.

## INTRODUCTION

Attractive thermo mechanical properties of electron beam - physical vapour phase deposited (EB-PVD) thermal barrier coatings (TBCs) are attributed to their unique microstructure. The primary columns present in these microstructures are separated by inter-columnar gaps oriented perpendicular to the substrate's plane, leading to excellent thermo-shock resistance of these coatings under thermal cyclic conditions. It is noteworthy that the inter-columnar gaps are oriented parallel to the heat flux, and thus, unfavourable for the capability of the TBCs as thermal insulators. The heat flux is principally transported through the solid column material being equilibrated by intra-columnar pores and voids between feather-arms.

During the EB-PVD coating process, primary columns start to grow on the substrate surface following a nucleation stage. The columnar growth occurs in a preferred direction, typically perpendicular to the plane of the substrate. Inter-columnar gaps and feather-arm features are created due to the shadowing effect of the neighbouring column tips which impede the vapour flux to reach the bottom of the valley between the columns[1]. Moreover, the formation and growth of the voids between feather-arms are influenced by the next factors:

(a) The sunshine-sunset shadowing effect of the neighbouring columns tips, resulting in different morphological sequences at the directions parallel and perpendicular to the plane of vapour incidence (PVI);

(b) The substrate temperature, that regulates the diffusion of atoms at the growing surface and

(c) The rotation speed, influencing the solid material growth and yielding the "banana" shape pores due to shadowing after each completed rotation movement.

Since the feather-arm gaps are located at the periphery of the columns, the column-tip shadowing effect will be enhanced during the growing process with the support of the other two mentioned factors (i.e. those given by b and c). Therefore, this phenomenon is significant at the ultimate edge of the columns, forming their conical cross section. For this reason, the feather-arm gaps can be designated as open intra-columnar pores created at the columns periphery due to lower vapour flux available for the complete solid material deposition. In addition, intra-columnar pores are created inside the columns at regions of highest vapour incidence angles (VIA), where the edge of the deposited vapour after each complete rotation overlaps with the initial growth generated by the next rotation's movement. They grow through the deposited material parallel to the feather-arms in an elongated "banana" shape following the altered direction of the vapour incidence angle (VIA) in a sunrise-sunset pattern.

Thermal exposure of the EB-PVD deposited TBCs results in morphological changes which consequently affect the properties such as thermal conductivity and Young's Modulus. These may be due to the occurrence of a series of thermal processes varying from formation of bridges between the columns, formation of sintering necks at contact points between the feather-arms, and eventually changes in pore geometry and sizes of the intra-columnar pores[2]. All these thermally activated processes generate surface area reduction and may follow in a similar way as the sintering processes[3].

Previous studies[4-6] indicate that the microstructural configuration of the EB-PVD PYSZ TBCs significantly contribute to the intrinsic thermal properties of the material. Thus, the quantitative information on the spatial and geometrical characteristics of the pores within these coatings is required to correlate those with their thermal properties. Sophisticated techniques such as USAXS and USANS have shown to be effective in thorough characterization of as-coated EB-PVD TBCs[7-9]. However, no study is known up-to-date which benefits from these techniques to determine the morphological alterations on ageing and to precisely identify the role of each morphological feature and alteration on the in-service stability of the thermal conductivity in EB-PVD manufactured PYSZ TBCs.

In this study, three different microstructures of EB-PVD produced PYSZ TBCs were manufactured by altering coating process parameters. Their morphological characterization was carried out in as-coated and aged (1100°C/100h) conditions via Ultra Small-Angle X-rays Scattering Method (USAXS). The measurements were carried out in two orthogonal directions per specimens due to the anisotropy of the pores. The resulting raw data were fitted employing a computer based model[10] capable to determine the mentioned characterization of the pores; i.e., volume, size, aspect ratio, shape and orientation of the inter-columnar gaps, gaps between feather-arms, and intra-columnar pores. Correspondingly, thermal conductivity of specimens under the same conditions was measured via Laser Flash Analysis Method (LFA).

## MATERIALS AND METHODS

### Materials and Processing

Partially Yttria Stabilized Zirconia (PYSZ) coatings were manufactured via EB-PVD process by employing "von Ardenne" pilot plant equipment having a maximum EB-power of 150 kW. Evaporation was carried out from a single evaporation source having the ingot dimensions of 62.5 mm diameter and 150 mm length. The chemical composition of the ingot was standard 7-8 wt.%$Y_2O_3$ stabilized $ZrO_2$. Deposition of the vapor phase on flat substrates was carried out under conventional rotating mode by mounting the substrates on a holder with its horizontal axis perpendicular to the evaporation source as described in[11] (i.e. perpendicular to PVI). During the coating process, the substrates were rotated at different speeds and heated to different temperatures (Table I). Table I lists the designation of the samples and the applied process parameters to manufacture these three investigated EB-PVD-morphologies.

**Table I.** EB-PVD coating conditions and designation for the three manufactured microstructures

| Morphology | Chamber Pressure (mbar) | Substrate Temperature (°C) | Rotation Speed (rpm) |
|---|---|---|---|
| Feathery | $8x10^{-3}$ | 850 | 30 |
| Intermediate | $8x10^{-3}$ | 950 | 12 |
| Coarse | $8x10^{-3}$ | 1000 | 3 |

For USAXS characterization, the EB-PVD PYSZ coatings of app. 400 μm thickness were deposited on Ni-basis substrates which were previously coated with a NiCoCrAlY bond coat. USAXS specimens were prepared by cutting and polishing the coatings into 200μm thickness slices. Two orthogonal slices (e.g. perpendicular and parallel to the PVI) per specimen were obtained through this process. For thermal conductivity characterization by Laser-Flash-Analysis (LFA), the coatings were deposited on 12.7 mm diameter discs of FeCrAl-alloys (without bond coat) in the same run. For LFA sample preparation, the FeCrAl-alloy substrates were chemically etched to obtain coatings in free-standing conditions. Finally, these free-standing coatings were additionally coated on both sides with a thin Pt-layer via sputter method to avoid laser penetration during the measurements. Subsequently, corresponding specimens were heat treated in air at 1100°C/100h using a heating rate of 5°C/min.

## Methods of Characterization

### Microstructural Characterization

The coating microstructures were characterized visually by using a Field-Emission Scanning Electron Microscope (FE-SEM, LEITZ LEO 982).

### Ultra-Small Angle X-ray Scattering (USAXS)

The effective pinhole-collimated USAXS instrument which is equipped with a data processing system and available at the UNICAT 33-ID of the Advance Photon Source, ANL, was employed for our measurements[8]. Effective-pinhole USAXS enables the characterization of anisotropic microstructures with nearly the same resolution previously available for isotropic materials. This instrument utilizes the advantages of Bonse-Hart[12] double-crystal diffraction optics to extend its range of the scattering vector Q ($Q = (4\pi/\lambda)(\sin\theta)$, where $\lambda$ is the wavelength of the incident X-rays and $2\theta$ is the scattering angle) to noticeably lower values ($1{,}2 \times 10^{-4}$ Å$^{-1}$ < $|Q|$ < $0{,}1$ Å$^{-1}$) by decoupling the resolution of the instrument from the primary beam size. This effective-pinhole configuration allows the measurement of the scattering vector (Q) in one direction (1D), which is perpendicular to the substrate's plane within the plane of the coating. Therefore, to determine the 2D distribution of the scattering intensities, rotation of the specimens in small increments of an azimuthal angle ($\alpha$) is required.

In order to obtain the complete characterization of spatial and geometrical anisotropic microstructures such as EB-PVD TBCs, two measuring methods are usually employed:

(1) The scattering intensity, I(Q) is measured as function of the azimuthal angle ($\alpha$) for a constant Q value (aniso-scans) to determine the principal max. and min. scattering intensities and their respective $\alpha$ values;

(2) The scattering intensity (I(Q)) is measured as a function of Q at the principal $\alpha$ values.

By combining the results obtained by these methods a quantitative map of the microstructural anisotropy as function of the pore sizes can be done[8].

Since the analyzed coatings enclose anisotropic stereometric characteristics at the azimuthal and polar spatial domains, each specimen was measured at two orthogonal directions (perpendicular and parallel to the plane of vapor incidence). Several (i.e. five) aniso-scans at different fixed Q-values were done, which allowed the characterization of the complete size range of the pores within the coatings. Additionally, the scattering intensity at important anisotropic azimuthal orientations was measured.

Scattering form-factor functions have been already derived for different shapes of scattering elements such as spheroids, rods, discs, networks, etc. and reported in previous studies [13-16]. Moreover, it is known that the scattering structure-factor function is dependent of local order, describing spatial correlations that may exist between scattering elements, e.g. monodispersed population of spherically shaped scatterers[17], arrays of parallel cylindrical shaped scatterers[18], or a fractal system[19]. Since real stereometric characteristics of every pore population enclose a certain deviation, the applied model considers Gaussian distribution for the calculation of the orientation and size values. It uses idealized particle shapes for which small-angle scattering can be reasonably well calculated (e.g. as ellipsoidal oblate and prolate, and spheres), and allows the optimization of the parameters by a last square fitting as given in[10]. The model allows the use of five independent pore populations composed of individual spheroid shapes with $R_0$, $R_0$ and $\beta R_0$ axes. In the employed coordinate system the x axis, and y and z axes lay perpendicular (in other words; parallel to the column axis), and parallel to the substrate's plane, respectively. Furthermore, the orientation of the pores in the space is described by its $\beta R_0$ axis with respect to the coordinate system by two independent angles α (azimuthal angle) and ω (polar angle). An anisotropic orientation model was applied in a study[7] which assumes the solution of the differential scattering cross-section as function of the orientation for each scattering pore population. Due to uncertainties in the calibration of the USAXS-specimens thickness, the total volumes of the pores were measured by Archimedes Method. Thus, in this work, the USAXS-model predicts the volume fraction of each pore population.

Laser Flash Analysis Method (LFA)

The thermal diffusivity of the Pt-coated free-standing specimens was measured employing a Netsch-LFA 427 instrument. The calculation of the thermal conductivity was calculated through the formula:

$$\lambda = \alpha \cdot \rho \cdot C_p \quad (5)$$

Where, λ is the thermal conductivity (W/m·K), in this case ρ represents the bulk density (gr/cm$^3$) of the free-standing coatings measured by Archimedes Method and Cp is the specific heat (J/g·K) measured by Differential Scanning Calorimeter (DSC).

RESULTS AND DISCUSSION

Microstructural observation of the coatings in the as-coated state displays the typical EB-PVD PYSZ morphology for all three intended microstructures. The main differences are being in the column diameter and in the feather-arm feature configuration. The microstructure "coarse" showed larger column diameters and the microstructure "feathery" more defined feather-arm features (Fig. 1). The resulting volume fractions of each pore population calculated with the USAXS-model were scaled to the corresponding total volume measured with Archimedes Method for each microstructure, which were 27.46% for the "feathery", 27.15% for the "intermediate", and 22.55% for the "coarse", respectively. As SEM micrographs also indicate, USAXS analysis deliver numerical data on the fact that basically all investigated microstructures show variations in their pore distribution and in their column density at the cross-sections, perpendicular and parallel to the plane of vapour incidence (PVI). At all three analysed coating, in the as-coated state, the inter-columnar gaps are slightly broader, and contain evidently higher volumes at the direction perpendicular to the plane of vapour incidence (PE-PVI) compared with that at the parallel direction (PA-PVI). This can also be observed in the micrographs shown in Fig. 1. As a matter of fact, some previous studies has delivered similar qualitative results solely relying on microstructural investigations[1].

For simplicity purposes in this context, we compared the quantitative data obtained on the cross-sections of three investigated microstructures perpendicular to PVI (PE-PVI). Fig. 2 shows the measured and modeled polar distribution of the scattered intensities versus the azimuthal angle (α) at different scattering vector (Q) values for as-coated state of the "coarse" microstructure (top row) and after ageing at 1100°C/100h (bottom row). The intensities drawn with symbols are fittings obtained by modeling. It has to be realized that the scattering intensities corresponding to the different mean opening dimensions (MOD) include scatterings with a specific range of similar sizes. The model uses Gaussian distribution averaging "size distribution" of 40-60%. Scattering intensities oriented at 90° and 270° azimuthal angles correspond to the inter-columnar gaps; while, the scattering intensities at 45-54° and 225-245°

azimuthal angles correspond to those from the feather-arms as well as intra-columnar pores which typically align behind the feather-arm features.

The results of the USAXS-modeling are given in Table II. The obtained quantitative USAXS results confirm clearly the anisotropic character of the porosity at EB-PVD TBCs, especially at the Q ranges between 0.00149 and 0.01285 (corresponding to a main opening dimension of 420nm and 48nm, respectively) (see Table II). These features display somewhat smaller growth-orientation angles (42°) at the microstructures "intermediate" and "feathery". The size distribution of the different pores was in some cases so broad, that these have to be modeled in two separated void populations. These result show that the thickness of the inter-columnar gaps at all investigated microstructures lie in the range of 602-670nm in as-coated conditions, making 4.08% - 8.38% of volume fraction. In the case of feather-arm openings, there are two representative dimensional ranges for the microstructures "intermediate" and "coarse, whereas only one dimension for the microstructure "feathery". Their volume fraction is a factor of four or higher at the microstructures "intermediate" and "coarse" (5.39% and 9.56%, respectively) than that at the microstructure "feathery" (1.40%). The measured coarser opening dimensions for the feather-arm features are given in the following after the schema [diameter/thickness (aspect ratio)]. It is found that these are 1.3/0.12μm (0.09) at the microstructure "intermediate", 1.6/0.15μm (0.09) at the microstructures "coarse" and 1.8/0.09μm (0.05) at the microstructure "feathery". Furthermore, the finer dimensional range for the "intermediate" and "coarse" is represented with 0.7/0.05μm (0.07) (see Table II).

Finally, the intra-columnar pores in the three investigated coatings show also two dimensional ranges. Although these can not be packed into specific dimensional groups being representative for all three coatings, nevertheless, it is distinctive that there is one group of intra-columnar pores yielding prolate ellipsoid shape (aspect ratio = 20) at all coatings. The dimension of those at the microstructure "coarse" (0.4/0.02μm) is approx. twice as large as those in the other two microstructures, counting solely for the smallest volume fraction (5.80%). It is noticeable that a higher volume fraction (19.82%) of the porosity at the microstructure "feathery" corresponds to these prolate ellipsoid shaped intra-columnar pores [diameter/thickness = 0.17/0.008μm]. Moreover, the intra-columnar pores at the microstructure "intermediate" fall in their dimension and volume fraction (0.2/0.01μm and 11.63%) between those of the other two microstructures. These pores are mostly located as aligned rows behind the voids between feather-arms and, thus, oriented at the same angles as the feather-arm voids (see Table II and Fig. 3). The second group of intra-columnar pores display a disc shape with an intermediate aspect ratio value. These are also oriented at similar angles as the feather-arms. The determination of the pore shapes has been carried out by fitting the modelling curves with those of the measured USAXS scattering curves, since each of the modelling shapes (oblate ellipsoid, sphere, and prolate ellipsoid) affects the shape factor P(Q) of the measured scattering intensities I(Q) by modifying the slope of fitted curves[20].

The USAXS results for the coatings in the as-coated state can be summarised as such that the "feathery" microstructure contains the finest dimension and highest volume of intra-columnar pores followed by the "intermediate" and finally by the "coarse" microstructure. Thus, due to their dimensions and aspect ratios, it can be assumed that, on thermal exposure, the intra-columnar pores of the first opening dimension are predestined to break into quasi-spherical pores due to the occurrence of sintering process[21].

By considering the pores as thermal insulators (i.e. insignificant radiation contribution), it can be stated that the spatial and geometrical distribution of all pore types at EB-PVD TBCs will contribute, according to their effectiveness, to the reduction of the thermal conductivity by interruption of phonon flow through the coating. This means that, as well as the finer dimensions, the higher volume of these cylindrical intra-columnar pores contributes significantly in thermal conductivity reduction of EB-PVD TBCs. This hypothesis is supported by the fact that this relationship defined by USAXS-modeling analysis agrees well with those experimentally determined thermal conductivity values of the three as-coated microstructures (see Fig. 4).

In the case of the microstructures "intermediate" and "coarse", larger intra-columnar pore sizes are anticipated and measured due to the applied lower rotation speeds during processing. Moreover, since these microstructures are manufactured at higher substrate temperatures, only a fraction of these pores are able to form at each rotation phase and will only survive until the next array of new pores are created. These both coatings consist of a lower volume fraction of intra-columnar pores which are heterogeneously distributed. Thus, it is plausible that their measured thermal conductivity values are higher.

After heat treatment at 1100°C for 100 hours, the sintering process becomes active, leading to morphological changes at all pore types[3]. Mass transfer occurs through bridging at the contact points between primary columns forming two dimensional groups of inter-columnar pores. This phenomenon is enhanced by the fact that the finer columns, which are cumulated mostly at the bottom zone of the coating and interrupted from growing through the coating thickness, tend to pull together and create finer channels between them. The volume fraction of such fine inter-columnar gaps is higher at the microstructures "feathery" and "intermediate" (15.58% and 14.08%, respectively) than those at the microstructure "coarse" (8.55%) (see Fig. 5).

On ageing, significant changes occur at the finer feather-arm regions and intra-columnar pores. The effect of sintering is obvious, considering the decrease in the USAXS-modeled aspect ratios which is controlled by the surface area reduction of the pores. The dimension group addressing the finer gaps disappears completely after ageing. These break into arrays of quasi-spherical pores at their inner pyramidal ends, leaving isolated openings with lower aspect ratio at the edge of the feather-arms. The secondary columns of feather-arm features appear to sinter into groups, leaving broader but shorter gaps between the feather-arms (see Fig. 5).

Finally, high aspect ratio cylindrical intra-columnar pores (e.g. fine "banana" shaped pores) which are created at each rotation phase and connect the two pore rows break into quasi-spherical pores (see Fig. 6). Moreover, on ageing, the volume fraction of these pores is drastically reduced, especially at the microstructure "feathery" from 19.82% in the as-coated state to 5.41% after heat-treatment. Consequently, a different equilibrium configuration of the pores results from the sintering process altering the effectiveness of the pores in interruption of the heat transfer through the coatings.

Considering the changes at the thermal conductivity of the coatings after heat-treatment at 1100°C/100h (Fig. 4) and in their morphological changes (Figs. 3 and 6), it can be clearly stated that the changes at the intra-columnar cylindrical pores, aligned behind the feather-arm features are responsible for the drastic thermal conductivity increase in the microstructure "feathery". These changes occur not only in their shape (from prolate ellipsoids to quasi-spheres) but also in their volume fraction, resulting in a decrease reaching to nearly a factor four (from 19.82% to 5.41%). Their significant influence can be attributed to the disappearance of heat-flux hindering paths and formation of isotropic features which facilitate heat transfer.

CONCLUSIONS

In this study employed USAXS-analysis supported with modeling was able to quantitatively and rather accurately determine the geometry and location of anisotropic voids within the three EB-PVD TBC microstructures in the as-coated and aged conditions. The variation of the EB-PVD process parameters produces evident differences in the spatial and geometrical characteristics of the porosity within the manufactured TBCs. Also the heat treatment of these at 1100°C/100h induce irreversible thermal activated processes (i.e. sintering), modifying the distribution and geometry of the pores. According to the results of the thermal conductivity measurements and those from the USAXS–modeling, it is discernible that the intra-columnar pores are the principal constructors of the heat-flux (phonons) through the coatings enhanced by their high volume concentration and distribution.

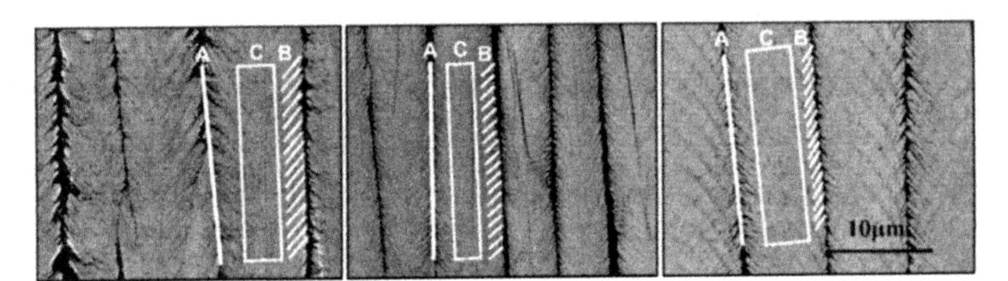


**Figure 1:** Scanning electron micrographs of EB-PVD TBCs cross sections at the direction perpendicular to the PVI showing the inter-columnar pores (A), pores between feather-arms (B) and intra-columnar pores (C): feathery (left), intermediate (middle), and coarse (right).

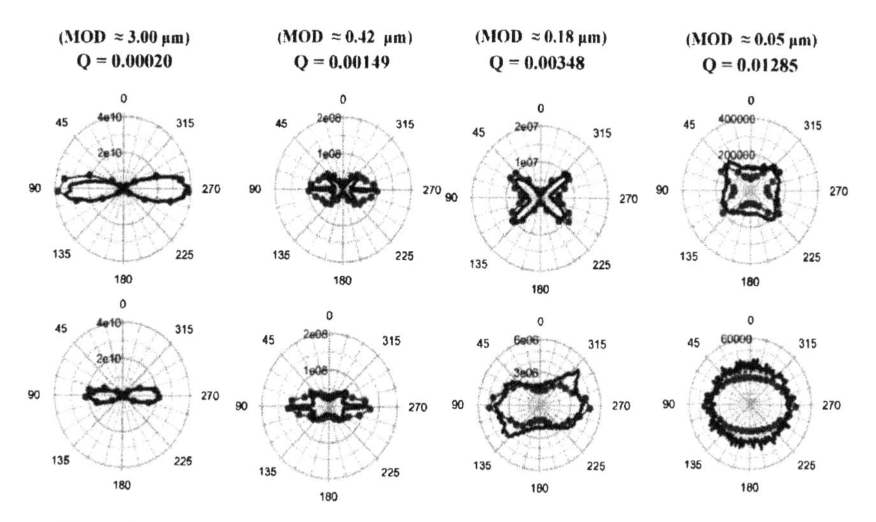


**Figure 2:** Measured (—) and modeled (-●-) polar distribution of the USAXS-scattered intensities versus the azimuthal angle ($\alpha$) at different Q values and the corresponding main open dimensions (MOD) for the "coarse" microstructure in cross-section perpendicular to PVI: (a) as-coated and (b) after heat treated conditions at 1100°C/100h.

**Table II:** Calculated values of spatial and geometrical characteristics of the porosity within the three analyzed EB-PVD TBCs in as coated and after ageing conditions via USAXS-modelling.

| Pore type as-coated "Intermediate" | Ellipsoidal Shape | Azimuthal angle | Diameter (nm) | Thickness (nm) | Aspect Ratio | Volume fraction |
|---|---|---|---|---|---|---|
| Inter-columnar-1 | Oblate | 82 | - | 673.00 | 0.05 | 8.38 |
| Between feather-arms-1 | Oblate | 42 | - | 118.30 | 0.091 | 1.21 |
| Between feather-arms-2 | Oblate | 42 | 705.70 | 45.87 | 0.065 | 4.18 |
| Intra-columnar-1 | Prolate | 42 | 13.00 | 260.00 | 20.00 | 11.63 |
| Intra-columnar-2 | Oblate | 42 | 33.50 | 15.10 | 0.45 | 1.74 |
| Pore type heat-treated "Intermediate" | Shape | Azimuthal angle | Diameter (nm) | Thickness (nm) | Aspect Ratio | Volume fraction |
| Inter-columnar-1 | Oblate | 84 | - | 469.18 | 0.05 | 8.65 |
| Inter-columnar-2 | Oblate | 84 | - | 124.80 | 0.26 | 5.43 |
| Between feather-arms-1 | Oblate | 45 | 580.00 | 174.00 | 0.30 | 3.08 |
| Intra-columnar-1 | Oblate | 45 | 120.52 | 84.36 | 0.70 | 4.65 |
| Intra-columnar-2 | Oblate | 45 | 44.00 | 35.20 | 0.80 | 5,32 |
| Pore type as-coated "coarse" | Shape | Azimuthal angle | Diameter (nm) | Thickness (nm) | Aspect Ratio | Volume fraction |
| Inter-columnar-1 | Oblate | 86 | - | 602 | 0.05 | 5.92 |
| Between feather-arms-1 | Oblate | 54 | - | 154.11 | 0.095 | 2.78 |
| Between feather-arms-2 | Oblate | 54 | 718.65 | 53.907 | 0.075 | 6.78 |
| Intra-columnar-1 | Oblate | 54 | 210.17 | 63.05 | 0.30 | 1.26 |
| Intra-columnar-2 | Prolate | 54 | 22.31 | 446.33 | 20.00 | 5.80 |
| Pore type heat-treated "coarse" | Shape | Azimuthal angle | Diameter (nm) | Thickness (nm) | Aspect Ratio | Volume fraction |
| Inter-columnar-1 | Oblate | 92 | - | 563.00 | 0.05 | 7.05 |
| Inter-columnar-2 | Oblate | 92 | - | 156.00 | 0.26 | 1.50 |
| Between feather-arms-1 | Oblate | 54 | 872.75 | 174.55 | 0.20 | 4.95 |
| Intra-columnar-1 | Oblate | 54 | 261.30 | 209.04 | 0.80 | 5.27 |
| Intra-columnar-2 | Oblate | 54 | 51.43 | 46.28 | 0.90 | 3.78 |
| Pore type as-coated "feathery" | Shape | Azimuthal angle | Diameter (nm) | Thickness (nm) | Aspect Ratio | Volume fraction |
| Inter-columnar-1 | Oblate | 84 | - | 602.45 | 0.07 | 4.08 |
| Inter-columnar-2 | Oblate | 84 | - | 61.70 | 0.14 | 2.14 |
| Between feather-arms-1 | Oblate | 42 | 1868.80 | 93.44 | 0.05 | 0.60 |
| Between feather-arms-2 | Oblate | 42 | 222.71 | 31.18 | 0.14 | 0.80 |
| Intra-columnar-1 | Prolate | 42 | 8.58 | 171.71 | 20.00 | 19.82 |
| Pore type heat-treated "feathery" | Shape | Azimuthal angle | Diameter (nm) | Thickness (nm) | Aspect Ratio | Volume fraction |
| Inter-columnar-1 | Oblate | 88 | - | 591.33 | 0.05 | 4.48 |
| Inter-columnar-2 | Oblate | 88 | - | 85.00 | 0.22 | 11.10 |
| Between feather-arms-1 | Oblate | 52 | 1467.36 | 161.41 | 0.11 | 1.14 |
| Intra-columnar-1 | Oblate | 52 | 123.45 | 98.76 | 0.80 | 5.31 |
| Intra-columnar-2 | Oblate | 52 | 37,36 | 29,88 | 0,80 | 5.41 |

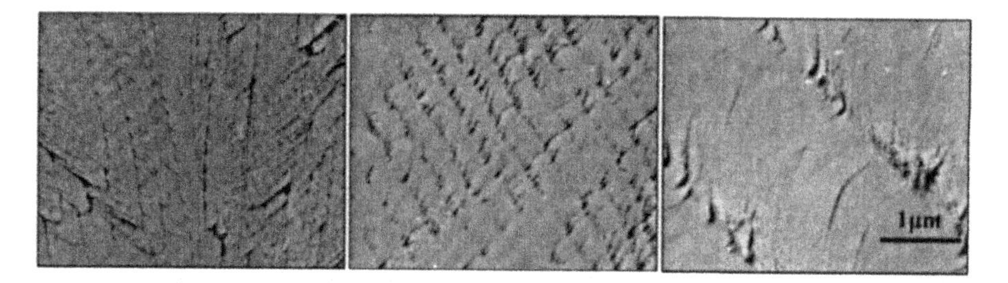


**Figure 3:** Scanning electron micrographs of EB-PVD TBCs cross sections at the direction perpendicular to the PVI showing the intra-columnar pores: feathery (left), intermediate (middle), and coarse (right).

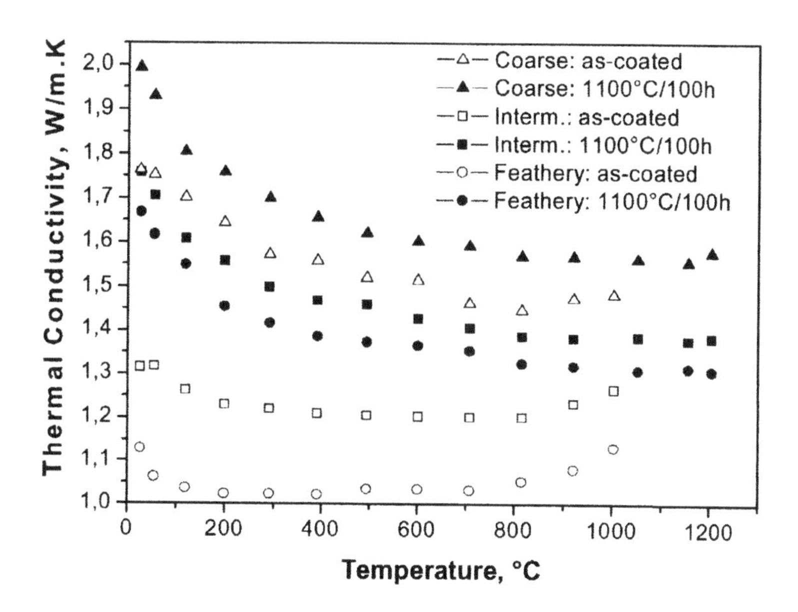


**Figure 4:** Thermal conductivities vs. measuring temperature values of the three analyzed microstructures: feathery, intermediate and coarse.

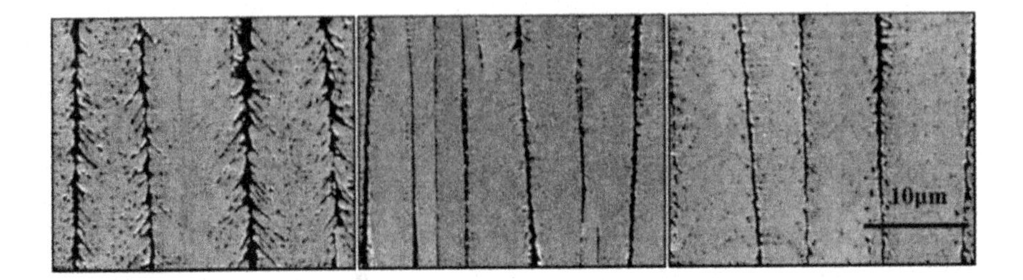


**Figure 5:** Scanning electron micrographs of heat treated (1100°C/100h) EB-PVD TBCs cross sections after heat treatment at the direction perpendicular to the PVI: feathery (left), intermediate (middle), and coarse (right).

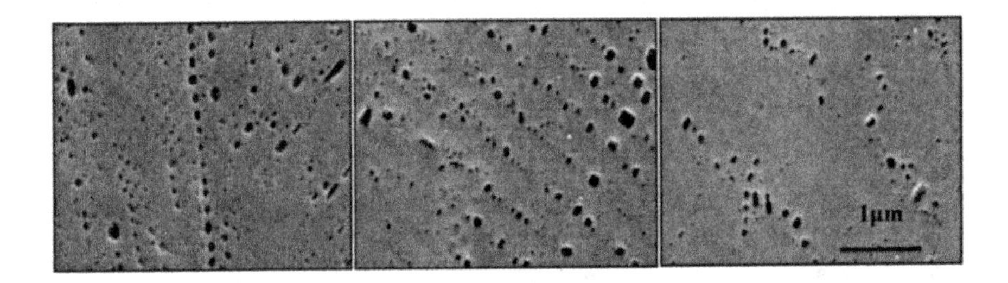


**Figure 6:** Scanning electron micrographs of heat treated (1100°C/100h) EB-PVD TBCs cross sections at the direction perpendicular to the PVI showing the intra-columnar pores: feathery (left), intermediate (middle), and coarse (right).

REFERENCES

[1]S.G. Terry, Evolution of microstructure during the growth of thermal barrier coatings by Electron-Beam Physical Vapor Deposition. Materials Department, University of California, Santa Barbara, 2001, p. 197.
[2]A.F. Renteria, B. Saruhan, U. Schulz, H.-J. Raetzer-Scheibe, Effect of Morphology on Thermal Conductivity of EB-PVD PYSZ TBCs, Surface and Coatings Technology, accepted for publication, (2006).
[3]A.F. Renteria, B. Saruhan, Effect of ageing on microstructure changes in EB-PVD manufactured standard PYSZ top coat of thermal barrier coatings, Journal of the European Ceramic Society, in Press, Vol.: 26, (2006).
[4]H.-J. Ratzer-Scheibe, U. Schulz, T. Krell, The effect of coating thickness on the thermal conductivity of EB-PVD PYSZ thermal barrier coatings. Surface and Coatings Technology In Press, Corrected Proof (2005).

[5]U. Schulz, C. Leyens, K. Fritscher, M. Peters, B. Saruhan-Brings, O. Lavigne, J.-M. Dorvaux, M. Poulain, R. Mévrel, M. Caliez, Some recent trends in research and technology of advanced thermal barrier coatings, Aerospace Science and Technology, 7, 73-80 (2003).
[6]C.G. Levi, Emerging materials and processes for thermal barrier coatings, Current Opinion in Solid State and Materials Science, 8, 77-91 (2004).
[7]T.A. Dobbins, A.J. Allen, J. Ilavsky, G.G. Long, A. Kulkarni, H. Herman, P.R. Jemian, Recent developments in the characterization of anisotropic void population in thermal barrier coatings using ultra-small angle x-rays scattering, in: Ceram. Eng. and Sci. Proc. (ed. by W.M. Kriven, H.-T. Lin), The American Ceramic Society, pp. 517-524 (2003).
[8]J. Ilavsky, A.J. Allen, G.G. Long, P.R. Gemian, Effective pinhole-collimated ultrasmall-angle x-ray scattering instrument for measuring anisotropic microstructures, Review of scientific instruments, 73, 1-3 (2002).
[9]A.J. Allen, Characterization of ceramics by x-ray and neutron small-angle scattering, J. Am. Ceram. Soc., 88, 1367-1381 (2005).
[10]J. Ilavsky, A.J. Allen, A. Kulkarni, T. Dobbins and H. Herman, Microstructure characterization of thermal barrier coatings deposits - practical models from measurements, in: E. Lugscheider, A. International, M. Park (Eds.), International Thermal Spray Conference, Basel, Switzerland, 2005.
[11]U. Schulz, S.G. Terry, C.G. Levi, Microstructure and texture of EB-PVD TBCs grown under different rotation modes, Materials Science and Engineering A 360, 319-329 (2003).
[12]U. Bonse, M. Hart, An X-ray interferometer with long separated interfering beam paths, Appl. Phys. Lett., 7, 99-100 (1965).
[13]A.J. Allen, J. Ilavsky, G.G. Long, J.S. Wallace, C.C. Berndt, H. Herman, Microstructural characterization of yttria-stabilized zirconia plasma-sprayed deposits using multiple small-angle neutron scattering, Acta Materialia, 49, 1661-1675 (2001).
[14]G. Porod, General theory, in: O. Glatter, O. Kratky (Eds.), Small-angle X-rays scattering, Academic Press, London, (1982).
[15]J.S. Pedersen, Form factors of block copolymer micelles with spherical, ellipsoidal and cylindrical cores, J. Appl. Cryst., 33, 637-640 (2000)
[16]G. Beaucage, Approximations leading to a unified exponential/power law approach to small-angle scattering, J. Appl. Cryst., 28, 717-728 (1995).
[17]A.J. Allen, S. Krueger, G. Skandan, G.G. Long, H. Hahn, H.M. Kerch, J.C. Parker, M.N. Ali, Microstructure evolution during sintering of nanostructured ceramic oxides, J. Am. Ceram. Soc.,9, 1201-1212 (1996).
[18]D. Marchal, B. Demé, Small- angle neutron scattering by porous alumina membranes of aligned cylindrical channels, International Union of Crystallography, pp. 713-717 (2003)
[19]D. Sen, A.K. Patra, S. Mazumder, S. Ramanathan, Pore morphology in sintered $ZrO_2$-$_8$% mol $Y_2O_3$ ceramic: a Small-Angle Neutron Scattering investigation, Journal of Alloys and Compounds, 340, 236-241 (2002).
[20]R.-J. Roe, Small-Angle Scattering, in: J.E. Mark (Ed.), Methods of X-ray and neutron scattering in polymer science, Oxford University Press, New York, pp. 155-209 (2000).
[21]J.S. Stoelken, A.M. Glaeser, The morphological evolution of cylindrical rods with anisotropic surface free energy via surface diffusion, Scripta Metallurgica et Materialia, 27, 449-454 (1992).

# SEGMENTATION CRACKS IN PLASMA SPRAYED THIN THERMAL BARRIER COATINGS

Hongbo Guo, Hideyuki Murakami and Seiji Kuroda
Materials Engineering Laboratory, National Institute for Materials Science (NIMS)
1-2-1 Sengen, Tsukuba
Ibaraki 305-0047, Japan

## ABSTRACT

Thick thermal barrier coatings (TBCs) containing segmentation cracks have seen a successful application in combustion chamber parts. In this work, rather thin TBCs were produced by spraying yttria stabilized zirconia (YSZ) coatings onto two kinds of bond coats: CoNiCrAlY and Pt-Ir. The effects of process parameters including substrate temperature on the density of segmentation cracks were studied. The thermal cycling performance and failure mechanisms of the segmented thin TBCs were also investigated. The coating sprayed at 1073 K contains a segmentation crack density of approximately 5 $mm^{-1}$. The segmented coatings significantly improved the thermal cycling lifetime as compared with the non-segmented coatings. The segmented TBC with Pt-Ir bond coat attained a lifetime of more than 5000 cycles (each cycle includes 3 minute heating up to the maximum temperature and 7 minute holding at the temperature), revealing an excellent thermal cycling performance.

## INTRODUCTION

Thermal barrier coatings (TBCs) perform the vital function of insulating gas turbine components, such as burner cans and turbine vanes or blades, which are subjected to excessive temperatures. The benefits of a TBC can be utilized in several ways: increased engine efficiency and combustion temperature, prolonged lifetime of parts, reduced transient stresses in parts, reduction of cooling air that results in higher cooling efficiency and lower emission.

Thermal barrier coatings are usually produced by electron beam physical vapor deposition (EB-PVD) or plasma spray. Compared to the EB-PVD TBC, the plasma sprayed coating has a better thermal insulation due to its multi-layered microstructure. Failure of plasma sprayed TBCs often occur at the interface between thermally grown oxide (TGO) layer and topcoat or within the topcoat due to the growth of TGO and thermal mismatch stresses [1-3]. Coating failures can decrease engine performance and accelerate substrate metal deterioration. Microstructure modifications of TBCs have been done in order to improve their thermal cycling performance. For traditional TBCs (usually<0.5 mm in thickness) sufficient durability is achieved by the use of porous and microcracked coatings [4]. For thick TBCs the thermal shock resistance can be significantly improved by inducing so-called segmentation cracks into the coatings as these cracks increase the compliance of coatings to substrates [5-7].

In TBC system, the bond coat is either MCrAlY (M=Ni, Co, etc) alloy or a Pt modified aluminide coating. To improve the protective performance of the PtAl coating, much attention has been focused on Pt-based binary alloy coatings, such as Pt-Ni and Pt-Rh [8,9]. Recently, a Pt-Ir coating was proposed [8,10,11], because Ir has the highest melting temperature (2716 K) among platinum group metals, excellent chemical stability, low oxygen diffusivity, and lower diffusivity into Ni-based alloys and lower cost than Pt.

In this paper, thin segmented TBCs are sprayed onto the MCrAlY and Pt-Ir bond coats, aiming at improving the thermal cycling performance of the thin TBCs. The effect of segmentation crack density on the thermal cycling lifetime of the thin TBCs will be studied. Also, the failure mechanisms of the thin TBCs will be investigated.

## EXPERIMENTAL

A Ni-based single crystal superalloy, TMS-82+ with <100> orientation, with the compositions as given in Table 1, and an Inconel 718 Ni-based superalloy were used as substrate materials. Co-32Ni-21Cr-8Al-0.5Y powder (Sulzer Metco 9954) and $ZrO_2$-8mass% $Y_2O_3$ powder (Sulzer Metco 204 NS) were chosen for spraying MCrAlY bond coat and yttria stabilized zirconia (YSZ) topcoat, respectively. The MCrAlY bond coat of around 200 μm were sprayed onto the Inconel 718 substrates by low pressure plasma spray (Plasma Giken Corp., Japan). Pt-Ir coatings of approximately 7~10 μm were deposited on TMS-82+ substrates at 873 K by DC magnetron sputtering from Pt-Ir target. The deposition parameters have been reported elsewhere [11,12]. The Pt-Ir coated specimens were embedded in an aluminum retort containing a mixture of $Al_2O_3$, Al, Fe and $NH_4Cl$ and aluminizing treatment was carried out at 1273 K for 5 h in flowing argon atmosphere.

Figs.1a and b show the micrograph of cross-section of Pt-Ir coated TMS 82+ specimen after aluminizing treatment and associated element distributions across the coating thickness, respectively. The top layer of around 25 μm basically consists of $PtAl_2$ and β-(Ni,Pt,Ir)Al phases and below this, an interdiffusion layer mainly comprising β-NiAl phase is formed on the substrate. YSZ coatings of around 400 μm were sprayed onto the CoNiCrAlY and Pt-Ir coated substrates in atmospheric plasma spraying using an SG 100 gun (Praxair, USA). Four sets of spray parameters, as shown in Table 2, were used for spraying YSZ coatings aiming at attaining different levels of segmentation crack densities. The choice of processing parameters for spraying segmented TBCs is based on the experimental details described in some literatures [13,14]. The main idea is that segmentation cracks were created by thermal tensile stresses during the deposition and hence, the heat input to the substrate should be enough high. Regarding to this, the substrate temperatures were varied by changing spray distance and plasma power. Condition 1 (C1), featured with a lower plasma power and a long spray distance as compared to other spray conditions, was used for spraying traditional non-segmented TBCs.

Table 1 Nominal chemical composition. wt%

| Element | Ni | Co | Cr | Mo | W | Al | Ti | Ta | Hf | Re |
|---|---|---|---|---|---|---|---|---|---|---|
| | bal | 7.8 | 4.9 | 1.9 | 8.7 | 5.3 | 0.5 | 6.0 | 0.1 | 2.4 |

Table 2 Spray parameters for YSZ topcoat (V: traverse speed of plasma gun; $T_s$: substrate temperature).

| No. | Power (KW) | Ar (slpm) | He (slpm) | Distance (mm) | Feed rate (g/min) | V (mm/s) | $T_s$ (K) |
|---|---|---|---|---|---|---|---|
| C1 | 23.1 | 50 | 27 | 120 | 20 | 150 | 593 |
| C2 | 26.4 | 50 | 27 | 100 | 20 | 150 | 773 |
| C3 | 34 | 50 | 27 | 80 | 20 | 150 | 923 |
| C4 | 40.8 | 50 | 27 | 60 | 20 | 150 | 1073 |

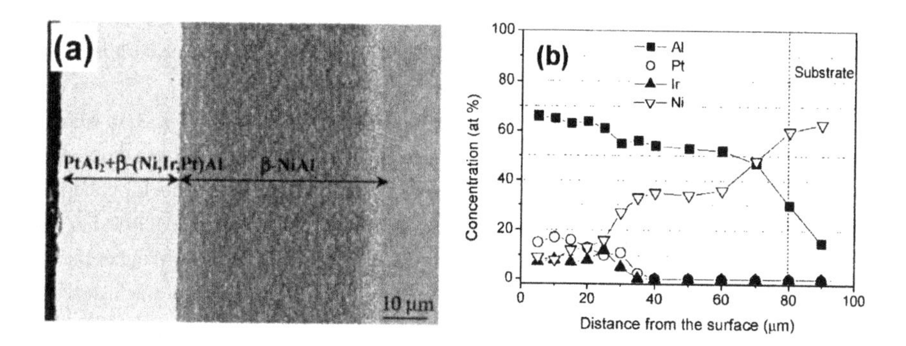


**Fig. 1** (a) SEM micrograph of cross-section and (b) concentration profiles of the elements of the Pt-Ir coated TMS-82+ substrate after aluminizing treatment.

Thermal cycling testing of disk-shaped specimens (ϕ25mm×3mm) was performed in a high temperature furnace equipped with an automatic mechanical system to move 4 specimens in and out of the furnace simultaneously. The coated specimens were placed in $Al_2O_3$ holders that connected to a water tank so as to attain a certain temperature gradient across the specimen thickness during heating. The furnace temperature during the testing was set to be 1200 °C. The specimens were heated to the maximum temperature within 3 min and held at the temperature for 7 min and then cooled by air jets for 3 min from the coating side. The coating surface and substrate temperatures of the specimens were measured by thermocouples fixed to the specimens. The lifetime of a TBC specimen is defined as the number of cycles the specimen underwent before more than 1/3 surface area of the coating spalled off from the substrate.

The microstructures of the coatings before and after thermal cycling testing were examined by optical microscopy (OM), scanning electron microscopy (SEM) equipped with energy dispersive spectroscopy (EDS), and the phases in the coatings identified by X-ray diffractometry (XRD).

## RESULTS AND DISCUSSION

### Microstructures of segmented thin TBCs

Figs. 2a and b show the micrographs of the TBCs sprayed under C1 condition and C4 condition. The coating sprayed under C1 condition is highly porous since larger voids are abundant for the coating. Compared to the coating at C1 condition, the coating sprayed at C4 condition is much denser and contains some segmentation cracks running perpendicular to coating surface. In this study, the segmentation cracks are defined as the vertical cracks penetrating at least half the coating thickness.

The segmentation crack density of the as-sprayed coatings is shown in Fig. 3, in which the crack density is calculated by dividing the number of segmentation cracks found in a cross-section with the length of the cross-section. The crack density of the coating sprayed under C1 condition is nearly zero and hence the coating is considered as traditional non-segmented TBC. For those coatings deposited at higher substrate temperatures, the crack densities are much higher than that of the coating at C1 condition, especially for the coating at C4 condition, a crack density of approximately 5 $mm^{-1}$ is achieved. Therefore, it can be concluded that the substrate temperature is an important factor in affecting the origin of segmentation cracks and a high substrate temperature gives rise to an increased segmentation crack density. The segmentation cracks initiate and propagate during the deposition phase, as a result of biaxial tensile stresses that arise from in-situ sintering. The mechanism for the development of segmentation cracks has been discussed in some literatures [15,16]. A good joining between lamellae is a key point for the propagation of segmentation cracks. Accordingly, high heat input to the substrate is one of most favorable factors in developing segmentation cracks because high substrate and particle temperatures promote the joining of adjacent lamellae.

Thermal cycling of the TBC with CoNiCrAlY bond coat

During thermal cycling tests, the maximum temperature at the coating surface is around 1393 K, while the substrate temperature is in a range of 1323 K to 1343 K. The lifetimes of the specimens are strongly dependent on the segmentation crack density, as shown in Fig. 4.

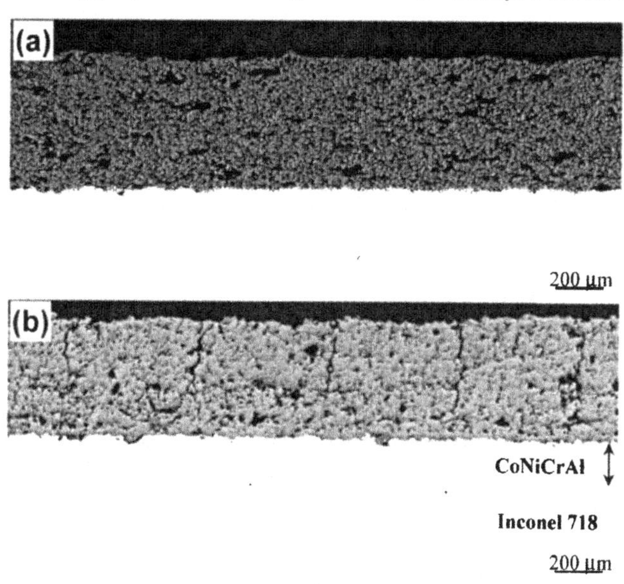


**Fig. 2.** Cross-sections of TBCs sprayed at C1 condition (a) and C4 condition (b).

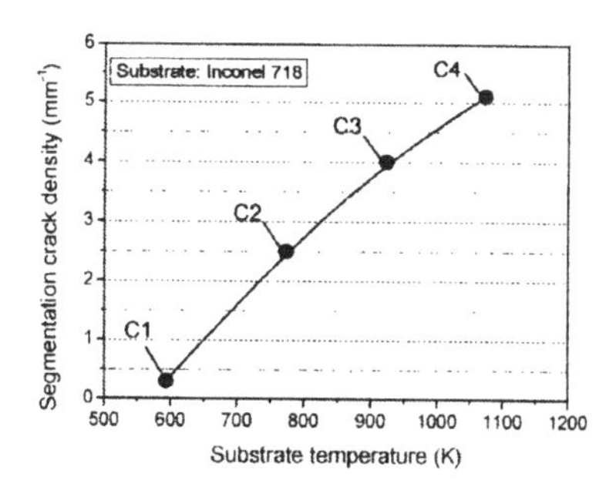


**Fig. 3.** Effect of substrate temperature on segmentation crack densities of TBCs with CoNiCrAlY bond coat.

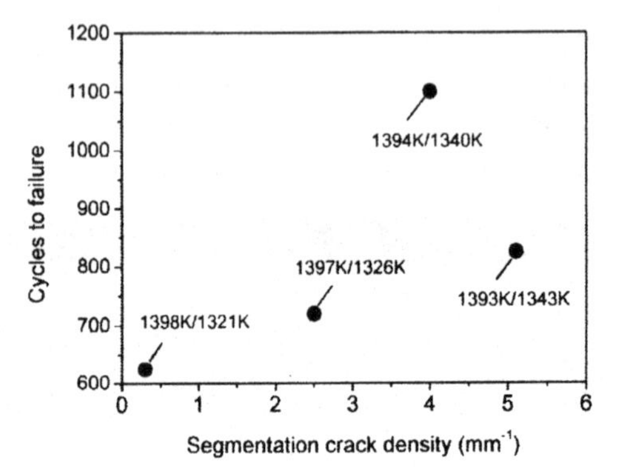


**Fig. 4.** Effect of segmentation crack density on thermal cycling lifetimes of the TBCs with CoNiCrAlY bond coat. The corresponding coating surface temperature (the right) and substrate temperature (the left) is also indicated.

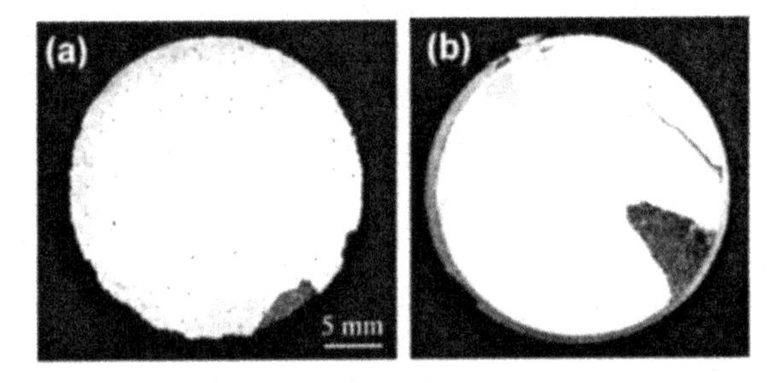


**Fig. 5.** Photographs of TBC sprayed under C1 condition cycled to 1398 K ($T_{surface}$) for 600 cycles (a) and TBC under C3 condition cycled to 1394 K ($T_{surface}$) for 1100 cycles (b).

The non-segmented TBC specimen has a lifetime of less than 650 cycles, while all of the segmented TBC specimens showed much longer lifetimes. A maximum lifetime of approximately 1100 cycles is achieved for the coating sprayed under C3 condition that has a crack density of around 4 $mm^{-1}$. It should be noted that the coating sprayed under C4 condition did not show the best thermal cycling performance, although the coating contains the highest crack density among the sprayed coatings. This can be explained by the processing parameters used for spraying the coating. For the coating sprayed under C4 condition, oxidation of the bond coat could occur due to quite high plasma power and short spray distance. On the other hand, too high a substrate temperature, up to 1073 K achieved during the spraying, could result in a high residual stress in the coating upon cooling down, which would degrade the bond strength between the bond coat and YSZ topcoat. Regarding to this, C3 condition is considered as the optimized condition for spraying TBC in terms of thermal cycling performance.

Both the non-segmented and the segmented TBC specimens show a spallation of large parts of ceramic coatings that occurred at the edge of the coatings, as shown in Fig. 5. This type of failure is primarily caused by the growth of TGO on the bond coat due to long time exposure at high temperature. The cross-sections of the above failed coatings were examined, as shown in Fig. 6. TGO layers were formed in the coatings and the TGO layer in the segmented coating (C3) is apparently thicker than that in the non-segmented coating (C4) due to relatively longer time exposure of the segmented coating. Some horizontal cracks propagated along the interface between bond coat and YSZ topcoat, indicating that the spallation observed in Fig. 5 close to the interface was induced by these cracks. On the other hand, the coatings after cycling is much thinner than the coatings before cycling and the coating surfaces even became rougher. This suggests that another kind of failure occurred in the YSZ coatings simultaneously during cycling by spallation or chipping close to the coating surface. Therefore, it can be concluded that the thin segmented TBCs reveal two kinds of failures, one of which occurred by spallation of large parts of coatings close to the YSZ/bond coat interface induced by the growth of TGO, and the other spallation or chipping close to the coating surface that was caused by thermal expansion mismatch stresses on cooling.

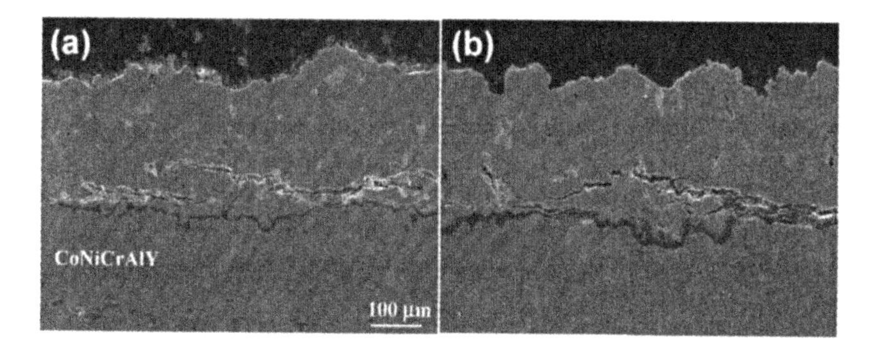


**Fig. 6.** SEM micrographs of cross-sections of TBCs sprayed under C1 condition after 600 cycles (a) and sprayed under C3 condition after 1100 cycles (b).

Thermal cycling of the TBC with Pt-Ir bond coat

Some YSZ coatings were sprayed onto the Pt-Ir coated TMS 82+ substrate under C3 condition and the thermal cycling performance of these sprayed TBC were also evaluated under the same testing condition as that used for the TBC with CoNiCrAlY bond coat. The segmentation crack density of the sprayed coating is measured to be around 2.9 $mm^{-1}$, which is a little lower than that of the TBC coated Inconel 718 substrate. The lower crack density could be attributed to the different coefficients of thermal expansion (CTE) between the two substrate materials. The Inconel 718 substrate has a larger CTE than the TMS-82+ substrate (in a temperature range of 0~100°C, $13\times10^{-6}$/°C for Inconel 718 and $11.6\times10^{-6}$/°C for TMS-82+). The coating surface and substrate temperatures were measured to be around 1401 K and 1338 K, respectively. The average lifetime of the TBC specimens exceed 5000 cycles, much longer than that of the TBC with CoNiCrAlY bond coat.

The failure of the TBC with Pt-Ir bond coat is similar to that of traditional TBC with MCrAlY bond coat, as shown in Fig. 7, which primarily caused by the growth of TGO layer. Fig. 8 shows the micrograph of finely polished cross-section of the TBC after 5200 cycles. The thickness of the interdiffusion zone increased to more than 100 μm (the thickness of the interdiffusion zone before testing is around 50 μm) due to the long time exposure at high temperature. A TGO of around 10 μm was formed on the bond coat, below which large horizontal cracks initiated, as shown in Fig. 8b. Besides, some large voids (as marked by arrow 2 in the Figure) are visible in the Pt-Al layer, which could be caused by spallation of some hard phases such as Ir-enriched phase during the polishing process. It should be noted that some black oxides (as marked by arrow 1) mainly consisting of α-Al2O3 are also present at the interface between the Pt-Ir layer and the interdiffusion layer. The formation of these oxides was related to internal oxidation. During the sputtering of Pt-Ir coating, there could be some residual oxygen left in the surface of substrate and during the thermal cycling, selective oxidation of Al occurred due to the low oxygen pressure. Some horizontal cracks initiated and propagated along the interface between Pt-Ir layer and interdiffusion layer partially due to the growth of the oxides. It can be inferred that the propagation of these horizontal cracks could finally result in coating debonding. In order to avoid the formation of these oxides, the sputtering processing and subsequent treatment such as annealing should be further optimized. Additionally, TCP phase (white points denoted by arrow 2) is also formed in the interdiffusion zone.

Fig. 9 shows the XRD pattern of the failed TBC specimen after spallation of YSZ topcoat. The β-(Ni,Pt,Ir)Al and $PtAl_2$ are the fundamental phases in the bond coat. Besides, small amount of α-$Al_2O_3$ phase and t'-$ZrO_2$ phase are also detected in the surface layer of the specimen. This confirms that the spallation of TBC specimen occurred by cracking at the interface between YSZ and Pt-Ir layer and the failure is caused by the oxidation of the bond coat.

Compared to the TBC with MCrAlY bond coat, the TBC with Pt-Ir bond coat exhibited a better thermal cycling performance. A possible explanation is that the failure of the thin TBCs is dominated by the oxidation of bond coat. The Pt-Ir bond coat has a better oxidation resistance than the MCrAlY bond coat [17] and therefore, a better thermal cycling performance.

In general, the segmented TBCs show a better thermal cycling resistance than the non-segmented coatings. An explanation of this behavior is that the segmentation cracks improved the strain tolerance of coatings. These cracks can easily open during tensile loading and close during compressive loading and hence, the thermal stresses were greatly relaxed. Another advantage of the segmented TBCs is the high temperature capability. The segmented coatings produced by "hot" spraying (high heat input to the substrate) are rather denser than traditionally sprayed coatings. Therefore, the segmented TBCs reveal lower sintering and shrinkage rate than traditional coatings that are essential for a good high-temperature capability.

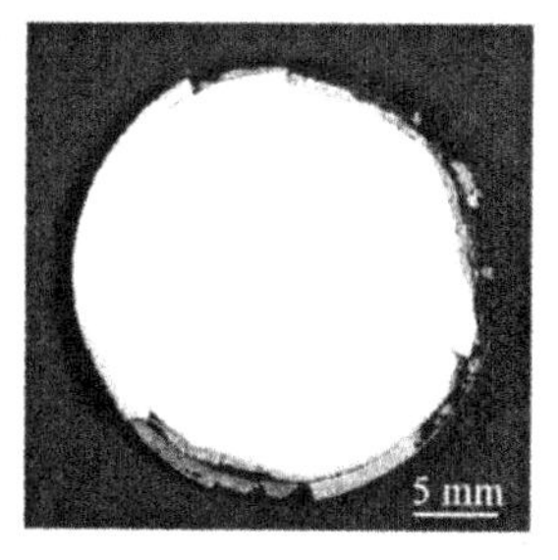


Fig. 7. Photograph of TBC with Pt-Ir bond coat cycled at 1401 K ($T_{surface}$)/1338 K ($T_{substrate}$) for 5200 cycles.

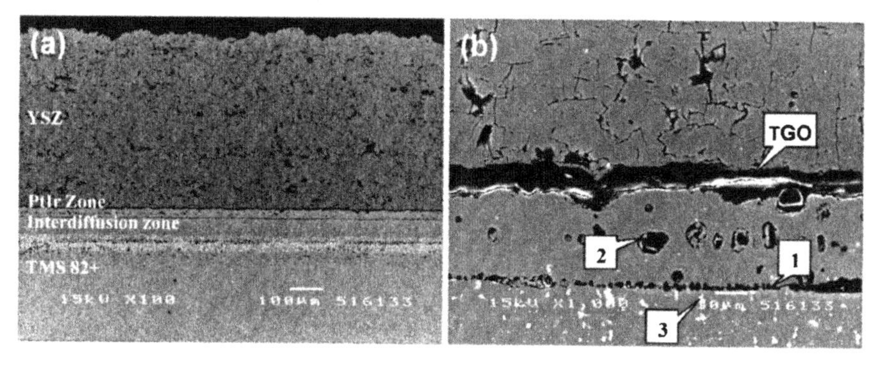


**Fig. 8.** (a) Back scattering electron (BSE) micrograph of TBC with Pt-Ir bond coat after 5200 cycles. (b) Higher magnification of (a).

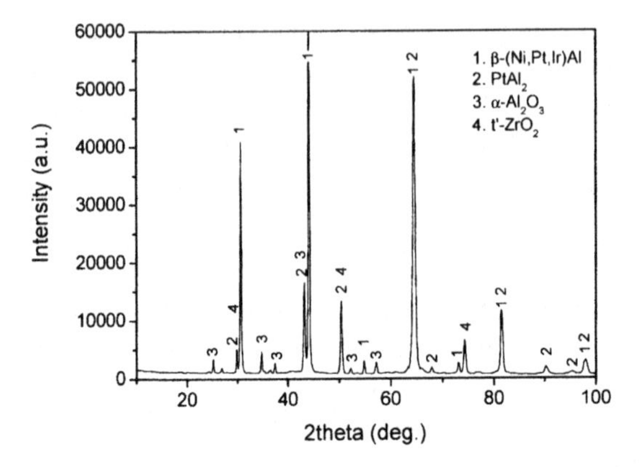


**Fig. 9.** XRD pattern of the Pt-Ir bond coated specimen after spallation of YSZ topcoat.

SUMMARY

Thin TBCs with different levels of segmentation crack density were sprayed onto the CoNiCrAlY and Pt-Ir bond coat. The substrate temperature played a crucial role in affect the origin of the cracks and the coating deposited at 1073 K attained a crack density of around 5 $mm^{-1}$. The segmented coatings showed longer thermal cycling lifetimes compared to the non-segmented coatings. Two kinds of failures occurred in the segmented coatings, one of which is chipping or spallation close to the coating surface, and the other spallation close to the interface between YSZ and bond coat primarily caused by the growth of TGO. The segmented TBC with Pt-Ir bond coat attained a lifetime of more than 5000 cycles, revealing an excellent thermal cycling performance.

ACKNOWLEDGMENTS

This work is supported by the Japan Society for Promotion of Science (JSPS) Fellowship program. Mr. M. Komatsu, Mr. M. Shibata and Mr. A. Yamaguchi from National Institute for Materials Science (NIMS) are gratefully acknowledged for their contributions in experiments.

REFERENCES

[1]D.R. Clarke, R.J. Christensen and V. Tolpygo, "The Evolution of Oxidation Stresses in Zirconia Thermal Barrier Coated Superalloy Leading to Spalling Failure", Surf. Coat. Technol., 94-5(1-3), 89-93 (1997).

[2]A. Rabiei and A.G. Evans, "Failure Mechanisms Associated with the Thermally Grown Oxide in Plasma-Sprayed Thermal Barrier Coatings", Acta Mater., 48(15), 3963-76 (2000).

[3]Nitin P. Padture, Maurice Gell and Eric H. Jordan, "Thermal Barrier Coatings for Gas-Turbine

Engine Applications", Science, 296, 280-84 (2002).
[4]R. Vaβen, F. Traeger and D. Stöver, "Correlation between Spraying Conditions on Microcrack Density and Their Influence on Thermal Cycling Life of Thermal Barrier Coatings", J. Therm. Spray. Technol., 13(3), 396-404 (2004).
[5]T.A. Taylor, D.L. Appleby, A.E. Weatherill and J. Griffiths, "Plasma-Sprayed Yttria-Stabilized Zirconia Coatings: Structure-Property Relationships", Surf. Coat. Technol., 43/44, 470-80 (1990).
[6]P. Bengtsson, T. Ericsson and J. Wigren, "Thermal Shock Testing of Burner Cans Coated with a Thick Thermal Barrier Coating", J. Therm. Spray. Technol., 7(3), 340-48 (1998).
[7]H.B. Guo, R. Vaβen and D. Stöver, "Thermophysical Properties and Thermal Cycling Behavior of Plasma Sprayed Thick Thermal Barrier Coating", Surf. Coat. Technol., 192(1), 48-56 (2005).
[8]Y. Zhang, J.A. Haynes, W.Y. Lee, I.G. Wright, B.A. Pint, et al., "Effects of Pt Incorporation on the Isothermal Oxidation Behavior of Chemical Vapor Deposition Aluminide Coatings", Met. Mater. Trans. A, 32(7), 1727-41 (2001).
[9]G. Fisher, P.K. Datta, J.S. Burnell-Gray, "An Assessment of the Oxidation Resistance of an Iridium and an Iridium/Platinum Low-Activity Aluminide/MarM002 System at 1100 Degree C", Surf. Coat. Technol., 113(3), 259-67 (1999).
[10]F. Wu, H. Murakami, Y. Yamabe-Mitarai, H. Harada, H. Katayama and Y. Yamamoto, "Electrodeposition of Pt-It Alloys on Nickel-Based Single Crystal Superalloy TMS-75", Surf. Coat. Technol., 184(1), 24-30 (2004).
[11]P. Kuppusami and H. Murakami, "A Comparative Study of Cyclic Oxidized Ir Aluminide and Aluminized Nickel Base Single Crystal Superalloy", Surf. Coat. Technol., 186, 377-88 (2004).
[12]P. Kuppusami, H. Murakami and T. Ohmura, "Behavior of Ir-24at.-% Ta Films on Ni Based Single Crystal Superalloys", Surface Engineering, 21(1), 53-59 (2005).
[13]H.B. Guo, S. Kuroda and H. Murakami, "Microstructures and Properties of Plasma Sprayed Segmented Thermal Barrier Coatings", J. Am. Ceram. Soc., 89 (4), 1432-39 (2006).
[14]H.B. Guo, S. Kuroda and H. Murakami, "Segmented Thermal Barrier Coatings Produced by Atmospheric Plasma Spraying Hollow Powders", Thin Solid Films, 506-507, 136-39 (2006).
[15]P. Bengtsson and J. Wigren, "Segmentation Cracks in Plasma Sprayed Thick Thermal Barrier Coatings ". in: P.J. Maziasz, I.G. Wright et al (Eds.), Gas Turbine Materials Technology: Conference Proceeding from ASM Materials Solutions, Rosemont, IL, 1999, p. 92.
[16]H.B. Guo, R. Vaβen and D. Stöver, "Atmospheric Plasma Sprayed Thick Thermal Barrier Coating with High Segmentation Crack Density", Surf. Coat. Technol., 186, 353-63 (2004).
[17]F. Wu, H. Murakami and H. Harada, "Cyclic Oxidation Behavior of Iridium-Modified Aluminide Coatings for Nickel-Based Single Crystal Superalloy TMS 75", Mater. Trans. JIM, 44(9), 1675-78 (2003).

# DESIGN OF ALTERNATIVE MULTILAYER THICK THERMAL BARRIER COATINGS

H. Samadi and T. W. Coyle
Centre for Advanced Coating Technologies,
University of Toronto,
Toronto, Ontario, M5S 3E4 Canada

## ABSTRACT

Increasing the combustion temperature in diesel engines is an idea which has been pursued for over 20 years. Increased combustion temperature can increase the power and efficiency of the engine, decrease the specific fuel consumption and CO emission rate. Ceramic thermal barrier coatings have been identified as the most promising approach to meeting these objectives. The most commonly used system is Yttria Partially Stabilized Zirconia (Y-PSZ). However, in contrast to the widespread use in aircraft and power generation turbine engines, Y-PSZ TBCs have not met with wide success in diesel engines. To reach the desirable temperature of 850-900°C in the combustion chamber, a coating with a thickness of at least 1mm is required. This introduces different considerations than in the case of turbine blade coatings, which are on the order of 100μm thick. The design of a multilayer coating employing relatively low cost materials with complementary thermal properties is described. Numerical models were used to optimize the thickness for the different layers to yield the minimum stress at the operating conditions while achieving the desired temperature gradient.

## INTRODUCTION

Diesel engines are commonly used in buses, trucks and, outside of North America, in passenger cars. In a diesel engine, the fuel is compressed to a high pressure at which automatically ignites and burns. The idea of decreasing heat transfer from the combustion chamber is based on the knowledge that only 30-40% of the entering fuel energy is converted to useful work on the output shaft [1]. In the past few decades, the use of advanced ceramics to insulate the combustion chamber to reduce heat rejection has been widely investigated. The goal is to have an engine with around 50% efficiency rather than the typical 33%. [2].

In the 1980s there was an effort to use thermal barrier coatings (TBCs) in diesel engines in pursuit of advantages including higher power density, fuel efficiency, and multi-fuel capacity due to higher combustion chamber temperature (900°C vs. 650 °C) [3,4]. Preliminary studies showed that using TBCs in diesel engine could increase engine power by 8%, decrease the specific fuel consumption by 15-20% and increase the exhaust gas temperature 200K [5]. At the same time, TBCs should protect the metallic substrate against the corrosive attack of fuel contaminants (Na, V, and S). However the main problem was still unsolved: durability.

Ceramic thermal barrier coatings (TBCs) have been in use for some time as protective coatings in gas turbine hot sections. The super alloy turbine blades cannot tolerate the high temperature (more than 1000°C) and corrosive environment experienced in current turbine engines for extended periods without protection. A thin layer of Yttria Partially Stabilized $ZrO_2$ (YSZ) with a thickness of 100- 200 μm, fulfills the requirements.

However, the service environment of the coating in the turbine is markedly different from the diesel engine. In the former, the service temperature is high (more than1100°C). The superalloy substrate's maximum service temperature is higher than 800°C. The thickness of coating is a few hundred microns and is applied to reduce the substrate temperature and protect it against oxidation, hot corrosion, thermo-mechanical fatigue and creep. Due to the high substrate temperature, oxidation of the bond coat plays a major role in coating failure. On the other hand, in the diesel engine the gas temperature, currently less than 650°C would ideally approach 900°C. The substrate temperature is limited to approximately 200°C, and therefore a thick coating (up to 1mm) is required which leads to a high thermal gradient . In a thick thermal barrier coating (TTBC) the bond coat temperature is too low for severe oxidation and creep [6].

In a thick thermal barrier coating, when the surface of the YSZ coating is heated, a compressive stress is developed in the surface. Stress relaxation through creep and sintering mechanisms may occur at the service temperature. Upon cooling, the stress at the surface may become tensile, initiating cracks (figures 1 and 2)[6]. Due to the mismatch in thermomechanical properties of the top coat, bond coat and substrate, these interfaces are sources for cracking and delamination [7].

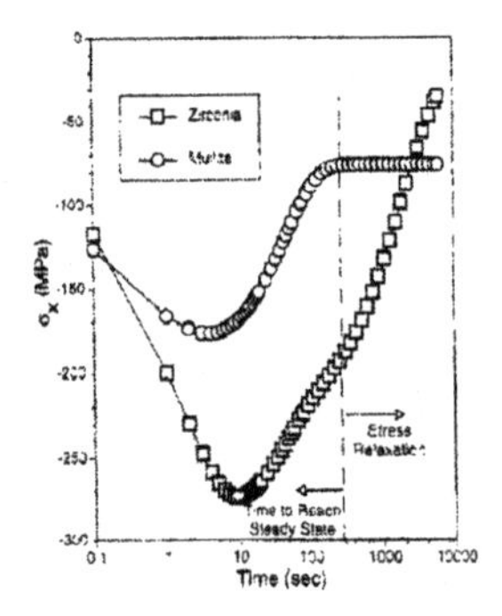


Figure 1. Surface Stress for Zirconia Mullite during Transient Heating (q= kWm$^{-2}$) (Reprinted from [6] with per from Elsevier).

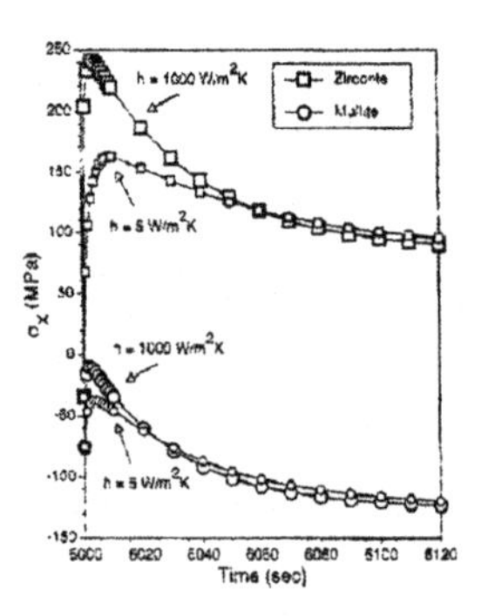


Figure 2. Surface Stress for Zirconia and Mullite during Transient Cooling (q=270 kWm$^{-2}$) (Reprinted from [6] with permission from Elsevier).

Another major difference between the two systems is that due to the on/off nature of diesel engine, TBC in a diesel engine is experiencing more thermal transients than a turbine blade. While being cooled the surface stress of the coating tends to become tensile leading to crack initiation and failure [6].

## RESIDUAL STRESS IN THERMAL BARRIER COATINGS

One of the most important parameters controlling the durability of plasma spray coating is the residual stress [8]. There are two types of residual stress in plasma spray coatings: macro and micro. Macro residual stresses exist in the body of component over a range much larger than the grain size. Micro residual stresses vary at or below in the scale of the grain size [9].

Macro residual stresses may originate from non-uniform heating and cooling operations and coefficient of thermal expansion mismatch between phases. Chemical residual stresses related to chemical reactions, precipitation, or phase transformations may also produce long range stresses [9]. Residual stress is introduced in the TBC system because of coefficient of thermal expansion mismatch between the top coat and the bond coat (or substrate) and also oxidation of the bond coat [10]. In thermal spraying, the cooling time for a single splat is in the order of a few milliseconds. Thermal contraction due to cooling of an individual splat to the temperature of the substrate (or the previously deposited solid layer) results in a tensile stress in the splat which is known as the "quenching" stress [11]. Analytical and numerical approaches have been used to model stress in multilayer system with unequal coefficients of thermal expansion and elastic moduli.

In a thick TBC, a low CTE is desirable for the hot surface to minimize stresses due to temperature gradients and sensitivity to thermal shock. However, a relatively high CTE is desirable to avoid a large CTE mismatch with the metallic substrate which would limit coating adhesion. The MCrAlY bond coat fulfills this need to some extend in YSZ TBC systems. A multilayer system may permit these opposing requirements to be satisfied.

Most traditional, high temperature refractory ceramic materials are found in the $Al_2O_3{\cdot}SiO_2{\cdot}MgO$ system. Among these oxides, some have been considered as alternatives to YSZ in TBCs. A general advantage of materials in this system is their low price relative to YSZ (Figure 3) [12].

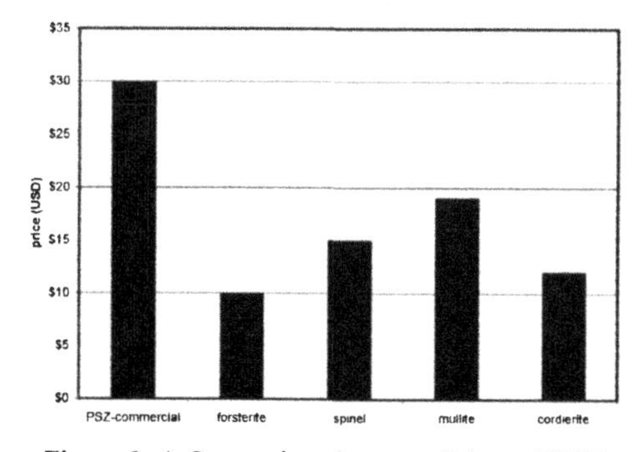


Figure 3. A Comparison between Prices of YSZ vs. Some Oxides in $Al_2O_3.SiO_2.MgO$ Phase Diagram

The multilayer system has been designed [12] consisting of forsterite, spinel and mullite. Table 1 contains the physical, mechanical and thermal properties of the substrate and coating materials. Forsterite has a CTE matching the steel substrate (11-12.5 × $10^{-6}$ $K^{-1}$ [13]). Mullite has a low CTE (4.5× $10^{-6}$ $K^{-1}$ [6]), which may minimize thermal strain at the hot face. These two phases are not stable in contact with one another at elevated temperatures, therefore an intermediate layer of spinel is needed to prevent reaction. Spinel exhibits a CTE between that of forsterite and mullite.

Table 1. Physical, thermal and mechanical properties of the materials being used [13, 14].

| Parameter | Stainless Steel | Forsterite | Spinel | Mullite |
|---|---|---|---|---|
| Thermal conductivity ( $Wm^{-1}K^{-1}$) | 44.5 | 3.5 | 8 | 2.5 |
| Density ($Kgm^{-3}$) | 7850 | 2600 | 3100 | 2800 |
| Heat capacity ($JKg^{-1}K^{-1}$) | 475 | 800 | 750 | 920 |
| Poisson's ratio | 0.28 | 0.24 | 0.29 | 0.25 |
| Young's modulus (GPa) | 205 | 87.7 | 137 | 30 |
| Thermal expansion coeff. (*e-6 $K^{-1}$) | 12.3 | 11 | 7.68 | 4.5 |

A finite-element based model of the multilayer system on a stainless steel substrate (1.5 mm thickness) has been developed. The thicknesses of the forsterite, spinel and mullite layers are 200, 50 and 640 μm, respectively. The stresses in the multilayer coating are compared to a conventional TBC with a MCrAlY bond coat (100 μm) and partially stabilized zirconia top coat (500 μm) on a 1.5 mm stainless steel which exhibits the same thermal insulating performance. To achieve this, the thickness of mullite in the multilayer system has been chosen such that both systems experience the same substrate-coating interface temperature when subjected to equal thermal flux on the surface of the TBC. Figure 4 illustrates the geometry and boundary conditions of both systems. The coatings were assumed to be isotropic and quenching stresses were not considered.

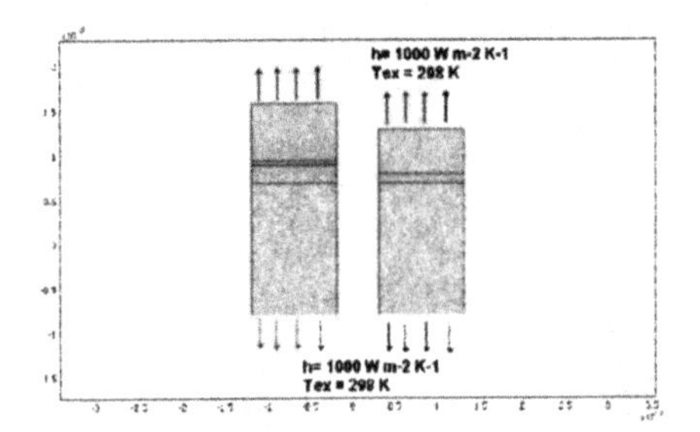


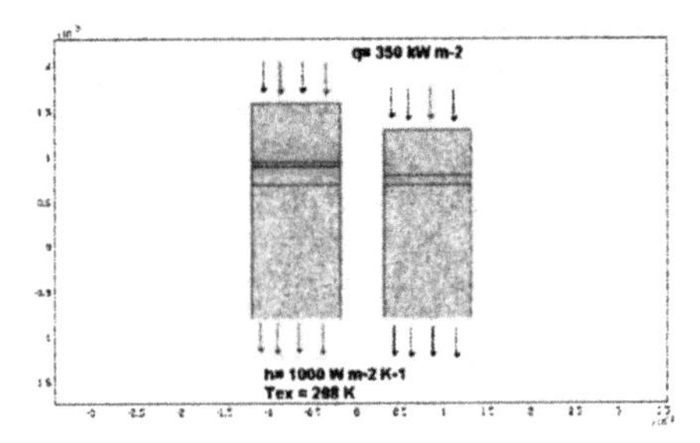


Figure 4. Geometry and boundary conditions of multilayer and duplex systems during a) cooling from the deposition temperature, and b) subsequent heating of ceramic surface.

Stresses were analyzed during cooling from the deposition temperature and during subsequent heating of the surface of the coating. In the first step, the whole system was considered to be initially at 650°C and stress free. It was then cooled from both sides assuming a heat transfer coefficient of 1000 $Wm^{-2}K^{-1}$ and an external temperature of 298 K. Subsequently a heat flux was imposed on the top coat (350 kW $m^{-2}$) while the back of the substrate was cooled with the same conditions as in the first step.

## RESULTS AND DISCUSSION

Figure 5 illustrates the surface temperature of both systems. Due to its higher thermal diffusivity and heat capacity, the surface temperature of the mullite layer changes more slowly than that of the zirconia layer and the final surface temperature under the imposed heat flux is 150K lower than that of zirconia. Considering the lower creep rate of mullite [6], stress relaxation and densification in the mullite should be reduced compared to that occurring in zirconia top coats.

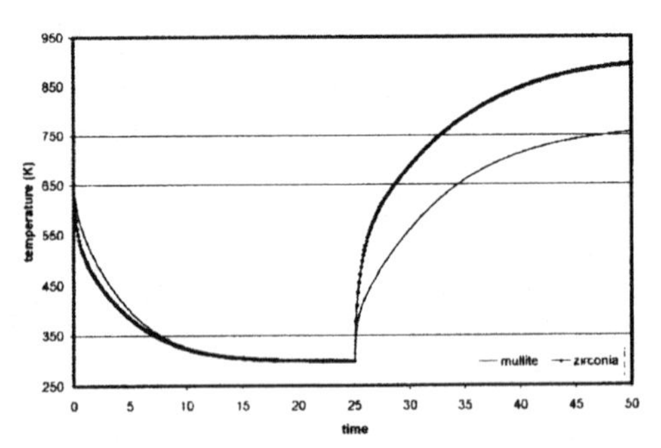


Figure 5. Surface temperature of both systems while cooling and heating.

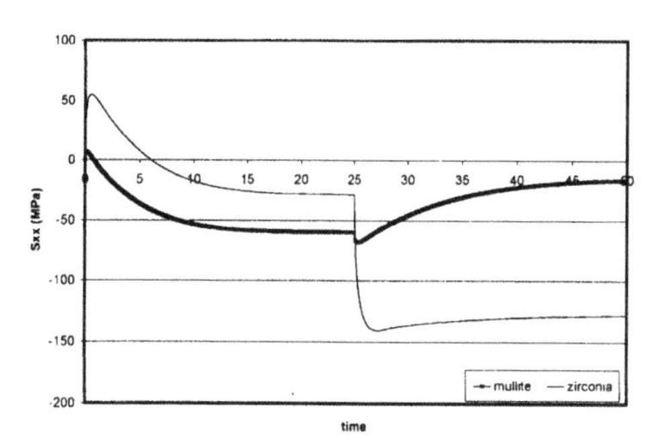


Figure 6. Surface stress ($\sigma_x$) of both systems.

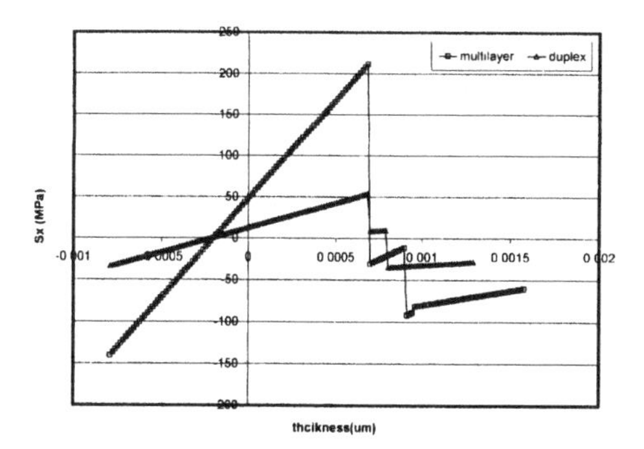


Figure 1. Surface Stress for Zirconia and Mullite during Transient Heating (q= 270

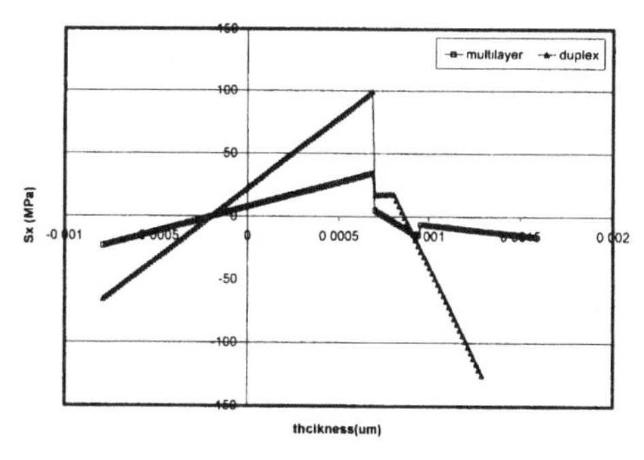


Figure 8. Stress distribution in both systems after the heating step

The in-plane surface stress, $\sigma_x$, of the systems is illustrated in figure 6. During the cooling stage, the surface of the zirconia first becomes tensile, and then as the underlying material begins to cool the stress gradually decreases, becoming compressive at room temperature. The surface of mullite reaches a much lower tensile stress at the beginning, and the residual stress at the end of the cooling stage is -60 MPa. After applying the heat flux, the stress in the surface of zirconia undergoes a sudden drop to about -140 MPa and the steady state stress of this coating is -130 MPa. Due to its lower CTE, no such a sudden drop is seen in mullite and the steady state stress is no greater than -15 MPa.

In figure 7, the stress distribution through the cross-section of both systems after cooling shows that in both cases, the ceramic coatings are in compression. During the heating step, a steady-state stress of 5 MPa was found for the forsterite layer at the interface of forsterite and the substrate, compared to 17 MPa for PSZ at the interface with the bond coat in the duplex coating (Figure 8).

Other than the small region of low tensile stress in the forsterite layer adjacent to the substrate, all three ceramic layers remain in compression through-out both the cooling and heating stages. In most thermally sprayed coatings, debonding and lack of cohesion is one of the main causes of failure; lower tensile stresses in the interfaces should make the coating more reliable. The multilayer system has this advantage.

## CONCLUSION

A new multilayer thermal barrier coating was modeled using the finite element method. Comparing the internal stress of this system with a conventional duplex TBC, it was found that:

- During cooling, the surface of the multilayer coating experienced higher compressive stresses.
- During heating, the steady state compressive stress at the surface of the multilayer coating is lower.
- During the cooling/heating cycle, the difference between maximum and minimum surface stresses in the multilayer system is much less than in the duplex system.

These results suggest that the multilayer coating would be a promising system for the operating conditions considered here. More detailed analysis of the transient temperature and stress distributions are underway.

ACKNOWLEDGEMENTS

Support for this research from the Auto 21, Network of Centres of Excellence and Iran Ministry of Science, Research and Technology is gratefully acknowledged.

REFERENCES

[1]C. A. Amann, The low heat rejection diesel Advanced Diesel Engineering & Operation, Ellis Harwood Ltd, UK (1988) p. 173-180.

[2]R. A. Churchill, J. E. Smith, N. N. Clark, R.A. Turton, "Low-Heat Rejection Engines – A Concept Review", SAE Technical Paper Series 880014.

[3]P. Ramaswamy, S. Seetharamu, K. B. R. Varma, N. Raman and K. J. Rao, "Thermomechanical Fatigue Characterization of Zirconia (8% Y2O3-ZrO2) and Mullite Thermal Barrier Coatings on Diesel Engine Components: Effect of Coatings on Engine Performance", Proc. Instn. Mech Engrs., vol. 214, part.C (2000) 729-742.

[4]D. Zhu, R. A. Miller, "Thermal Barrier Coatings for Advanced Gas Turbine and Diesel Engines", NASA/TM—1999-209453.

[5]T. Hejwowski, A. Weronski, "The Effect of Thermal Barrier Coatings on Diesel Engine Performance", *Vacuum*, 65 (2002) 427-432.

[6]K. Kokini, Y. R. Takeuchi, B. D. Choules," Surface Thermal Cracking of Thermal Barrier Coatings Owing to Stress Relaxation: Zirconia vs. Mullite", *Surf. & Coat. Tech.*, 82(1996) 77-82.

[7]S. Rangaraji, K. Kokini, "Interface Thermal Fracture in Functionally Graded Zirconia-Mullite-Bond Coat Alloy Thermal Barrier Coating", *Acta Mater.*, 51(2003) 251-267.

[8]S. Kuroda, T. W. Clyne, "The Quenching Stress in Thermally Sprayed Coatings", *Thin solid films* **200** (1991) 49-66.

[9]F. A. Kandil, J. D. Lord, A. T. Fry, P. V. Grant, "A Review of Residual Stress Measurement Methods- A Guide to Technique Selection", NPL Report MATC (A)04, Project CPM 4.5 (2001).

[10]B. G. Nair, J. P. Singh, M. Grimsditch, "Stress Analysis in Thermal Barrier Coatings Subjected to Long-term Exposure in Simulated Turbine Conditions", *J. Mater. Sci.* **39** (2004) 2043-2051.

[11]Y. C. Tsui, T.W. Clyne, "An Analytical Model for Predicting Residual Stresses in Progressively Deposited Coatings, Part I: Planar Geometry", *Thin Solid Films*, **306** (1997) 23-33.

[12]R. Soltani, H. Samadi, E. Garcia, T.W. Coyle , "Developemt of Alternative Thermal Barrier Coatings for Diesel Engines", SAE World Congress, Detroit, MI, Paper No. 2005-01-0650 (2005).
[13]H. Wang, H. Herman, "Thermomechanical Properties of Plasma-Sprayed Oxides in the MgO-Al2O3-SiO2 System" *Surface & Coatings Tech.* **42** (1990) 203-216.
[14] Materials/ Coefficients Library, Comsol Multiphysics Software, Ver. 3.2 (2005).

# CREEP BEHAVIOUR OF PLASMA SPRAYED THERMAL BARRIER COATINGS

Reza Soltani, Thomas W. Coyle, Javad Mostaghimi
Centre for Advanced Coating Technologies, University of Toronto
40 St George Street
Toronto, Ontario, Canada

## ABSTRACT

Engineers have always been concerned about the creep/sintering properties of thermal barrier materials, especially thermally sprayed zirconia coatings, which in the past few years have received exceptional attention. Under service conditions, the surface of these coatings is under compressive stress as a result of the high surface temperatures and steep thermal gradient. Stress relaxation occurs through mechanisms such as creep and sintering. Upon cooling, a new stress distribution develops, which may introduce cracks and reduce service life. In this work coatings from two different zirconia feedstocks (nanostructured and hollow sphere particles, HOSP™) with 6-8Wt% $Y_2O_3$ as stabilizer, were prepared by air plasma spraying. By altering the process parameters of spraying, different amounts of porosity and non-melted particles were incorporated into the deposits. The creep strain and creep rate of these coatings were measured in four point flexure under a range of load and temperature levels. Results show that in spite of having almost the same rate of creep in the secondary stage (equal stress exponents, n), the total creep strain of coatings produced from nanostructured feedstock is lower than the coatings produced by the hollow sphere feedstock.

## 1. INTRODUCTION

Plasma sprayed thermal barrier coatings for diesel engines and turbine blades are being developed, however the scenario for diesel engines is different than that for turbines. As operating temperatures approach 1000°C the presence of relatively high compressive stresses at the coating surface lead to stress relaxation by creep and/or sintering. Shrinkage at high temperatures followed by crack initiation and propagation during cooling will cause spallation of the coating, consequently shortening the lifetime of the deposit. Creep/sintering behaviour of zirconia and its plasma sprayed form have been investigated by several researchers.[1-10]

At service temperatures sintering may heal microcracks and eliminate some porosity, creating a denser structure which could be an advantage in reducing oxygen transport to the bond coat. Unfortunately upon cooling micro cracks reappear. Eventually the coating becomes sufficiently damaged that spalling occurs. The thickness of the coating is another important factor; nowadays depositions of more than 1mm in thickness are required in some applications to increase protection and reduce heat transfer. The thicker the coating, the greater the thermal gradients present during service, which results in higher stresses within the cross section of the material.[3]

Therefore the development of a new top coat structure which shows more resistance to sintering is a topic of active research. Bimodal structured thermal barrier coatings can provide the improved performance of an alternative material without the necessity of redesigning the entire component. Researchers have shown that bimodal nanostructured coatings can exhibit better properties than conventional coatings of the same composition.[11-13]

In the current work, partially stabilized zirconia (PSZ) coatings were deposited by air plasma spraying. Two different types of powder feedstock were employed, a powder consisting of hollow spherical particles and a nanostructured particle powder. The influence of the powder type on the thermo-mechanical behaviour of the coatings was investigated.

## 2. EXPERIMENTAL PROCEDURE

Two commercially available 6-8 wt% $Y_2O_3$-PSZ powders were used to deposit coatings by air plasma spraying. A nano-structured powder (Nanox S4007, Inframat Corp., CT, USA) was deposited using the SG-100 (Praxair, Concord, NH, USA) plasma torch. This powder consists of porous agglomerated particles 15-150 μm in diameter made up of crystallites on the order of 200 nm. A powder consisting of hollow spherical particles (204B-NS, Sulzer-Metco, NY, USA) was deposited with the same gun. The particle size of this HOSP$^{TM}$ powder (hollow oven spherical particles)was 45-75 μm.

A range of deposition parameters was used to produce coatings to get structures with different levels of porosity and non-melted particles. These parameters as well as mean velocity and temperature of in flight particles have been given elsewhere.[19] Substrates were carbon steel plates, with dimensions of 50x50x4 mm (WxLxH). The thickness of the coating was more than 3 mm. Samples then were cut to 3x4x40 mm (HxWxL) and then substrates were detached from the coatings. The tensile and compressive surfaces of the specimens, where the upper and lower rollers of flexure fixture contact, then were ground to assure the surfaces were flat, smooth, and parallel.

The powder and coating microstructures were examined by scanning electron microscopy (SEM Hitachi S-4500, Japan) using low voltage. The chemical composition of the powders was obtained from flame photometry (FP) and inductively coupled plasma (ICP). Image analysis (Clemex Vision Professional, Clemex, QC, Canada) was used to measure the average porosity of each sample; details were published elsewhere.[19] Creep tests were conducted employing a SiC four-point bend test fixture (with 20 and 40 mm inner and outer spans, respectively) inside a box furnace at elevated temperatures and under a range of applied loads.

## 3. RESULTS

Microstructure: prior to conducting creep tests, SEM investigations of feedstock particles and coatings deposited using 204B-NS and Nanox powders was conducted, Figs.1 and 2. Thin shell particles increase the probability of having fully melted particles resulting in very thin final splats in coating.

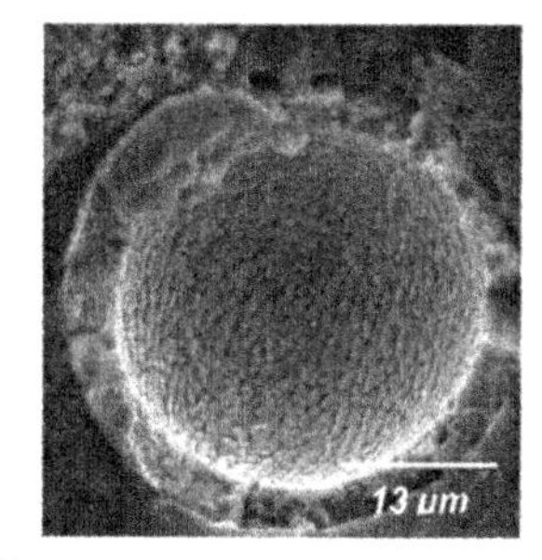


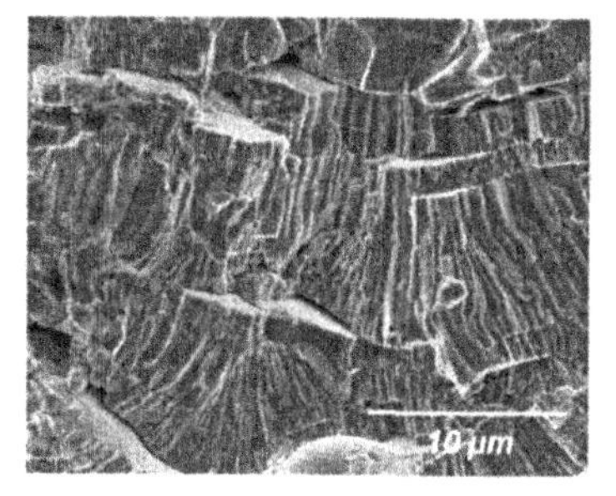


**Fig. 1.** Typical SEM image of 204B-NS; left: feedstock particle and right: polished and etched cross section of 204B-NS coating.

Feedstock of Nanox particles is an agglomeration of nano particles in spherical shapes; some of these nano species will be retained in coating as is clear in Figure 2.

Porosity and non-molten nano particle percentages in deposits were investigated by image analysis, Table I. The most significant process parameters were argon and hydrogen flow rate, current and type of powder feeder (internal/external).

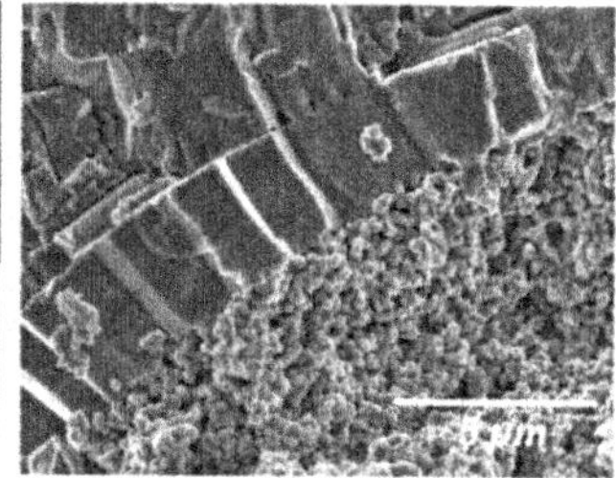

**Fig.2.** SEM image of Nanox S4007; left: feedstock particles and right: fractured cross section showing nano particles next to big columnar grains of an adjacent splat.

In this work the Nanox sample which showed the lowest thermal diffusivity is compared with the 204B-NS deposit.

Table I: Porosity and non molten percentages of Nanox and 204B-NS coatings.

| Powder | Plasma power (KW) | Feeder | Velocity m/sec | Temp. °C | Thermal diffusivity $m^2$/Sec | Image Analysis % | |
|---|---|---|---|---|---|---|---|
| | | | | | | Porosity | Non molten |
| Nanox | 17 | Ext. | 130 | 2533 | 5.09E-7 | 17 ± 1 | 4.5 ± 1 |
| 204B-NS | 25 | Int. | 144 | 2809 | 6.12E-7 | 11 ± 1 | --- |

Int.: Internal, Ext.:External

Creep tests were performed at a range of temperatures (800,1000 and 1200 °C) and stresses for 2 days. Creep strain and strain rate of samples were calculated and plotted versus time, details can be found elsewhere[14]. At least two specimens of each sample were tested and results showed a very good repeatability.

Fracture images of 204B-NS samples, under 30 N applied force at 1000 °C, illustrate phenomena such as crack healing and inter/intra splat grain growth due to material transport by diffusion, Fig.3. Under these circumstances, SEM investigations show that inter-splat grain growth is more significant than intra-splat joining of grains. It may be because of a preferential growth direction of the columnar grains created during solidification and the stable geometrical shape of grains in a splat.

Figure 4 shows SEM images of some nano particles after two and five days under 30 N bending load at 1000°C. Little sintering is apparent after two days but after five days extensive sintering could be observed in many areas.

Figure 5 shows the results of creep tests conducted under a range of stresses for Nanox and 204B-NS coatings at 1000 °C. With increasing applied stress, creep strain rises but there is not any remarkable difference in strain rates after the first day of testing, showing little dependency of strain rate on stress in the quasi-stationary stage.

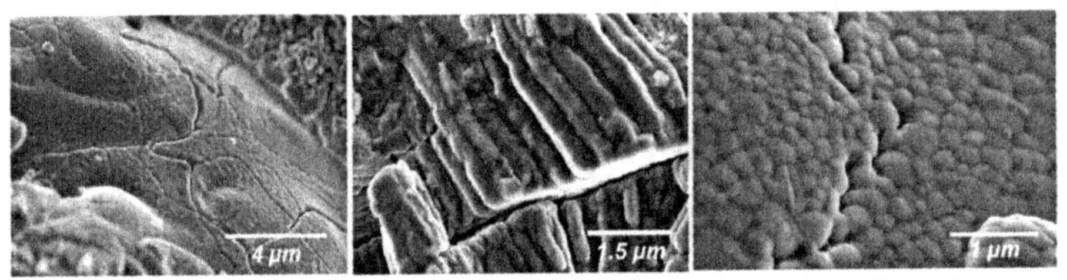


**Fig. 3.** SEM image of 204B-NS splats after two days at 1000 °C under 30 N flexure load; left: surface of splat shows intra splat grain growth, middle: inter splats grain growth, right: micro-crack healing.

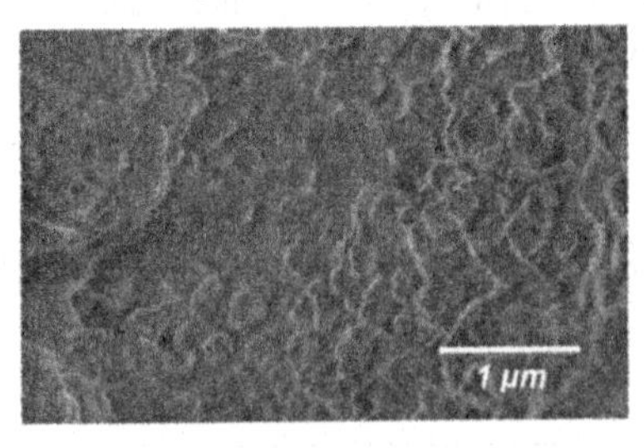


**Fig. 4.** SEM image of non molten nano particles after 5 days at 1000 °C under 30 N flexure

Figure 5 shows the results of creep tests conducted under a range of stresses for Nanox and 204B-NS coatings at 1000 °C. With increasing applied stress, creep strain rises but there is not any remarkable difference in strain rates after the first day of testing, showing little dependency of strain rate on stress in the quasi-stationary stage.

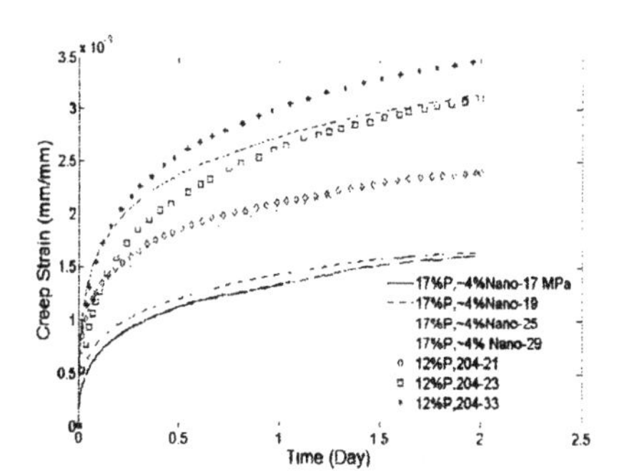


**Fig. 5.** Creep strain of 204B-NS and Nanox under different stresses at 1000 °C

Plotting strain rate against stress in a log-log scale will provide the stress exponent in a power law equation, Fig. 6. Creep exponents of Nanox and 204B-NS samples are 1.1 and 1.3, respectively, very close to 1.

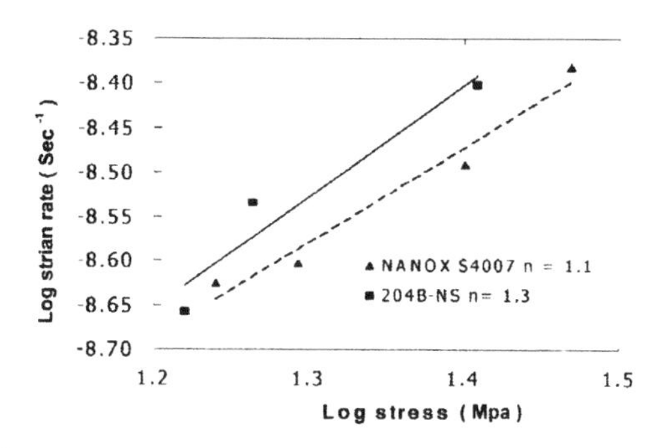


**Fig.6.** Stress versus strain rate of 204B-NS and Nanox samples at 1000 °C .

Activation energy is another concern in creep properties; lower activation energy generally means a lower resistance to creep.

Thermal barrier coatings experience compression stresses at high temperatures. If relaxation mechanisms such as creep and/or sintering eliminate this compressive stress, upon cooling to room temperature tensile stresses will develop in the coating which could in turn lead to cracking of the coating; consequent delamination reduces the service life of the deposit. Therefore a high activation energy in thermal barrier deposits is advantageous. In order to determine the apparent activation energy, creep tests were conducted for both materials under the same stress and at temperatures 800, 1000 and 1100°C, Fig.7.

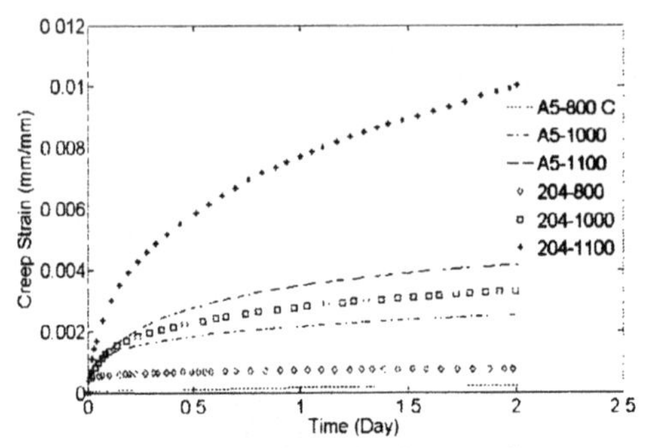


**Fig. 7.** Creep strain of 204B-NS samples under 25 MPa flexure stress at different temperatures.

Plasma sprayed thermal barrier coatings posses a long primary creep stage which is not only a function of temperature and stress but also is time dependent. Generally the strain rate of these samples can be expressed by following equation :

$$\dot{\varepsilon} = A\{ \frac{\sigma^n}{D^p} \}\exp\{ \frac{-Q}{RT} \}\ t^{-s} \tag{1}$$

Equation (1) was used to plot natural log of strain rate versus 1/T, where $\dot{\varepsilon}$ denotes creep rate ($Sec^{-1}$), A, a material constant ($Sec^{-1}$), $\sigma$ the normalized applied stress, n the stress exponent, D grain size, p grain size exponent, Q the apparent activation energy of creep (J/mol), R the gas constant (J/mol K) and T the absolute temperature (K). Slope of the fitted line will give Q. Figure 8 shows results for two samples of Nanox and 204B-NS, having apparent activation energies of 154 and 190 KJ/mol, respectively.

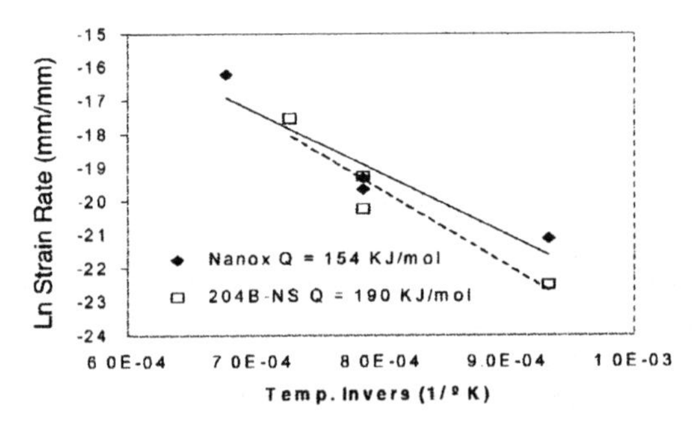


**Fig. 8.** Activation energy of 204B-NS and Nanox under 30 MPa flexure stress.

## 4. DISCUSSION

The amount of deformation in a creep test,™, is a function of stress $\sigma$, time t , temperature T and structure S[15]:

$$™ = f(\sigma, t, T, S) \qquad (2)$$

Since deformation is strongly dependent on the constitution of the material, a structure term is required which involves both macro and micro structural features .[26] In this work the effect of some of these structural features on creep behaviour of zirconia coatings has been examined.

Compositional analysis was performed to investigate the levels of several common impurities in the two feedstocks. No major differences between the two materials were found (Table II). Particular components such as silica and alumina, which may create a glassy phase at grain boundaries, were present in almost the same amounts in both powders. Therefore compositional differences do not account for the observed differences in creep behavior.

The area percentage of non-melted nano-particles does not exceed 5% in any sample, excluding porosity from the calculated area, Table I. Therefore the discrete areas of nano particles could not have a significant effect on the creep/sintering behaviour.

Considering the lamella structure of thermally sprayed coatings, similar composition and a negligible content of non melted nano particles, the thickness of splats and the location and shape of porosity seem to be the main features influencing creep/sintering processes.

Table II: Chemical composition of the two feedstock powders; nanostructured and 204B-NS.

| | Nanox(%) | 204B-NS(%) | Technique |
|---|---|---|---|
| $Al_2O_3$ | 0.66 | 0.66 | ICP |
| CuO | 0.007 | 0.008 | ICP |
| CaO | 0.025 | 0.026 | ICP |
| $Fe_2O_3$ | 0.006 | 0.009 | ICP |
| $HfO_2$ | 1.74 | 1.52 | ICP |
| $P_2O_5$ | 0.04 | 0.04 | ICP |
| $SiO_2$ | 0.12 | 0.11 | ICP |
| $TiO_2$ | 0.06 | 0.14 | ICP |
| $Y_2O_3$ | 6.43 | 7.30 | ICP |
| Na2O | 0.03 | 0.02 | FP |

There are several parameters which affect the final thickness of splats. The velocity and temperature of the in-flight particles are particularly important. Table I shows velocity and temperature of in-flight particles for each coating. The following equations show correlations between the flattening ratio and these parameters. Temperature appears through the kinematic viscosity, which is exponentially dependant on temperatue[16]:

$$R_f = \alpha ( V_p . D_p / \nu )\beta \tag{3}$$

$$\nu = A \exp (B / KT) \tag{4}$$

Where $R_f$ is flattening ratio, $V_p$ Velocity, $D_p$ particle diameter, $\nu$ kinematic viscosity, K Boltzmann's constant, T absolute temperature, $\alpha$, $\beta$, A and B are all constants. Based on equations 3 and 4, by increasing the velocity, temperature, and mean diameter of in-flight particles, the flattening ratio increases and therefore the thickness of splats decreases. The effect of particle size is the least significant among these. Thinner splats result in a higher density of splat interfaces in a given thickness of coating.

Table I shows that the 204B-NS sample exhibited a higher velocity and temperature of in-flight particles than the Nanox coating. Empirical constants for equation 3 obtained by Yoshida[17], Madejski[18] and Solonenko[19] indicate that the average thickness of 204B-NS splats would be more than 20% less than in the Nanox deposit. Extensive SEM examination of splat thickness in chemically etched samples ($H_3PO_3$ at 250°C) , see Fig. 9, showed good agreement with these predictions.

The lower thickness of splats in 204B–NS coatings results in an increase of about 25% in the linear density of splat interfaces in a given thickness. The inter-splat boundaries and their associated laminar pores are well recognized as a mechanical weak link in thermal spray coatings. Sliding along splat boundaries has been suggested to play a significant role in the inelastic deformation of thermal spray coatings.[1] According to Fig. 5, 204B-NS coatings show a creep strain which is more than 50% higher than the Nanox coatings, supporting the premise that the inter-splat boundaries play a controlling role in the creep behavior.

A large primary creep strain and a low creep activation energy for plasma sprayed zirconia coatings were observed (Figs.5 and 7). The creep rate continuously decreases during the test though no steady-state regime was observed. The large decrease of creep rate with time can be explained by microstructural changes in the coating during the creep test. At the beginning of the creep test the total contact area between splats is very small.[20] At the elevated temperatures experienced during the creep test, the flat surfaces of the columnar grains at the splat surface gradually develop spherical caps as grain boundary grooves form by surface diffusion. Wherever columnar grains in adjacent splats are separated by a small distance, the spherical caps can touch and start necking, as shown schematically in Figure 10. These inter-splat bridges pin the splats together, increasing the contact area between splats.

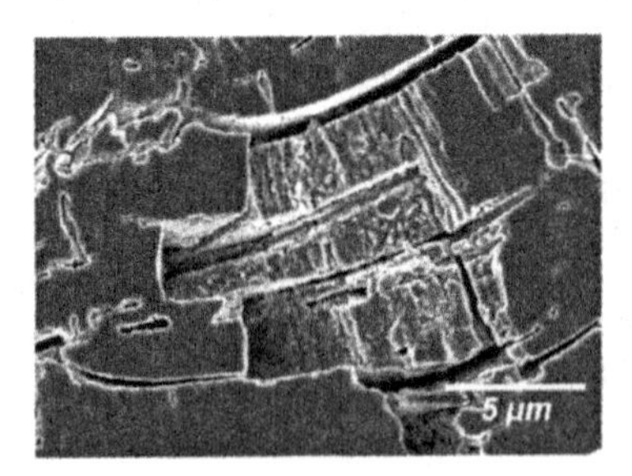


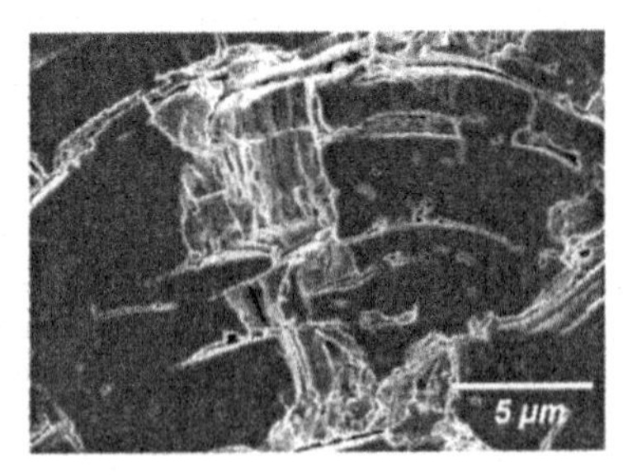


**Fig. 9.** Typical SEM images of Nanox (left) and 204B-NS (right) showing thickness of splats.

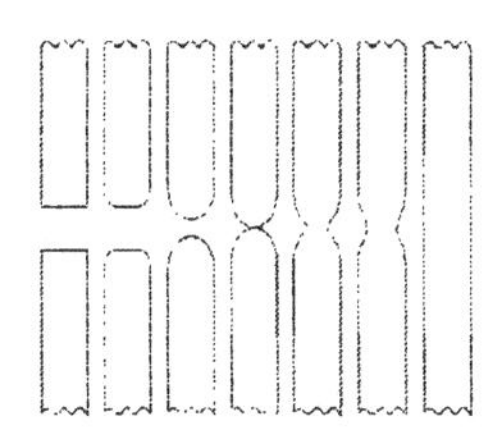

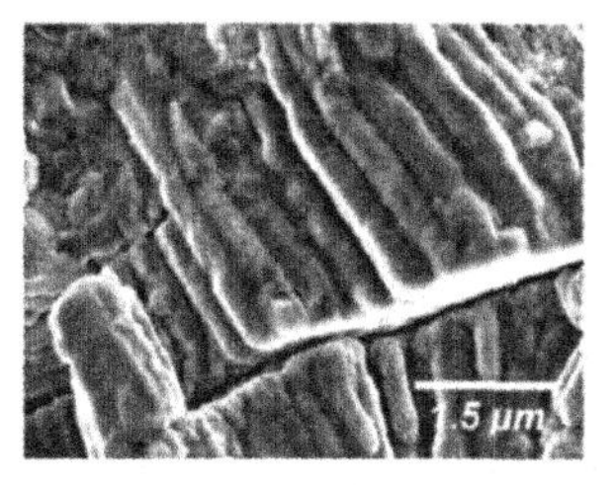


**Fig. 10.** Schematic and SEM images of inter-splat grain bridging.

5. CONCLUSION

At all temperatures and stress levels the bimodal coating showed lower creep strain and strain rate than 204B-NS coatings. Stress exponents were close to one, showing very little dependency of creep rate on applied stress. The apparent activation energy was 154 KJ/mol for Nanox samples and 190 KJ/mol for 204B-NS. The splat interfaces were found to be the key factor affecting the creep behavior of these coatings. The formation of bridges across splat interfaces was observed during the creep tests, and is believed to account for the steady decrease in creep rate with time.

6. REFERENCES

[1]D.Zhu and Miller R. A. , "Determination of creep behaviour of thermal barrier coatings under laser imposed temperature and stress gradients"; NASA Technical Memorandum 113169, ARL-TR-1556(1997).

[2]D.Zhu and Miller R.A., "Sintering and creep behaviour of plasma-sprayed zirconia –and hafnia-based thermal barrier coatings"; Surface and Coatings Technology, 108-109,114-120 (1998).

[3]Ed. F. Rejda, D.F. Socie and T. Itoh, "Deformation behavior of plasma-sprayed thick thermal barrier coatings": Surface and Coatings Technology, 113,218-226 (1999).

[4]A. H. Chokshi , "Diffusion creep in oxide ceramics", Journal of the European Ceramic Society 22, 2469-2478 (2002).

[5] A. H. Chokshi, "Diffusion, diffusion creep and grain growth characteristics of nanocrystalline and fine-grained monoclinic, tetragonal and cubic zriconia": Scripta Materialia, 48,791-796, (2003).

[6]R.Schaller, M. Daraktciev, "Mechanical spectroscopy of creep appearance in fine-grained yttria-stabilized zirconia"; Journal of the European Ceramic Society, 22, 2461-2467,( 2002).

[7]M.J. Adnrews, M. K. Ferber and E. Lara-Curzio, "Mechanical properties of zirconia –based ceramics as functions of temperature" ; Journal of the European Ceramic Society, 22, 2633-2639,(2002).

[8]K.Kokini, Y. R. Takeuchi and B.D. Choules, "Surface thermal cracking of thermal barrier coatings owing to stress relaxation: zirconia vs. Mullite"; Surface and Coating Technology, 82, 77-82 (1996).

[9]L. Vasylkiv, Y. Skka, and V. V .Skorokhod, "Low-temperature processing and mechanical properties of zirconia and zirconia-alumina nanoceramics"; J. Am. Ceram. Soc., 86[2],299-304 (2003).
[10]M.Daraktchiev and R. Schaller, "High-temperature mechanical loss behaviour of 3 mol% yttria-stabilized tetragona zirconia polycrystals (3Y-TZP)"; Journal of Phys. Stat. Sol. 2, 293-304 (2003).
[11]Shaw,L., Goerman, D., Ren,R. and Gell, M., "The dependency of microstructure and properties of nanostructured coatings on plasma spray conditions"; Surface and Coating Technology, 130, 1-8, (2000).
[12]Gell, M., Jordan, E.H., Sohn, Y.H. and Gberman, D., "Development and implementation of plasma sprayed nanostructured ceramic coatings"; Surface and Coating Technology, 146-147,48-54, (2001).
[13]Chen,H. and Ding , C.X., "Nanostructured zirconia coating prepared by atmospheric plasma spraying", Surface and Coatings Technology, 150, 31-36, (2002).
sprayed nanostructured zirconia coatings", Surface and coating technology, 135, 166-172, (2001).
[14]R.Soltani, T.W.Coyle, J.Mostaghimi, "Thermo-physical property of bimodal structured plasma sprayed TBCs", To be published.
[15] W.D.Kingery, H.K.Bowen and D.R.Uhlmann, "Introduction to Ceramics",2$^{nd}$ Ed., John Wiley &Sons, New York,1975.
[16] P.Fauchais, M.Fukumoto, A.Vardelle, M.Vardelle, "Knowledge concerning splat formation: An invited review", Journal of Thermal Spray Technology, 13 (3), 2004, p. 337-360.
[17] J. Madejski, "Solidification of droplets on a cold surface", Int. J.Heat Mass Transfer, 19, 1976, p. 1009-1013.
[18] T. Yoshida, T. Okada, H. Hamatani, and H. Kumaoka, "Integrated fabrication process for solid oxide fuel cells using novel plasma sprayingv" , Plasma Sources Sci. Technol., 1(3), 1992, p. 195-201.
[19] O.P. Solonenko, A.V. Smirnov, V.A. Klimenov, Y.G. Butov, and Y.F.Ivanov, "Role of interfaces in splat and coatings structure formation", Physical Mesomechanics, 2(1-2), 1999, p. 113-129.
[20] R. McPherson, "A Review of microstructure and properties of plasma sprayed ceramic coatings", Journal of Surface and Coatings Technology, 39/40 ,1989, p. 173- 181

# CORROSION RIG TESTING OF THERMAL BARRIER COATING SYSTEMS

Robert Vaßen, Doris Sebold, Gerhard Pracht, Detlev Stöver
Institut für Werkstoffe und Verfahren der Energietechnik
Forschungszentrum Jülich GmbH
Jülich, 52425, Germany

## ABSTRACT

A burner rig test facility was built up which allows the testing of protective coatings as thermal and environmental barrier coatings under harsh environmental conditions. In the experimental set-up liquids as salt solutions or kerosene are injected into a gas burner flame, accelerated and heated or burned in the combustion gases and finally deposited on the sample surface. As the samples are cooled by compressed air a thermal gradient can be established across the coatings allowing typically a surface temperature of 1150-1250 °C and a bond coat temperature of about 950 -1000°C.
YSZ based thermal barrier coating systems have been cycled in the new rig using different corrosive liquids as pure water, NaCl and $Na_2SO_4$ solutions. Depending on the concentration of the corrosive media the lifetime of the coatings were reduced up to a factor of more than 100. Microstructural evaluation as well as X-ray diffraction analysis will be presented and the reasons for the early failure of the coatings discussed.

## INTRODUCTION

Thermal barrier coating systems (TBCs) are widely used in land-based or aero gas turbines to improve their performance. The isolative layer can provide a significant reduction of the temperature of the metallic substrate which results in an improved component durability. Alternatively, an increase of efficiency can be achieved by allowing an increase of the turbine inlet temperatures [1].
TBC systems consist typically of two layers, a so-called bond coat layer, and an isolative, ceramic topcoat. The bond coat is in most cases an alloy and has two major functions. It improves the bonding between the substrate and the topcoat and it protects the substrate from corrosion and oxidation. Two types of bond coats are frequently used, a (platinum-) aluminide based one and a so-called MCrAlY with M being Ni and/or Co. The choice of the adequate bond coat depends on the used deposition technique for the topcoat. Electron beam physical vapour deposition (EB PVD) and atmospheric plasma spraying (APS) are the most frequently used techniques. The meanwhile as standard material established oxide is the 6-8 wt.-% $Y_2O_3$ stabilised zirconia (YSZ, [2]) which is frequently used in aero and stationary gas turbines from the beginning of the eighties [3, 4, 5, 6]. The stabilizing agent is essential for the performance of the TBCs to avoid the phase transition from tetragonal to monoclinic which is observed in pure zirconia.
All of the different components, substrate, bond coat, and topcoat, interact with each other or the environment to a more or less extent and/or they undergo detrimental changes due to thermo-mechanical treatments during operation.
This study is focused on the influence of corrosive media on the performance of TBCs. Corrosive media can be introduced from the environment e.g. in aviation engines from sand particles during landing and take-off. In addition, also the fuel contains corrosive constituents, the most harmful are vanadium, sulphur and sodium. The concentration can vary in wide ranges. While for aviation fuel the sulphur content is limited to 0.05 wt.%, it may reach 4 wt.% in heavy oil for stationary gas turbines [7].

A large number of investigations have been performed in the past to study the influence of the corrosive media in laboratory tests. An older review of the results has been made by Bürgel and Kvernes [8] and a more recent one by Jones [9], showing the most important reactions of the corrosive media with the TBCs. Under severe corrosive environments it is found that the stabilizing agent reacts with the corrosive media leading to a destabilizing of the coating. Two examples are:

$$ZrO_2(Y_2O_3) + V_2O_5 \rightarrow ZrO_2\ \text{(monoclinic)} + 2YVO_4 \quad (1)$$

$$ZrO_2(Y_2O_3) + 3SO_3(+Na_2SO_4) \rightarrow ZrO_2\ \text{(monoclinic)} + Y_2(SO_4)_3\ \text{(in } Na_2SO_4 \text{ solution)} \quad (2)$$

Also new stabilizing agents as $Sc_2O_3$ or new ceramics have been studied [10, 11]. A general improvement of the corrosion behaviour of these materials compared to YSZ was not found. To the authors knowledge no material was identified up to now which shows largely improved corrosion resistance compared to YSZ for most of the relevant testing conditions and corrosive media.

The distinct influence of the testing conditions also raises the question under which conditions corrosion tests should be performed. Typically, the testing is performed under isothermal conditions in a furnace with testing temperatures up to about 1100°C, typically between 800 and 1000°C. In earlier investigations sometimes also burner rigs are used for corrosion testing [12, 13]. The surface temperature in these studies was also well below 1100°C at about 982°C, the substrate was at about 842°C. In this study a major influence of the bond coat composition on the performance of the TBC systems was found. Typically higher Cr and Al contents improved the lifetime of the coatings. These results show that the corrosion of the bond coat and the attack of the formed thermally grown oxide play a major role. Typically in the temperature range above 800°C type I hot corrosion is expected. Reactions of the formed $Al_2O_3$ scale with $SO_3$ containing atmospheres can be as follows [14], with (3) being dominant at high temperatures (hot corrosion type I):

$$Al_2O_3 + O^{2-} \rightarrow 2\ AlO_2^- \quad \text{for low } SO_3 \text{ pressures} \quad (3)$$

$$Al_2O_3 + 3SO_3 \rightarrow Al_2(SO_4)_3 \quad \text{for high } SO_3 \text{ pressures} \quad (4)$$

In addition, under type I conditions, the sulphide formation of metals (Cr, Ni) at the slag/metal interface play an important role:

$$M + SO_2 \rightarrow \text{M-oxide} + \text{M-sulphide} \quad (5)$$

In the present investigation corrosive media are injected into the flame of a burner rig. With the used setup temperature profiles can be adjusted which are considered as relevant for modern gas turbines (i.e. bond coat temperatures in the range of 900 to 1000°C and surface temperature of about 1200°). Hence a rather realistic testing of TBC systems under corrosive conditions should be possible.

## EXPERIMENTAL

The investigated thermal barrier coating systems have been produced by plasma spraying with two Sulzer Metco plasma-spray units. Vacuum plasma spraying with a F4 gun was used to deposit a 150 µm NiCo21Cr17Al13Y0.6 bond coat (Ni 192-8 powder by Praxair Surface Technologies Inc., Indianapolis, IN) on disk shaped IN738 superalloy substrates. The

diameters of the substrates used for thermal cycling tests were 30 mm, the thickness 3 mm. At the outer edge a radius of curvature of 1.5 mm was machined to reduce the stress level.
The ceramic top coats with a thickness of about 350 μm were produced by atmospheric plasma spraying (APS) using a Triplex I gun. During the manufacture of the thermal cycling specimens also steel substrates were coated. These coatings were used to characterize the as-sprayed condition. The porosity level was measured using free - standing coatings with mercury porosimetry given values of about 12 vol.%. The Argon and Helium plasma gas flow rates were 20 and 13 standard liter per minute (slpm), the plasma current was 300 A at a power of 20 kW.
Corrosion tests were performed in a gas burner rig setup, in which the disk shaped specimens were periodically heated up to the desired surface temperature of about 1200 °C in approximately 1 min by a natural gas/ oxygen burner. After 5 min heating the specimens were cooled for 2 min from both sides of the specimens by compressed air. The surface temperature was measured with an infrared pyrometer operating at a wavelength of 9.6 - 11.5 μm and a spot size of 5 mm, while the substrate temperature was measured using a NiCr/ Ni thermocouple inside a hole drilled to the middle of the substrate. More details on the standard thermal cycling rigs can be found in [15]. In this investigation the central gas nozzle was used to introduce atomized liquids containing the corrosive media into the flame. In this paper only the results on water based salt solutions will be presented. Other media as kerosene are also possible.
In Figure 1 a photo of the rig in operation is shown. The injection of water droplets into the flame led to a reduce stability of the flame and therefore larger variations in the temperature profile. In order to evaluate this effect also a pure water injection (without additional corrosive media) was examined.

Fig. 1 Flame of the corrosion rig during injection of $Na_2SO_4$ solution.

The test conditions used with respect to the corrosive media are given in Table 1. The concentrations of the corrosive media have been calculated from the injected quantities of water-based solutions and the mean methane and oxygen gas flows during thermal cycling. It turned out that the gas flows of the burner gases had to be varied in rather wide ranges to adjust similar temperature profiles. This fact led to the rather large differences in the corrosive media concentration in Table 1 column 4. As the injection nozzle is located in the center of the flame the profile of the corrosive media concentration has a maximum in the centre. This concentration profile has not been considered for the calculation of the concentration in Table 1.

The test was stopped when obvious degradation of the coatings (large delamination) occurred during the thermal cycling. This definition of failure results in an uncertainty in the lifetime data especially if partial delamination of the coatings occurs. However, as seen below the resulting error does not effect the interpretation of the results.
Sprayed specimens were vacuum impregnated with epoxy, and then sectioned, ground and polished. As for the preparation water was used as a media, water-soluble corrosion products might be removed by the preparation process. Cross-sections of the coatings were examined by optical microscopy and scanning electron microscopy (Ultra 55, Zeiss). The surface of the tested samples were analysed by X-ray diffraction using a Siemens D5000 facility at a wavelength of 1.5406 Å.

| Sample # | Corrosive media | Mass flow of corrosive solution [g/min] | Appr. concentration of corrosive media (Cl/S) in relation to $CH_4$ [%] |
|---|---|---|---|
| A | Non | - | - |
| B | Water | 1.4 - 2.8 | - |
| C | 1% NaCl in water | 1.4 | 0.433 |
| D1 | 0.1 % $Na_2SO_4$ in water | 1.4 | 0.010 |
| D2a | 1.0 % $Na_2SO_4$ in water | 1.4 | 0.108 |
| D2b | 1.0 % $Na_2SO_4$ in water | 1.4 | 0.122 |
| D2c | 1.0 % $Na_2SO_4$ in water | 1.4 | 0.039 |
| D2d | 1.0 % $Na_2SO_4$ in water | 1.4 | 0.098 |
| D3 | 10 % $Na_2SO_4$ in water | 1.4 | 0.588 |

Table 1 Corrosive testing conditions.

## RESULTS AND DISCUSSION

In Table 2 and Figure 1 the temperatures and the cycles to failure of the thermal cycling tests are given. The listed temperatures are mean values calculated from the temperature recordings during the heating phase. The standard deviation of the bond coat temperature is about 50 K, for the surface temperature it is about 75 K. For the used test conditions the bond coat temperature is about 40 K higher than the substrate temperature. The results refer to single samples, D2a to D2d show results of samples cycled under similar conditions.

| Sample # | Substrate temperature [°C] | Surface temperature [°C] | Cycles to failure | Time at temperature [h] |
|---|---|---|---|---|
| A | about 950 | about 1200 | about 5000 | 333 |
| B | 965 | 1253 | 4402 | 293 |
| C | 944 | 1225 | 936 | 62 |
| D1 | 960 | 1242 | 1243 | 83 |
| D2a | 938 | 1241 | 547 | 36 |
| D2b | 923 | 1162 | 614 | 41 |
| D2c | 981 | 1195 | 150 | 10 |
| D2d | 948 | 1235 | 843 | 56 |
| D3 | 942 | 1156 | 21 | 1.4 |

Table 2 Results of thermal cycling with corrosive media (conditions of samples A to D are explained in Table 1).

The lifetime of the sample cycled with pure water injection (sample B) is similar to the results of coatings cycled without additional injection (sample A). The stated last value is a mean value taking into account several experiments in the indicated temperature regime. More details are given in [16]. Obviously the use of water does not significantly influence the lifetime of the coatings. In contrast, the cycling with corrosive media led to a significant reduction of lifetime for all investigated conditions.

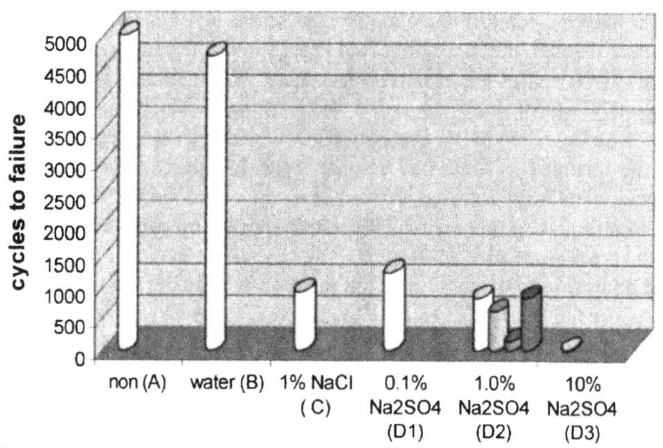


Figure 1 Cycles to failure for the used corrosion conditions (Table 2).

Cycling with NaCl media (sample C)

Sample C cycled with 1% NaCl injection showed an early failure with a lifetime reduction of about a factor of 5 compared to thermal cycling without corrosive media. The sample shows a massive spallation of the coating and also a yellowish and greenish appearance (Figure 2). In contrast, sample B cycled with water appears still white.

The XRD pattern (Figure 3) shows peaks corresponding to following three major phases, the tetragonal YSZ phase, the fcc phase of the Ni-base superalloy and $\alpha$– alumina phase. There is no indication of the formation of significant amounts of monoclinic phase. These results correspond to the results of Mitamura et al. who did not find a reaction between 3YSZ and NaCl + KCl in the temperature range between 850 and 1050°C [17].

Figure 2 Photos of samples after thermal cycling, left with water injection (sample B in Table 1+2), right with 1% NaCl injection (sample C).

SEM-images of the metallographic cross-section of the sample are shown in Figure 4 a,b). In the left image a long crack in the TBC is found which runs close to the bond coat / TBC interface. In the right micrograph the interface is shown with a higher magnification. At three locations EDX analysis has been performed (Fig. 4, c-e). At location 1 mainly the elements Al, Y, Cr, Co, Ni and Cr were found. At location 2 mainly Al, Cr and less Y showed up. At the third location peaks from Al, Ni, Co, Zr, relatively much Cr and clear indications of Na and Cl were visible. Although at other locations clear evidence of Na and Cl were found, it is assumed that the alumina scale is attacked by the liquid NaCl (melting point 800°C) which can easily penetrate through the interconnected cracks to the thermally grown oxide. Although it is known that $Al_2O_3$ scales are relatively resistant to Cl-containing gases [18] the given conditions obviously significantly degraded the alumina scale. Once the alumina scale is removed elements as Cr, Ni and Co can easily react with the Cl-species.

In addition to the changes of the TGO also at some locations an infiltration of the micro cracks by Cr-species has been detected. An explanation of this finding can be the formation of liquid or even volatile $CrCl_2$ species which are transported by capillary forces or via the gas phase through the crack network.

A more detailed analysis and discussion of the reaction will be performed in the future.

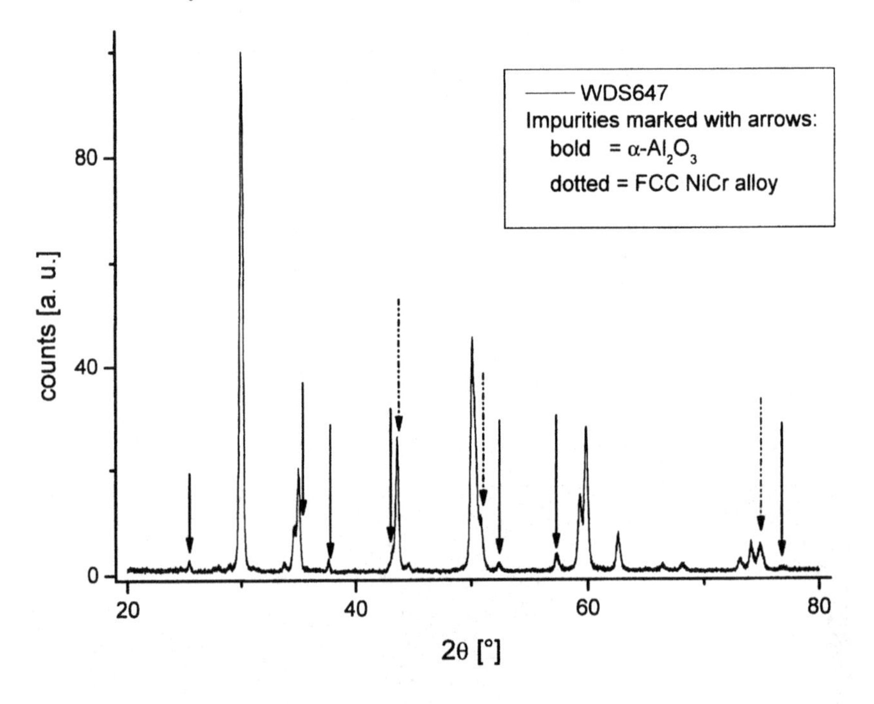


Figure 3 X-ray diffraction patterns of samples cycled with NaCl injection (sample C, Table 1). Peaks not indexed can be ascribed to YSZ (tetragonal phase).

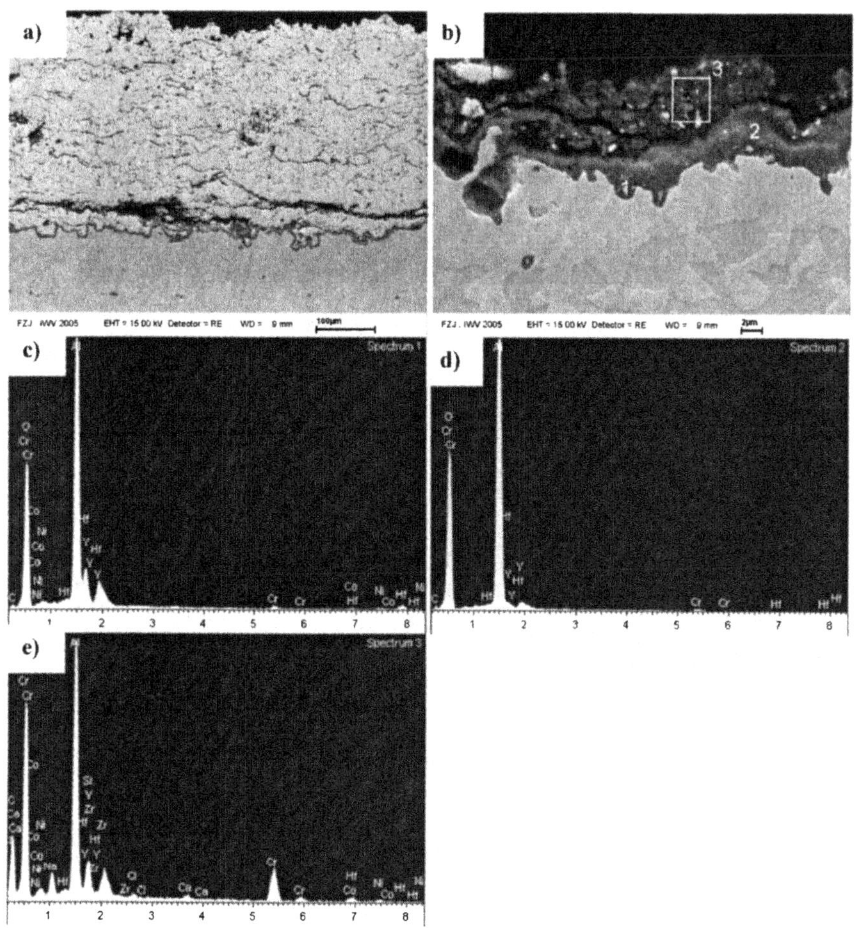


Figure 4 SEM images (a, b) of metallographic cross-sections as well as EDX results (c, d, e) at the different locations given in b) of the sample cycled with NaCl addition.

Cycling with $Na_2SO_4$ media (samples D)
Similar to the results of the thermal cycling experiment with NaCl addition also the $Na_2SO_4$ injection led to an early failure of the samples for all investigated concentrations (see Table 2). Photos of the samples after cycling are shown in Figure 4. Similar to the NaCl injection also here different colourations of the surface of the samples are visible. The circular spall pattern of sample D1 can be attributed to the rather circular concentration profile of the corrosive media due to the central injection.
For the $Na_2SO_4$ three different concentrations have been used. Clearly, an increased concentration of corrosive media led to a reduction of lifetime. Especially the use of solutions with the highest concentration (10%) resulted in a very early failure of the coatings after about 21 cycles (D3).

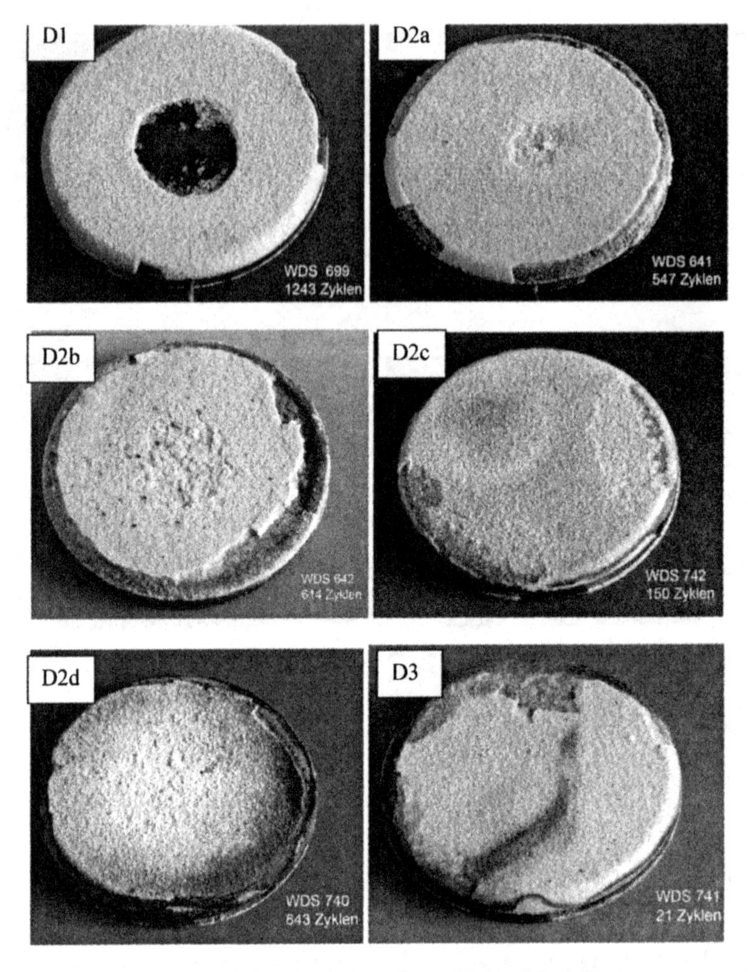


Figure 5 Photos of the samples after thermal cycling with $Na_2SO_4$ injection corresponding to the samples D1, D2a – d, and D3 from top left to bottom right.

In Fig. 6 the XRD patterns for samples cycled with the 0.1 and 1% $Na_2SO_4$ injection are plotted. All patterns reveal that the tetragonal YSZ phase is the dominant phase. There are only minor peaks at about $2\theta = 28°$ which can be attributed to the existence of monoclinic $ZrO_2$. The corresponding amount of monoclinic phase is similar to the one in the as-sprayed condition [19], i.e. the cycling under the investigated conditions did not significantly destabilize the YSZ as might be expected according to reactions as given in (2). Also in [11] YSZ underwent little change in a sulphur containing environment ($SO_2$ + $Na_2SO_4$ + $MgSO_4$) even after treatment for 360 h at 900°C. Obviously, the higher testing temperature (>1150°C for the TBC surface) used in the present investigation did not significantly accelerate the reaction of sulphur containing species and YSZ.

Besides the YSZ phase also additional phases can be identified by the XRD patterns. For the samples D2a and D2b $Na_2SO_4$ is clearly visible, which is expected due to its injection in the gas flame. For D1 further phases are found. Looking at Figure 5 it is visible that a part of the

coating spalled off completely. Hence, the layers formed on the bond coat are contributing to the XRD patterns. Consequently, the thermally grown oxide consisting of spinel and alumina layers are found. In addition, also the nickel substrate is visible.
In addition to the XRD analysis metallographic sections have been prepared and analysed by optical and in the case of sample D2a by electron microscopy. Two examples of optical micrographs are shown in Figure 7.
In both micrographs cracks are found within the TBC close to the bond coat. While in D2a the crack is still rather short, a long crack is observed in D3. In addition, also large dark areas are found in the TBC, which will be discussed below.
A better insight in the development of the TBC system during thermal cycling in corrosive media is possible with a SEM analysis of the coatings. Figure 8 shows SEM images of sample D2a (1% $Na_2SO_4$). In the two top images it is obvious that the dark "TGO" in the central part is much thicker than at the outer rim of the sample. This can be correlated to the central injection of the corrosive media. In fact, the "TGO" is not simply an alumina scale but much more complex as seen in Fig 8 c.

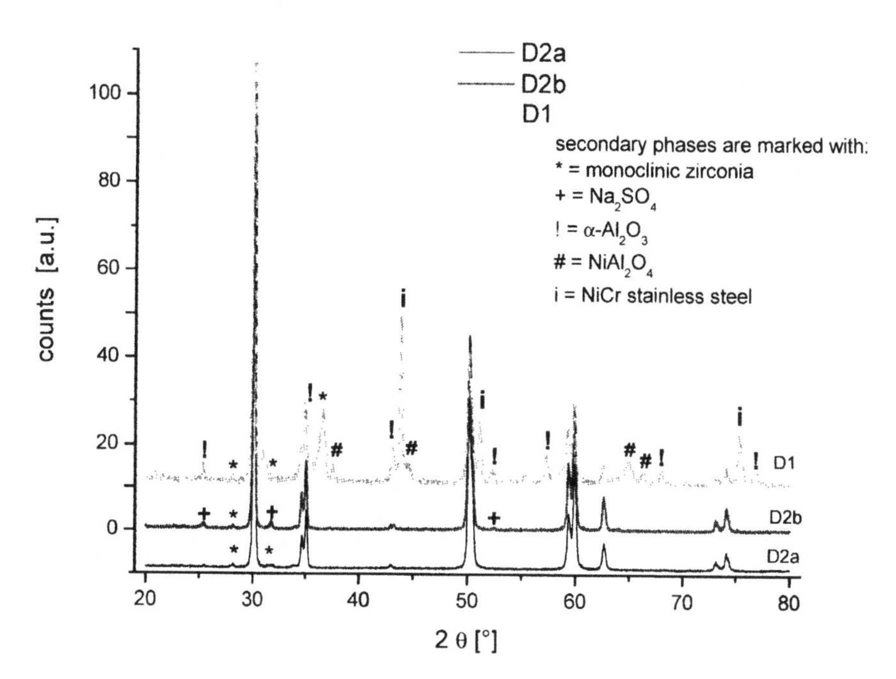


Figure 6 X-ray diffraction patterns of samples cycled with $Na_2SO_4$ injection (samples D1, D2a, D2b, Table 1). Peaks not indexed can be ascribed to YSZ (tetragonal phase).

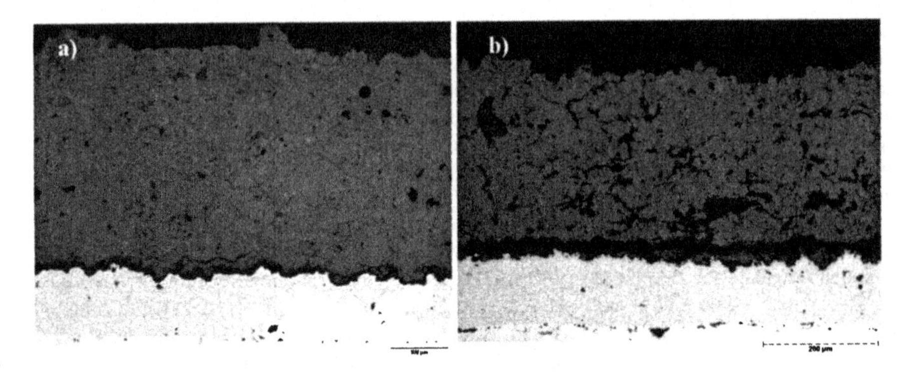


Figure 7 Optical micrographs of sample cycled with the injection of 1% (D2a, a) and 10 % $Na_2SO_4$ (D3, b).

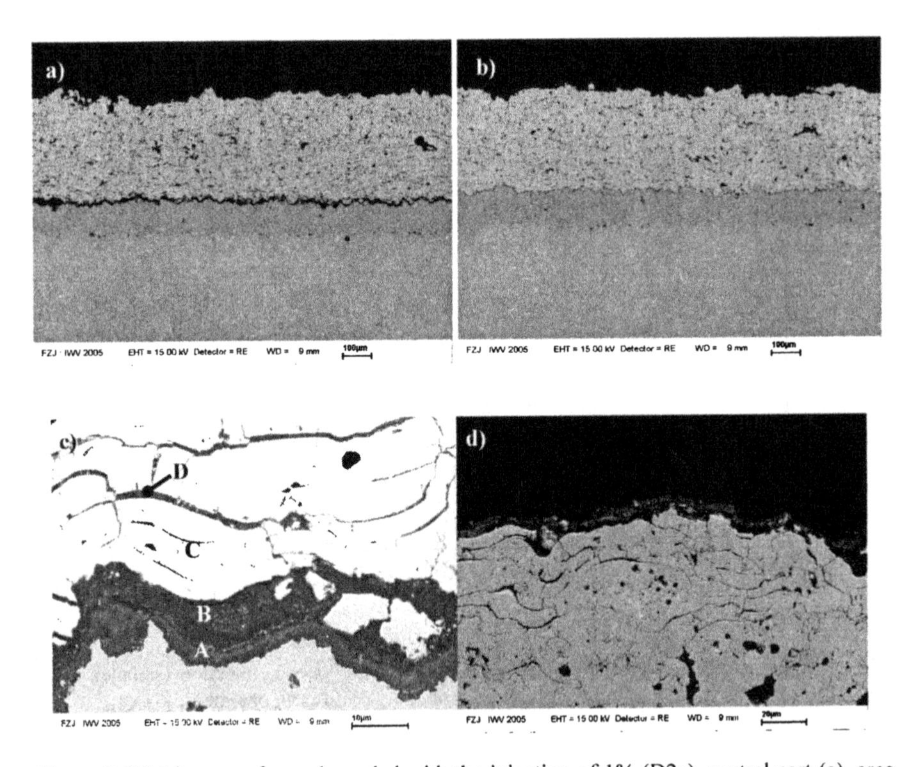


Figure 8 SEM images of sample cycled with the injection of 1% (D2a), central part (a), area close to the outer rim (b), central part at bond coat TBC interface (c), surface region of the TBC (d).

| Location | Al | Cr | Ni, Co | Na, S | Zr, Y |
|---|---|---|---|---|---|
| A | ++ | | | | |
| B | ++ | ++ | + | ++ | |
| C | | | | | ++ |
| D | ++ | ++ | | ++ | ++ |

Table 3 Qualitative results of an EDX analysis of the micrograph in Fig. 8c at the given locations A-D. Only major elements are shown. Oxygen is present at all locations.

A qualitative EDX analysis of the regions in Figure 8c is shown in Table 3. Close to the bond coat still a part of the alumina scale is found. On top of this scale, complex corrosive products have been formed (region B in Fig. 8c) which consist of Al/Cr and Na containing compounds with some additions of Ni and Co. According to equation (3) the formation of $NaAlO_2$ can lead to the basic fluxing of the alumina scale. After removal of the protective scale at different locations the reaction of $Na_2SO_4$ and oxygen with the Ni base bond coat will start leading to the observed corrosion products. Both oxygen and sulfur is observed in the EDX analysis. So, it can not be determined by the present investigation whether mixed oxide, sulfates or sulfides of the various elements as Al, Cr, Ni are present. In [20] complex oxides have been found together with some remaining alumina scale after hot corrosion tests with $Na_2SO_4$ at 950°C. However, in this investigation only low levels of sulfur were found in the oxide areas. This might be related to the use of pure oxygen in this investigation.
An indication of internal sulfide formation of the bond coat, often found in type I hot corrosion, according to reaction (5) was not found in the samples.
At location C only YSZ could be detected, no indication of TBC decomposition was found. Within the micro cracks (location D) often corrosion products have been observed. Probably capillary forces led to the filling of the micro cracks at high temperatures by the liquid corrosion products.
Sample D2c showed a significantly lower lifetime than the other samples cycled with the same $Na_2SO_4$ concentration. This might be related to the relatively high substrate temperature which can lead to an increased melt formation and penetration of corrosive products into the TBC. A detailed analysis of the influence of the temperature profile will be made in the future.
Finally, the sample tested with the highest amount of $Na_2SO_4$ (D3c) will be analyzed by SEM. In Figure 7 b already large dark areas have been detected in this sample. In the SEM image this areas can be identified as pores (Figure 9a). It is possible that these pores have been filled with water soluble corrosion products which have been removed during metallographic sample preparation. Clearly the size and the amount of the pores is much larger than in the as-sprayed condition. In addition, the appearance of the pore surface is unusual for plasma-sprayed coatings (Figure 9 b). The surface is highly fragmented indicating severe attack of the former rather straight surfaces. The described findings suggest a considerable attack of the YSZ by $Na_2SO_4$ at the given high concentration. Under this conditions failure took place within the ceramic and not by fluxing of the oxide layer on the bond coat. The oxide layer seems to be intact and rather thin (Figure 9a) corresponding to the short time at high temperature.

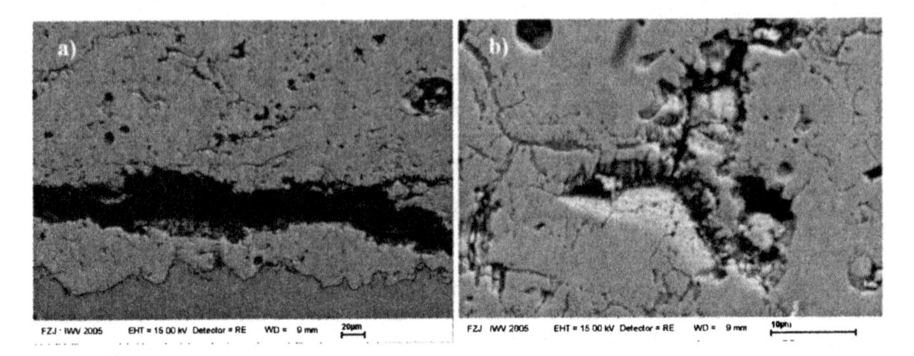


Figure 9 SEM-images of a metallographic section of a sample cycled with 10% $Na_2SO_4$ (D3c).

Further work will be performed in the near future to achieve a better insight in the corrosion mechanisms. In addition, additional corrosive media as kerosene and V containing compounds will be tested.

## CONCLUSIONS

The results of a new burner rig allowing corrosion testing of TBC systems under rather realistic conditions have been presented. The injection of NaCl and $Na_2SO_4$ led to a reduction of lifetime by more than a factor of 100 for the highest concentration (0.6 % S). Failure of the coatings was in most cases related to fluxing of the TGO oxide by the corrosive species. Only for the highest S-concentration failure of the ceramic topcoat has been observed.

## ACKNOWLEDGEMENT

The authors would like to thank Mr. K.H. Rauwald and Mr. R. Laufs (both IWV1, FZ Jülich) for the manufacture of the plasma-sprayed coatings and Mrs. A. Lemmens for the thermal cycling of the specimens. The authors also gratefully acknowledge the work of Mrs. H. Moitroux (IWV1), Mr. P. Lersch (IWV2), Mrs. S. Schwartz-Lückge (IWV1) and Mr. M. Kappertz (IWV1) who supported the characterization of the samples by photography, XRD, optical microscopy, and sample preparation.

## REFERENCES

1 P. Hancock, and M. Malik, Materials for Advanced Power Engineering Part 1, D. Coutsouradis et al. (eds.), Kluwer Academic Publishers, Dordrecht, 1994, 658-704.
2 S. Stecura, Advanced Ceramic Materials, 1 [1] (1986) 68-76.
3 S. Bose, J. DeMasi-Marcin, J. of Thermal Spray Technology 6 [1] (1997) 99-104.
4 W.A. Nelson, R.M. Orenstein, J. of Thermal Spray Technology, 176 (1997) 176-80.
5 J. Wigren, L. Pejryd, Proc. of the $15^{th}$ Int. Thermal Spray Conf., ASM International, Ohio, USA, 1998, 1531-1541.
6 D.R. Clarke and C.G. Levi, Annu. Rev. Mater. Res. 33 (2003) 383-417.
7 B.R. Marple, J. Voyer, C. Moreau, D.R. Navy, "Corrosion of Thermal Barrier Coatings by Vanadium and Sulfur Components," Materials at High Temperatures, 17 (3) (2000) 397-412.

8 R. Bürgel, I. Kvernes, "Thermal Barrier Coatings," High Temperature Alloys for GasTurbines and Other Applications 1986, W. Betz et al. , ed. D. Reidel Publiching, 1986, p.327-356.
9 R.L. Jones, "Some Aspects of the Hot Corrosion of Thermal Barrier Coatings," J. of Thermal Spray Technology, 6 (1) (1997) 77-84.
10 M. Yoshiba, K. Abe, T. Arami, Y. Harada, "High-Temperature Oxidation and Hot Corrosion Behavior of Two Kinds of Thermal Barrier Coating Systems for Advanced Gas Turbines," J. of Thermal Spray Technology, 5 (3) (1996) 259-68.
11 B.R. Marple, J. Voyer, M. Thibodeau, D.R. Nagy, R. Vassen, "Hot Corrosion of Lanthanum Zirconate and Partially Stabilized Zirconia Thermal Barrier Coatings," J. of Engineering for Gas Turbines and Power, July 2005 Vol 127, p1-9.
12 P.E. Hodge, R.A. Miller, M.A. Gedwill, "Evaluation of the Hot Corrosion Behaviour of Thermal Barrier Coatings," Thin Solid Films, 73 (1980) 447-453.
13 I. Zaplatynsky, "Performance of Laser-Glazed Zirconia Thermal Barrier Coatings in Clyclic Oxidation and Corrosion Burner Rig Tests," Thin Solid Films, 95 (1982) 275-284.
14 P. Kofstad, High Temperature Corrosion, Elsevier Appl. Sci., London, 1988.
15 F. Traeger, R. Vaßen, K.-H. Rauwald, D. Stöver, "A Thermal Cycling Setup for Thermal Barrier Coatings," Adv. Eng. Mats., 5, 6 (2003) 429-32.
16 F. Traeger, M. Ahrens, R. Vaßen, D. Stöver, "A Life Time Model for Ceramic Thermal Barrier Coatings," Materials Science and Engineering, A358 (2003) 255-65.
17 T. Mitamura, E. Kogure, F. Noguchi, T. Iida, T. Mori, Y. Matsumoto, "Stability of Tetragonal Zirconia in Molten Fluoride Salts," Advancces in Ceramics, Vol 24°, Science and Technolgy of Zirconia III, S. Somiya, N. Yamamoto, H. Yanagida, eds., Amercan Ceramic Society, 1988, p. 109-118.
18 R. Bürgel, Handbuch Hochtemperatur-Werkstofftechnik, Friederich Vieweg-Verlags-gesellschaft, Braunschweig, 1998.
19 J.-E. Döring, R. Vaßen, D. Stöver, R.G. Castro, „Particle Properties Tailor Coating Microstructure, Porosity and Phase Composition," Proceedings of the 2003 International Thermal Spray Conference, Orlando, FL, USA, 5-8 May 2003, edited by B.R. Marple, C. Moreau, ASM International, Materials Park, OH, 2003, pp. 1197-1204.
20 C. Leyens, I.G. Wright, B.A. Pint, „Hot Corrosion of an EB-PVD Thermal Barrier Coating System at 950°C," Oxidation of Metals, 54, 5/6 (2000) 401-424.

# THERMAL PROPERTIES OF NANOPOROUS YSZ COATINGS FABRICATED BY EB-PVD

Byung-Koog Jang, Norio Yamaguchi and Hideaki Matsubara
Materials Research and Development Laboratory,
Japan Fine Ceramics Center (JFCC)
2-4-1 Mutsuno, Atsuta-ku, Nagoya, 456-8587, Japan

## ABSTRACT

This study investigates the specific heat, thermal diffusivity and thermal conductivity of $ZrO_2$-4mol% $Y_2O_3$ coating layers fabricated by electron beam physical vapor deposition (EB-PVD) in the temperature range between room temperature to 1000°C. The laser flash method and differential scanning calorimeter are used to measure the thermal diffusivity and specific heat of the coated samples, respectively. EB-PVD coatings reveal the porous columnar microstructure. Coatings are found to contain nano sized pores as well as micron sized pores. The thermal conductivities and thermal diffusivities of EB-PVD coatings decreased with increasing measuring temperature.

## INTRODUCTION

Thermal barrier coatings (TBCs) have received a large attention because they increase the thermal efficiency of gas turbine engines by increasing the gas turbine inlet temperature and reducing the amount of cooling air required for the hot section components. Among the various coating processes for producing TBCs, electron beam physical vapor deposition (EB-PVD) is widely used because it has several advantages in comparison with plasma sprayed coatings, including high deposition rate, use of high melting point oxides and excellent thermal shock resistance behavior due to porous columnar microstructure of the coatings [1-3].

The thermophysical property of coatings is one of the most important properties for obtaining superior TBCs. Specially, low thermal conductivity is one of the most important properties for obtaining superior TBCs. It is well known that thermal conductivity of materials closely depends on microstructural properties such as porosity, pore architecture and morphology. In particular, the thermal conductivity of a coatings film is sensitive to the deposition method, microstructural morphology, density and coating composition [4-7]. The purpose of this work is to investigate the influence of temperature on thermal diffusivity and thermal conductivity of nanoporous 4mol%$Y_2O_3$-$ZrO_2$ coatings fabricated by EB-PVD.

## EXPERIMENTAL

The coatings on zirconia substrates were deposited by EB-PVD (electron beam-physical vapor deposition) using commercially available 4 mol% $Y_2O_3$ stabilized zirconia targets. The substrates were first preheated at 900 ~ 1000°C in a preheating vacuum chamber using graphite heating element. The substrates were then moved to the coating chamber for deposition. An electron beam evaporation process was conducted in a coating chamber under a vacuum level of 1 Pa using 45 KW of electron beam power. The target material was heated above its evaporation temperature of 3500°C, and the resulting vapors were condensed on rotating substrates. Deposition was generally conducted in condition of rate of 5 μm/min and substrate rotation of 5 rpm. The coating thickness is about 200~300 μm. The substrate temperature was 950°C.

The thermal diffusivity was determined by the laser flash method using a thermal analyzer (Kyoto Densi, LFA-501). The laser flash method involves heating one side of the sample with a laser pulse of short duration and measuring the temperature rise on the rear surface with an infrared detector. The thermal diffusivity is determined from the time required to reach one-half of the peak temperature in resulting temperature rise curve for the rear surface as illustrated in Fig.1.

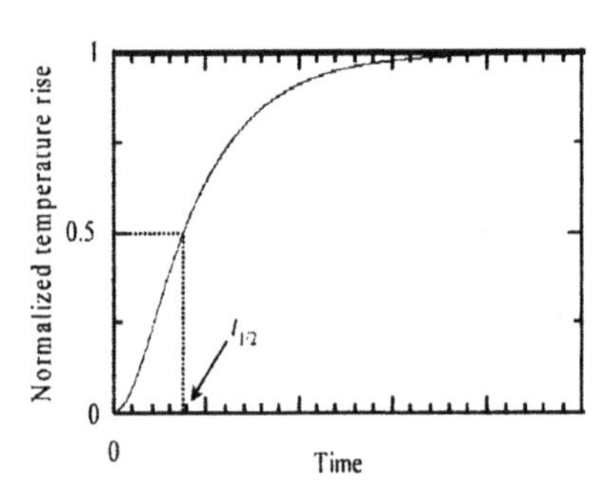


Fig. 1. Temperature response as a function of time at the opposite side of a coated specimen after laser pulse heating.

All the measurements of the coated samples of 10 mm diameter were carried out between 25°C and 1000°C in 200 degree intervals in a vacuum chamber. Because of the translucency of the specimens to the laser, the specimens were sputter-coated with a thin layer of silver and colloidal graphite spraying to ensure complete and uniform absorption of the laser pulse prior to thermal diffusivity measurement. Specific heat measurements of the coated samples of 5mm diameter and 300 μm thickness were made with differential scanning calorimeter (Netzsch, DSC 404C) using sapphire as the reference material. Measurements were made between room temperature and 1000°C in 200°C increments in an argon gas condition.

The thermal conductivity of the coatings was then determined using

$$k = \alpha C \rho \qquad (1)$$

where $k$ is the thermal conductivity, $\alpha$ is the thermal diffusivity, $C$ is the specific heat, and $\rho$ is the density of the coatings, respectively. Quantitative analysis of the pore size distributions in the coated layers was performed using a mercury porosimeter. The microstructure of the coated samples was observed by SEM.

## RESULTS AND DISCUSSION

The typical microstructure of 4mol%$Y_2O_3$-$ZrO_2$ coating deposited by EB-PVD is shown in Fig. 2. The top surfaces of EB-PVD coatings consist of square-pyramidal or cone-like grains. The coatings clearly reveal a columnar microstructure with all columnar grains oriented in the same direction. i.e., perpendicular to the substrate. The columnar microstructure characteristic of EB-PVD imparts the low thermal conductivity and excellent strain tolerance to the materials.

Fig.2. SEM micrographs of microstructure of EB-PVD coatings of 4mol%$Y_2O_3$-$ZrO_2$: (a) surface view and (b) side view.

X-ray diffraction patterns of top surface of coating layers were obtained as shown in Fig. 3. The spectrum contains (200) and (400) diffraction and no other phase were observed in coated surface. It is clear that the primary growth direction of the coating layers is dominant along the (200) with maximum diffraction intensity. The analysis of XRD patterns indicates that the observed phase in all coatings materials is the tetragonal phase.

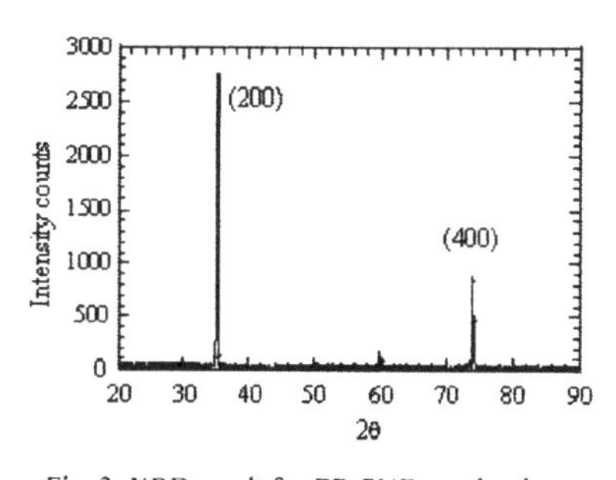


Fig. 3. XRD result for EB-PVD coating layer.

The formation mechanism of the porous column by EB-PVD can be explained as follows. Porous structures in coatings are formed during non-equilibrium condensation of the vapor phase. One of the main mechanisms of porous structure formation is based on the so-called "shadowing" effect. The porous columns can be formed by a flux shadowing mechanism. During nucleation and subsequent growth of various crystallographic faces of the nuclei at different rates., consolidation of grains occurs on the condensation surface.

The faces and micro protrusions, growing at a maximum growth speed, screen the adjacent regions of the surface from the surface of the vapor flow. This results in formation of inner micro-voids in the shadow region. The shadowing effect is also determined by the angle of the vapor flow incidence on the deposited surface. During coating procedure. the vapor flow is a non-uniform, as it is disturbed by the substrate rotation, resulting in various vapor incident angles. The shadowing can be significantly enhanced so that second phase particles form and grow on the deposited surface. Consequently, the shadowing effect contributes to the formation of a columnar coating layer with a highly porous structure.

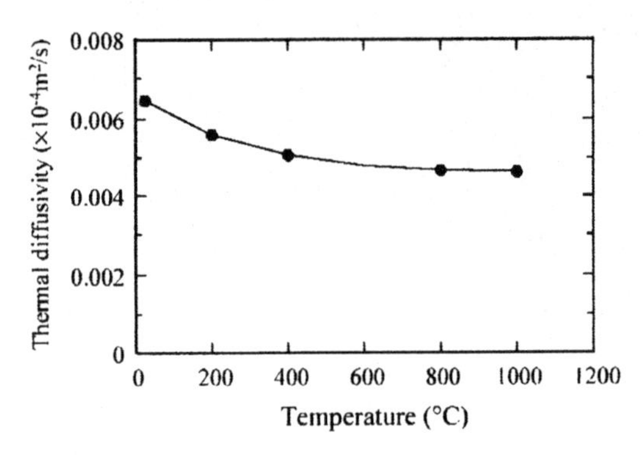


Fig. 4. Thermal diffusivity vs temperature of coatings specimens of 4mol%$Y_2O_3$-$ZrO_2$.

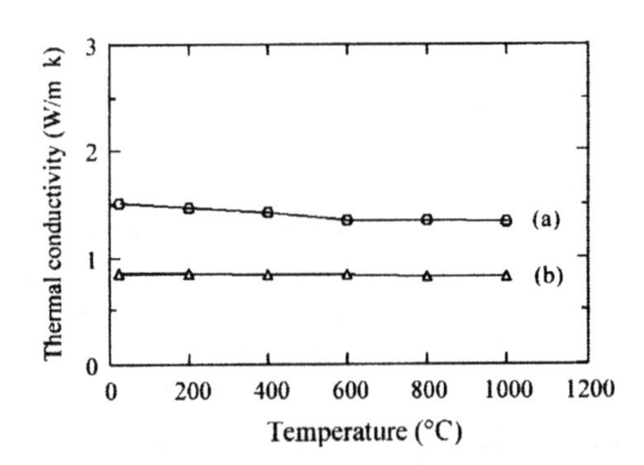


Fig. 5. Thermal conductivity vs temperature of coatings specimens of 4mol%$Y_2O_3$-$ZrO_2$; (a) EB-PVD coatings and (b) plasma sprayed coatings.

The specific heat increases between 0.45~0.62 kJ/(kg K) as the temperature increase up to about 600°C and show the almost constant above 600°C. These values are similar to that of plasma sprayed TBCs [8]. The thermal diffusivity of the coated specimens is plotted in Fig. 4 as a function of temperature. It shows that thermal diffusivity of the specimens decrease with increasing the temperature. Fig. 5 shows the thermal conductivity of the coated specimens as a function of temperature. However, thermal conductivity shows a little decrease as the temperature increase.

The reason of smooth decrease of thermal conductivity was attributed to the results of competitive compensation between decrease of thermal diffusivity and increase of specific heat with increasing the temperature. Thermal conductivity of plasma sprayed coatings [9] is significantly lower than that of EB-PVD coatings at given temperature in the Fig. 5. This reason is the difference of pores morphology as well as the microstructure, that is, EB-PVD coatings and plasma sprayed coatings are columnar and splat lamellar microstructure, respectively [10].

We considered the two reasons for the detailed explanation on the decrease of thermal conductivity. As first reason, lower thermal conductivity of plasma sprayed coatings in comparison with EB-PVD coatings is caused by the high porosity of plasma sprayed coatings, resulting in the enhancement of phonon scattering at pores. As second reason, the splat-boundaries between splat lamellar grains for plasma sprayed microstructure are thought to be contribution for increase of phonon scattering because they can intersect the heat flow path. In addition, the flat-pancake shaped pores perpendicular to the temperature gradient in plasma sprayed coatings, lead to the greater decrease in thermal conductivity than that for EB-PVD coatings.

Fig. 6 shows SEM micrographs of columnar gap and the elongated pores of the coated specimen. Gap type-pores between columnar grains are observed which extend from the substrate to the coating surface, as shown in Fig.6 (a). The gap type-pores evolve with the aligned columnar structure.

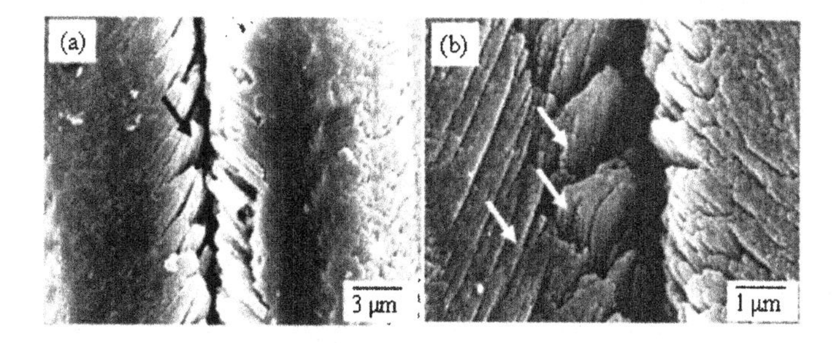


Fig. 6. SEM micrographs of microstructure of EB-PVD coatings of 4mol%$Y_2O_3$-$ZrO_2$: (a) columnar gap and (b) elongated nano pores at feather-like grains.

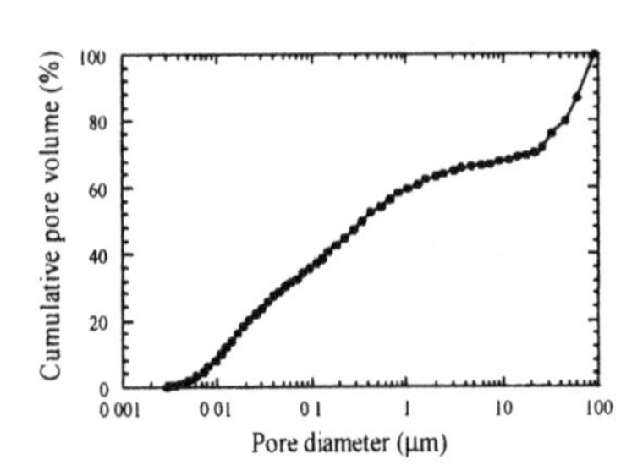


Fig. 7. Pore size distribution for EB-PVD coating layer by mercury porosimeter.

On both sides of the columnar grains, a pronounced dendrite structure, including many micro-pores as well as the elongated nano pores was formed, as shown in Fig.6 (b). The quantitative analysis of pore distribution is necessary to estimate the influence of porosity on thermal conductivity in coating layers. Fig. 7 indicates the result of pore distribution for coating layer with 200 μm coating thickness. According to pore distribution analysis, the volume portion of the nano pores (for consideration of size < 100 nm) on total porosity is approximately 38% and an average pore size is about 300 nm.

It seems that such nano pores exist mainly around feather-like columns as well as intracolumnar pores at inside of columns as shown in Fig. (6). The nano pores can contribute to phonon scattering. The effect of porosity on thermal conductivity can be explained as follows. The thermal conductivity can be usually reduced by decreasing the mean free path due to the phonon scattering. The phenomenon of phonon scattering occurs when phonons interact with lattice defects. Such defects include vacancies, dislocations, pores, grain boundaries, and atoms of different masses. The total mean free path by phonon scattering can therefore be reduced by introduction of pores as lattice imperfections. Thus, the present samples with higher porosity above 20% exhibit lower thermal conductivity. In particular, the elongated nano pores (Fig. 6) at both side of dendrite column are thought to be contribution for increase of phonon scattering because they can intersect the heat flow path due to the inclined arrangement of the elongated nano pores.

This can be explained from the theory of thermal conductivity by phonons where the thermal conductivity is limited by intrinsic phonon scattering. The thermal conductivity ($K_p$) due to lattice vibration can be described by following expression:

$$K_p = \frac{1}{3}\int_0^{\omega_D} C(\omega)v(\omega)l(\omega)d\omega \qquad (2)$$

where $C(\omega)d\omega$ represents the contribution to the lattice specific heat in the frequency range $d\omega$ at $\omega$, $v(\omega)$ is the sound velocity and $l(\omega)$ is the phonon mean-free path. The integration is performed between zero and $\omega_D$ (the latter of which is the Debye frequency). From equation (2), if the phonon mean-free path by increase of phonon scattering at pores can be decrease, it can be concluded that thermal conductivity is decreased.

## CONCLUSIONS

Thermal properties of 4 mol% $Y_2O_3$-stabilized zirconia coatings fabricated by EB-PVD have been investigated in the temperature range between 25°C to 1000°C. The EB-PVD 4mol%$Y_2O_3$-$ZrO_2$ coatings had a nanoporous columnar microstructure with gaps between columnar grains. The thermal diffusivity and thermal conductivity of EB-PVD coatings showed the decreasing tendency with increasing temperature. The thermal conductivity of EB-PVD coatings was found to be higher than those of plasma sprayed coatings. It could be concluded that a presence of pores in coating layers mainly leads to reduced thermal conductivity by the reduction of mean free path.

## ACKNOWEDGMENTS

The authors acknowledge the financial support of the New Energy and Industrial Technology Development Organization (NEDO), Japan.

## REFERENCES

[1]J. Singh, D.E. Wolfe and J. Singh, "Architecture of thermal barrier coatings produced by electron beam-physical vapor deposition (EB-PVD)," *J.Mater.Sci.*, **37**, 3261-3267 (2002).

[2]C.G.Levi, " Emerging materials and processes for thermal barrier systems," *Current Opinion in Solid State and Mat. Sci.*, **8**, 77-91 (2004).

[3]D.D. HASS, P.A.PARRISH and H.N.G. WADLEY, "Electron Beam Directed Vapor Deposition of Thermal Barrier Coatings," *J.Vac. Sci. Technol.*, **16**, 3396-3401 (1998)

[4]B.K.Jang and H. Matsubara, "Influence of Rotation Speed on Microstructure and Thermal Conductivity of Nano-Porous Zirconia Layers Fabricated by EB-PVD" *Scripta Mater.*, **52**, 553-558 (2005).

[5]J. R. Nicholls, K. J. Lawson, A. Johnstone and D. S. Rickerby, "Methods to reduce the thermal conductivity of EB-PVD TBCs," *Surface and Coatings Technol.*, **151-152**, 383-391 (2002).

[6]U. Schulz, "Phase Transformation in EB-PVD Yttria Partially Stabilized Zirconia Thermal Barrier Coatings during Annealing," *J. Am. Ceram. Soc.*, **83**, 904-910 (2000).

[7]B.K.Jang, M. Yoshiya and H. Matsubara, "Influence of Number of Layers on Thermal Properties of Nano-Pore Dispersed Zirconia Coating Layer Fabricated by EB-PVD" *J. Jpn. Inst. Metals.* **69,** 101-105(2005)

[8]B. Alzyab, C.H. Perry and R.P.Ingel, "High-Pressure Phase Transitions in Zirconia and Yttria-Doped Zirconia," *J. Am. Ceram. Soc.*, **70**, 760-765 (1987).

[9]R.B. Dinwiddie, S.C. Beecher, W.D. Poter, B.A. Nagaraj, ASME-96-GT-282.

[10]P.S. Anderson, X.Wang, P. Xiao, "Effect of Isothermal Heat Treatment on Plasma-Sprayed Yttria-Stabilized Zirconia Studied by Impedance Spectroscopy," *J. Am. Ceram. Soc.* **88**, 324-330 (2005).

# OXIDATION BEHAVIOR AND MAIN CAUSES FOR ACCELERATED OXIDATION IN PLASMA SPRAYED THERMAL BARRIER COATINGS

Hideyuki Arikawa, Yoshitaka Kojima
Hitachi Research Laboratory, Hitachi, Ltd
Hitachi, 317-8511, Japan

Mitsutoshi Okada, Takayuki Yoshioka and Tohru Hisamatsu
Materials Science Research Laboratory, Central Research Institute of Electric Power Industry
Yokosuka, 240-0196, Japan

## ABSTRACT

The growth of thermally grown oxide (TGO) can influence the durability of thermal barrier coatings (TBCs). In order to clarify its main causes and to find a method to restrain it, we examined an influence of surface treatment on a bond coat and its yttrium (Y) content on TGO growth. Shot peening was applied on the bond coat surface of TBC specimens, and high temperature oxidation behavior in air was compared with that of the TBC specimens without shot peening. As a result, in the specimen without shot peening, wart-like oxide was observed on a continuous oxide layer, which was formed between the bond coat and top coat. The wart-like oxide was caused due to unmelted particles of the bond coat. And in the continuous oxide layer, Y-rich granular microstructure was found. On the other hand, by means of shot peening on the bond coat surface, unmelted particles were removed, and the formation of the wart-like oxide was restrained. The growth of the continuous oxide layer, moreover, was also decreased because the Y-rich microstructure in the oxide was reduced. And the TBC specimen with the bond coat eliminating Y-content (Y-free bond coat) was prepared, and its oxidation behavior was examined by the high-temperature oxidation test in air. In the TBC specimen with Y-free bond coat, whether or not shot peening was applied, the growth of the continuous oxide was restrained since Y did not exist in the oxide. The possibility that Y-free bond coat could restrain TGO growth was found.

## INTRODUCTION

Turbine inlet temperatures (TIT) of gas turbines have been increased in order to improve thermal efficiency of combined cycles using them, and the temperatures have reached around 1500°C recently. In order to withstand such critical thermal conditions, Co-base or Ni-base superalloys with excellent high-temperature strength are employed in hot-gas-path parts, and, moreover, thermal barrier coatings (TBCs) are applied for better heat resistance.

In general, TBC consists of top coat (TC) and bond coat (BC). Yttria-partially stabilized zirconia (YSZ) is used for TC, and MCrAlY metals (M is cobalt and/or nickel), which are also used for corrosion-resistant coatings, are employed for BC[1].

TBC is exposed in high-temperature oxidizing environment during the operation of gas turbines, and the oxide layer at the boundary between TC and BC, which is called thermally grown oxide, TGO, grows due to the oxidation of BC. It is considered that the growth of the oxide layer may accelerate the delamination of TBC[2-6]. Thus, it is one of important issues to restrain the growth of the oxide layer in order to improve the reliability of the hot-gas-path parts and to increase their lives.

It is reported that the oxide layer formed in TBC grows faster than that on MCrAlY coating[2]. It is considered that the oxide growth is influenced by the formation of wart-like oxide consisting of BC compositions such as Ni and Co[2] and/or the segregation of yttrium in the oxide layer[3, 4]. As for the formation of the wart-like oxide, it is reported that they are caused by unmelted particles formed in spraying process, and that surface polish and chromate processing can restrain its growth. On the other hand, as for the segregation of yttrium, its influence on the accelerated oxidation has not been clarified yet. It can be concluded that the cause of the faster growth of the oxide layer in TBC has not been fully elucidated yet.

In this paper, the authors examined the main causes for the growth of the oxide layer at the boundary between TC and BC in TBC in order to find a method to restrain its growth. The oxidation behavior was observed when the unmelted particles were removed from the BC surface and when yttrium was eliminated from the chemical composition of BC. And, in each case, the observation was also performed when top coat was not applied. And then, we examined the effects to remove the unmelted particles and to eliminate yttrium on the restraint of the oxide growth at the boundary.

EXPERIMANTAL PROCEDURE

Figure 1 illustrates schematic representation of specimen shape and coatings. Substrate material was Inconel738LC, and its size and shape was 20mm×70mm×3mm. 100μm thick of CoNiCrAlY bond coat (Co-32.0Ni-21.2Cr-8.0Al-0.5Y (wt %)) was applied on the substrate plates by means of low pressurized plasma spraying, and after that, diffusion heat treatment was carried out. In order to examine the influence of surface treatment on BC, any treatment was not performed on one specimen, and shot peening was applied on the other. And 200μm thick of YSZ top coat ($ZrO_2$-8wt%$Y_2O_3$) was applied on one side of each specimen, and top coat was not

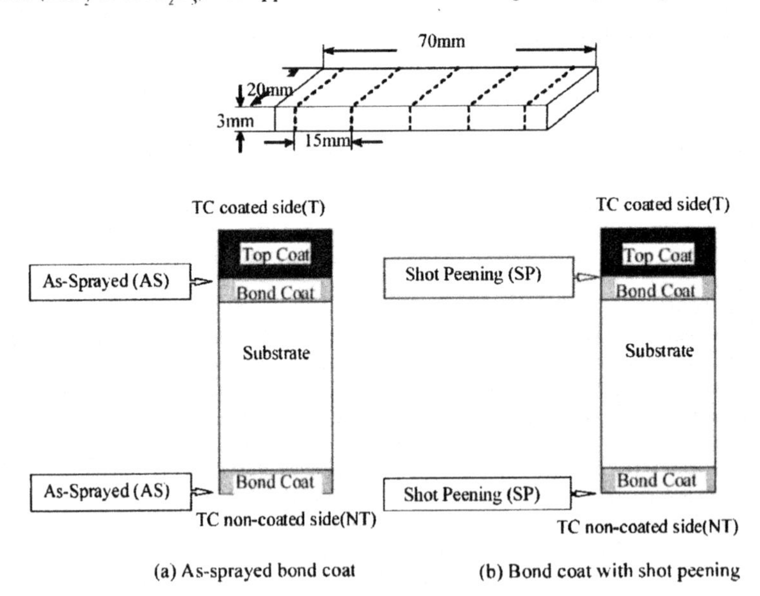


Fig. 1 Schematic representation of specimen shape and coatings

overlaid on the other. The influence of TC, therefore, was examined by comparing the sides with TC and without TC. In this paper, "AS" and "SP" represent as-sprayed BC and BC with shot peening, respectively. And "T" and "NT" represent the side with TC and the one without TC, respectively. For example, the TC coated side of the specimen with as-sprayed BC is described as "AS-T", and the TC non-coated side of the specimen with shot-peened BC was described as "SP-NT".

The CoNiCrAl metal powder was prepared in order to examine the influence of yttrium in BC on the oxidation behavior. The powder consists of the chemical composition where yttrium was eliminated from CoNiCrAlY metal, and the analysis of the powder showed its chemical composition of Co-31.7Ni-20.7Cr-7.8Al (wt%). The coated specimens were produced in the same condition as above. "YF" indicates that the BC without yttrium (Y-free BC) was applied. According to the analysis by an electron probe micro-analyzer (EPMA), the bond coat did not contain yttrium.

One plate was cut into four specimens after coatings were applied, and then they were used for oxidation tests.

The oxidation tests were carried out in air at 950°C. A test time was defined as the one held at the test temperature. The specimens were taken out from a furnace after certain oxidation time. And they were cut and polished, and then its microstructure in the cross section was analyzed. For the microstructural analysis, scanning electronic microscopy (SEM) and electron probe micro-analyzer (EPMA) were used.

The heat cycle test was conducted in air by means of an electric furnace. One heat cycle consists of a rapid heat-up from room temperature to 1100°C for 15minutes, a 10hours-hold at 1100°C, and a 15 minutes cooling to room temperature by a fan. TBC failure was defined as the cycle when 20% of whole area of top coat was delaminated, and then, the thermal cycling was stopped.

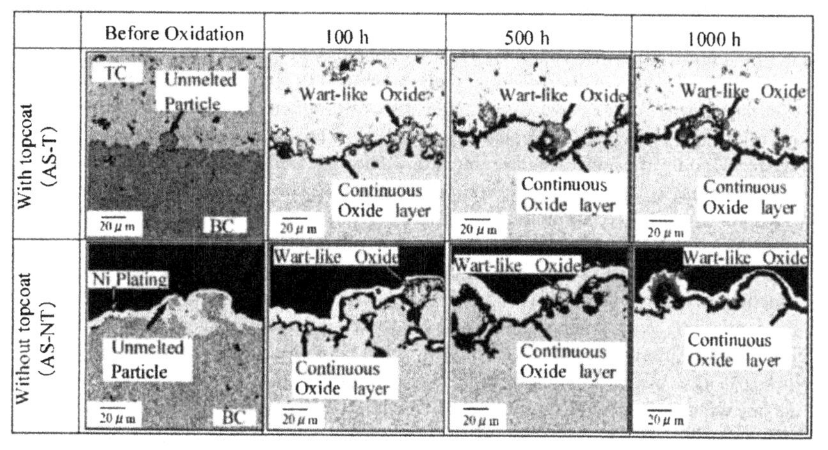


Fig. 2 Cross-sectional morphologies of the specimens with as-sprayed bondcoat after oxidation test at 950°C in air

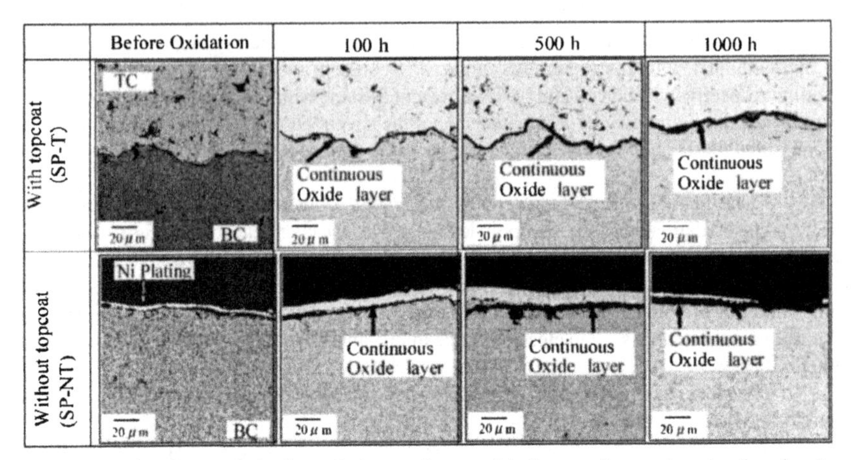


Fig. 3 Cross-sectional morphologies of the specimens with shot peening treatment on bondcoat surface after oxidation test at 950°C in air

RESULTS AND DISCUSSION

Influence of surface treatment for bond coat on oxide morphology

Figure 2 indicates cross-sectional morphologies of the specimens with as-sprayed bond coat. The upper side and the bottom side in the figure show the specimen with TC (AS-T) and the one without TC (AS-NT), respectively. In the specimen without TC, Ni-plating was performed so as not to damage the oxide layer in the process of cutting and grinding. In the specimen with TC, an oxide layer was observed along the BC/TC boundary, and it is defined as continuous oxide layer. The continuous oxide layer consisted mainly of alumina according to EPMA analysis. On the continuous oxide layer, moreover, wart-like oxides were formed locally. It has been reported that the wart-like oxide was found particularly in TBC. It has been also pointed out that the oxide results from unmelted particles left on the BC/TC boundary in spraying process. The unmelted particles were found also in our specimen with as-sprayed BC before the test. According to EPMA analysis, the wart-like oxide consisted of the oxides of aluminum and chromium in the external scale of the particle and of the metals of cobalt and nickel in its center after 100hours of the oxidation test. After 500hours, the metals in the center almost disappeared and the whole particle changed into the mixed oxide of aluminum, chromium, cobalt and nickel. The above analysis ensures that the wart-like oxide results from the unmelted particle of BC formed in coating spraying. Its formation process can be described as follows. At first, the unmelted particle, which was not contacted with BC, was oxidized, and mainly alumina was formed. And then, since aluminum in the particle was consumed for oxidation, the mixed oxide of cobalt and nickel was formed. In the specimen without TC, the continuous oxide layer and wart-like oxide were also observed on as-sprayed BC surface.

Figure 3 indicates cross-sectional morphologies of the specimens with shot peening on BC surface. The upper side and the bottom side in the figure show the specimen with TC (SP-T) and the one without TC (SP-NT), respectively. In the specimen without TC, Ni-plating was performed as well as in Figure 2. The BC surface treatment restrained the formation of the wart-like oxide with TC as well as without TC, and only the continuous oxide layer was formed. Even after 1000hours of the oxidation test, the wart-like oxide was not observed.

Whether or not TC was applied, the wart-like oxide was formed in the specimen with as-sprayed BC, and it was restrained in the specimen with BC surface treatment. That is because the BC surface treatment removes the unmelted particle that is the cause of the wart-like oxide.

Influence of top coat on the thickness of continuous oxide layer

In order to examine the influence of TC on the oxidation behavior, the continuous oxide layer and the wart-like oxide were distinguished, and only the thickness of the former was considered. That is because the formation mechanism of the wart-like oxide is different from that of the continuous oxide layer.

The thickness of the continuous oxide layer is defined as follows. The area of the continuous oxide layer was measured from the cross-sectional micrograph. And, dividing the area by the length of the BC/TC boundary, the thickness was obtained. The average of 10 fields of view with a magnitude of 500times was used as a thickness value of certain test condition. The average square root of the unbiased variance of the thickness was about 0.4μm in all the specimens. The oxide was observed in the specimen even before the oxidation test. It was formed during the thermal treatment after spraying process. Its thickness, however, was too small to be measured by the above method.

Figure 4 indicates the relationship between the thickness of continuous oxide layer and time. Close and open symbols represent the specimens with TC and those without TC, respectively. As shown in the figure, when the surface condition was same, significant difference in the thickness of the continuous layer was not found between the specimens with TC and without TC. This result, therefore, reveals that the chemical reaction of BC with TC and the change of the partial pressure of oxygen due to overlaid TC do not have an influence on the thickness growth of the continuous oxide layer to the extent of these oxidation tests.

Influence of surface treatment for bond coat on the growth of continuous oxide layer

The influence of the surface treatment for BC on the thickness of the continuous oxide

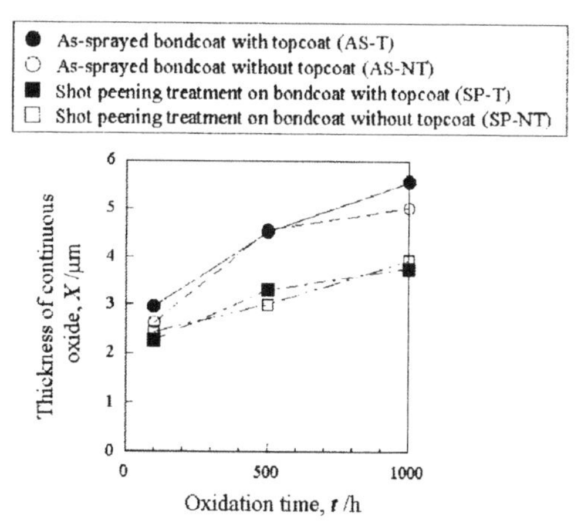


Fig. 4 Relationship between thickness of continuous oxide layer and time in oxidation tests at 950°C in air

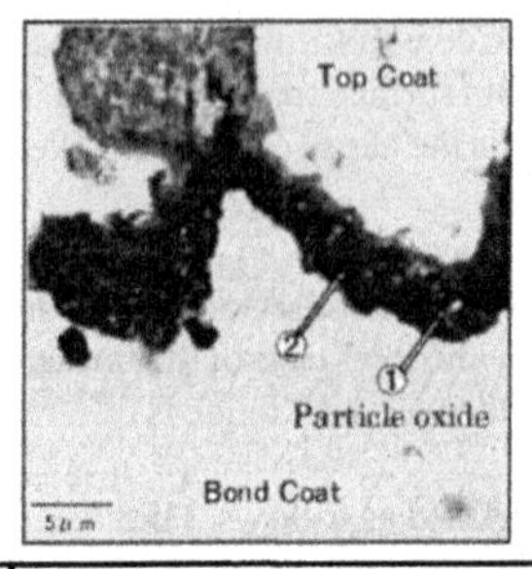


| | Concentrations (mass%) | | | | | | |
|---|---|---|---|---|---|---|---|
| | Al | O | Y | Co | Ni | Cr | Zr |
| ① | 43.8 | 45.5 | 9.1 | 0.6 | 0.6 | 0.4 | 0.01 |
| ② | 47.5 | 46.0 | 1.0 | 2.0 | 2.1 | 1.3 | 0.04 |

(a) Cross-sectional micrograph and results of quantitative analysis by EPMA

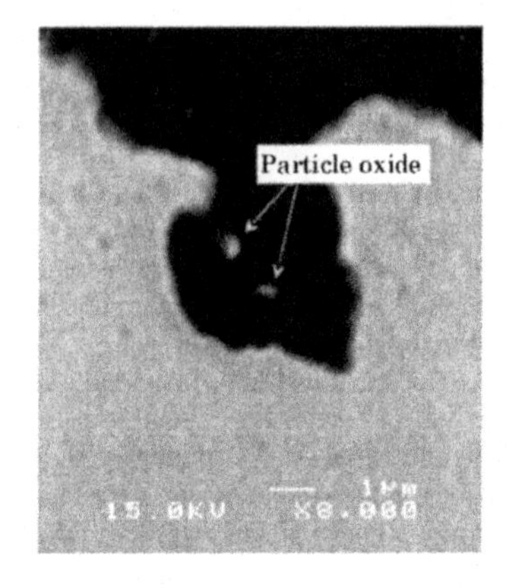


(b) Morphology of the boundary between continuous oxide layer and bond coat

Fig. 5 Analytical results of the continuous oxide layer in TBC specimen with as-sprayed bond coat (AS-T) after oxidation test at 950°C for 1000h in air

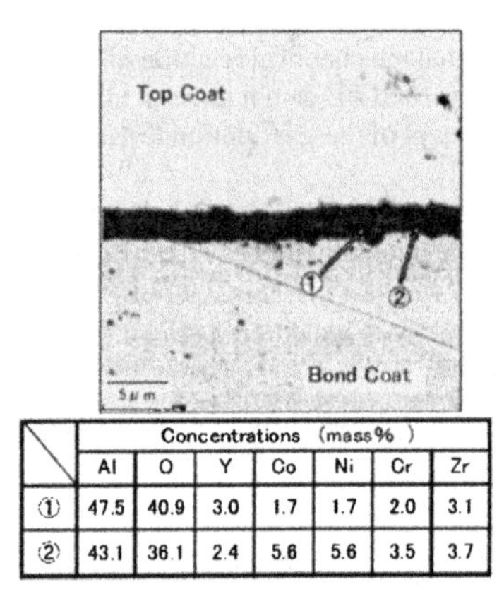


| | Concentrations (mass%) | | | | | | |
|---|---|---|---|---|---|---|---|
| | Al | O | Y | Co | Ni | Cr | Zr |
| ① | 47.5 | 40.9 | 3.0 | 1.7 | 1.7 | 2.0 | 3.1 |
| ② | 43.1 | 36.1 | 2.4 | 5.6 | 5.6 | 3.5 | 3.7 |

Fig. 6 Analytical results of the continuous oxide layer in TBC specimen with shot peening treatment on bond coat surface (SP-T) after oxidation test at 950°C for 1000h in air

layer was also shown in Figure 4. Whether or not TC is applied, the thickness of the continuous oxide layer in the specimen with the BC surface treatment was thinner than that in the specimen with as-sprayed BC. This result clarifies that the BC surface treatment not only restrains the formation of the wart-like oxide as described above, but also reduces the thickness of the continuous oxide layer.

The continuous oxide layer was observed in more detail and analyzed by EPMA in order to examine the cause of the difference in the thickness of the continuous oxide layer due to the BC surface treatment. Figures 5 and 6 show analytical results of the continuous oxide layer in the specimen with as-sprayed BC and of that with the BC surface treatment, respectively. Both specimens were oxidized at 950°C for 1000h in air.

White granular microstructure was scattered inside the continuous oxide layer in the specimen with as-sprayed BC (AS-T) as shown in Figure 5. The chemical compositions were analyzed by EPMA as for the granular microstructure ( ① in Figure 5 (a)) and the other part in the continuous oxide layer ( ② in Figure 5(a)). And the microstructure of ① was identified as yttrium-rich oxide. As shown in Figure 5(b), in some local parts, the continuous oxide layer surrounded the granular microstructure and the layer grew inside BC.

On the other hand, the white granular microstructure was hardly observed in the specimen with the BC surface treatment (SP-T) as shown in Figure 6. EPMA analysis could not find the Y-rich microstructure indicated in Figure 5.

Zirconium was detected in the EPMA analysis of figures 5 and 6. The authors estimate that this results from the chemical composition of TC.

The BC surface treatment causes the difference in the microstructure of the continuous oxide layer, and this leads to the difference in its thickness. It is suggested that yttrium-rich granular oxide is the origin of the oxidation and that it results in the growth of the boundary oxide layer since there is a possibility that the diffusion rate is comparatively high in the yttrium-rich oxide or in the boundary between the Y-rich oxide and alumina[5,7]. Our research also ensures that a number of Y-rich granular oxides were included in the continuous oxide layer of the specimen with as-sprayed BC and that this granular oxide was the origin of oxidation. The acceleration mechanism of oxidation due to yttrium such as how yttrium is distributed in bond coat and how it forms granular oxide, however, has not been clarified yet, and it is necessary to be investigated more in detail. On the other hand, in the specimen with the BC surface treatment, the formation of alumina with smaller amount of impurities such as Y-rich oxide restrains the growth of the continuous oxide layer. As for its cause, it is pointed out that the surface treatment produces lattice defects at the vicinity of surface and that it accelerates the oxidation of aluminum and the growth of protective alumina[6], but it is necessary to be examined further.

Influence of Yttrium in bond coat on the growth of continuous oxide layer

Figure 7 indicates cross-sectional morphologies of TBC specimens with Y-free BC after

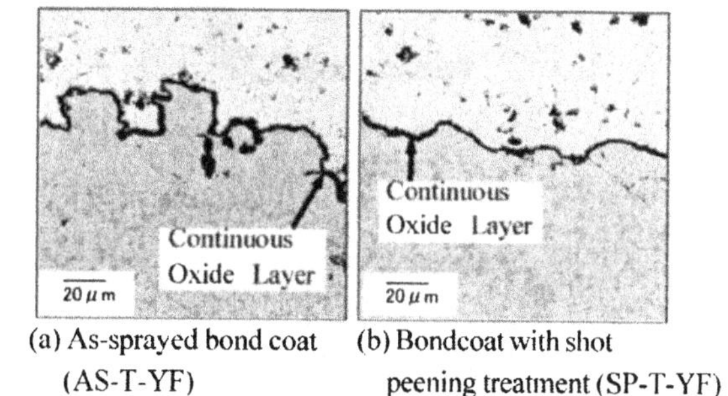


(a) As-sprayed bond coat (AS-T-YF) (b) Bondcoat with shot peening treatment (SP-T-YF)

Fig. 7 Cross-sectional morphologies of TBC specimens with Y-free bondcoat after oxidation test at 950°C for 1000h in air

oxidation test at 950°C for 1000h in air. Figures 7(a) and (b) show the TBC specimen with as-sprayed BC (AS-T-YF) and the one with BC surface treatment (SP-T-YF), respectively. In the Y-free BC specimen with the surface treatment (SP-T-YF), only the continuous oxide layer of alumina was formed along the boundary of BC/TC as well as in the Y-included BC (CoNiCrAlY coated) specimen with the surface treatment (SP-T) shown in Figure 3. In the specimen with as-sprayed BC (AS-T-YF), even though some parts were estimated to be unmelted particles, the continuous oxide layer was formed along the rough boundary, and the wart-like oxide including cobalt or nickel was not observed.

The relationship between the thickness of continuous oxide layer and time in TBC specimens with Y-free BC was indicated in Figure 8. In the specimens with Y-free BC (AS-T-YF

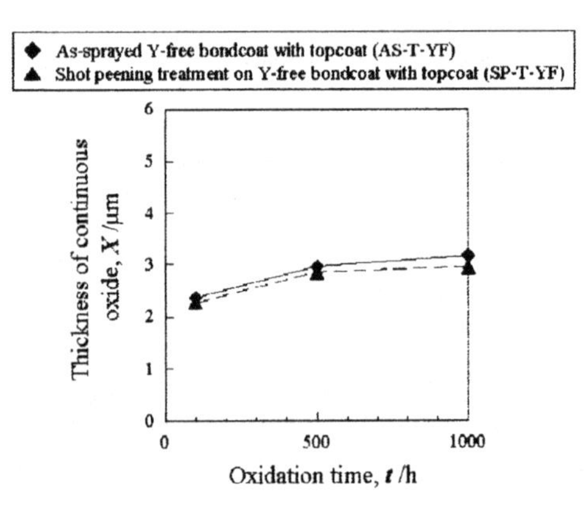


Fig. 8 Relationship between thickness of continuous oxide layer and time in TBC specimens with Y-free bondcoat in oxidation test at 950°C in air

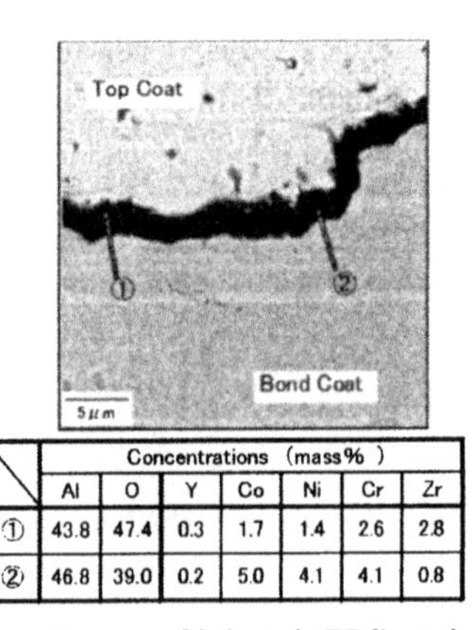


| | Concentrations (mass%) | | | | | | |
|---|---|---|---|---|---|---|---|
| | Al | O | Y | Co | Ni | Cr | Zr |
| ① | 43.8 | 47.4 | 0.3 | 1.7 | 1.4 | 2.6 | 2.8 |
| ② | 46.8 | 39.0 | 0.2 | 5.0 | 4.1 | 4.1 | 0.8 |

Fig. 9 Analytical results of the continuous oxide layer in TBC specimen with as-sprayed Y-free bond coat (AS-T-YF) after oxidation test at 950°C in air

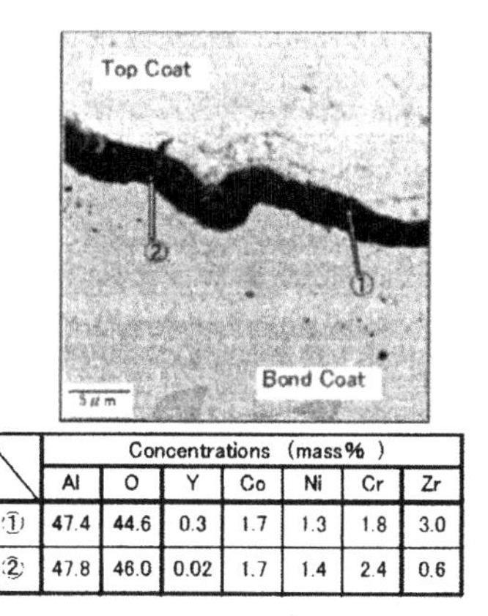


| | Concentrations (mass%) | | | | | | |
|---|---|---|---|---|---|---|---|
| | Al | O | Y | Co | Ni | Cr | Zr |
| ① | 47.4 | 44.6 | 0.3 | 1.7 | 1.3 | 1.8 | 3.0 |
| ② | 47.8 | 46.0 | 0.02 | 1.7 | 1.4 | 2.4 | 0.6 |

Fig. 10 Analytical results of the continuous oxide layer in TBC specimen with Y-free bond coat and shot peening treatment (SP-T-YF) after oxidation test at 950°C in air

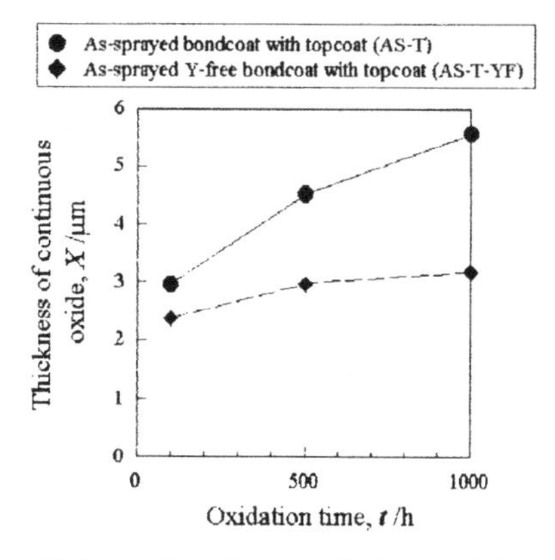


Fig.11 Relationship between thickness of continuous oxide layer and time in TBC specimens with CoNiCrAlY bondcoat and Y-free CoNiCrAl bondcoat in oxidation test at 950°C in air. Bond coat is as-sprayed.

and SP-T-YF), whether or not the BC surface treatment was carried out, the thicknesses of the continuous oxide layer were almost same. Figures 9 and 10 show analytical results of the continuous oxide layer in the specimen with as-sprayed Y-free BC and in the one with the surface treatment on Y-free BC, respectively. Whether or not the BC surface treatment was applied, granular microstructure was not included in the continuous oxide layer as observed in Figure 5(b). In both specimens, the shape of the boundary between the continuous oxide layer and BC was relatively flat, and the local growth of the oxide into BC shown in Figure 5(b) was not found. According to EPMA analysis on the specimens, moreover, the continuous oxide layer was alumina, and Y-rich microstructure was not found. Subtle content of yttrium was detected in the continuous oxide layer in Figures 9 and 10. That is because the thickness of the oxide layer was so small that the electronic beam of EPMA detected the yttrium of top coat or because yttrium might diffuse from TC. The EPMA analysis also comfirmed that Y-free BC did not contain yttrium even after the oxidation test.

Figure 11 shows the relationship between the thicknesses of the continuous oxide layer and time in the specimens with as-sprayed Y-included BC (AS-T) and with as-sprayed Y-free BC (AS-T-YF). In the specimen with Y-included BC, since the continuous oxide layer including yttrium was formed as described in the previous paragraph, thicker oxide layer grew. On the other hand, in the specimen with since Y-free BC, alumina with relatively smaller amount of impurities was formed, and its growth rate was smaller.

Figure 12 indicates the relationship between the thicknesses of continuous oxide layer and time in the specimens with Y-included bond coat and (SP-T) and with Y-free one (SP-T-YF) where the BC surface treatment was performed. Since the continuous oxide layer with smaller content of yttrium was formed in the specimens with the BC surface treatment, its thickness was almost same in the specimen with Y-included BC as in the specimen with Y-free BC. In the long-term oxidation test, the thickness was thinner in the specimen with Y-free BC than in the one with Y-included BC.

In the specimen with Y-free BC, whether or not the BC surface treatment was applied, yttrium was not included in the continuous oxide layer and its thickness was thinner. This clarifies that yttrium accelerates the growth of the continuous oxide layer. Since yttrium was not included by eliminating it from the chemical composition of BC, it was supplied from BC and the yttrium in TC does not have an influence on the oxidation.

Figure 13 summarizes the factors of the accelerated oxide growth in TBC. The oxidation at the boundary of TBC is accelerated by following factors; ① the formation of the wart-like oxide, and ② yttrium in bond coat. The formation of the wart-like oxide results from unmelted particles on BC surface, and the surface treatment can remove them. But, since the rough boundary is indispensable to adhere BC and TC, it is necessary to develop other surface treatment that not only

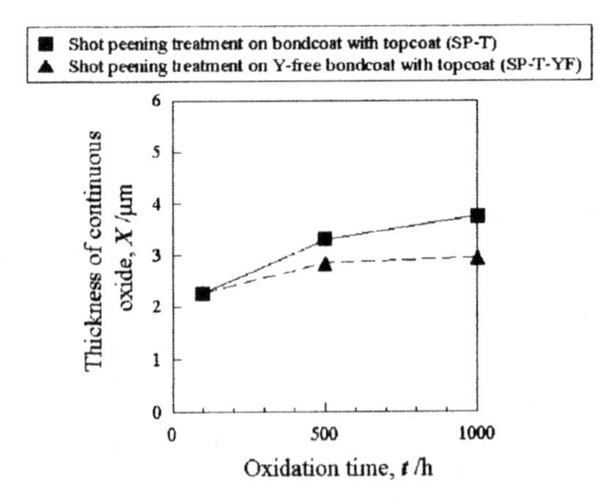


Fig. 12 Relationship between thickness of continuous oxide layer and time in TBC specimens with CoNiCrAlY bond coat and Y-free CoNiCrAl bond coat in oxidation test at 950°C in air. Shot peening was carried out on bond coat surface.

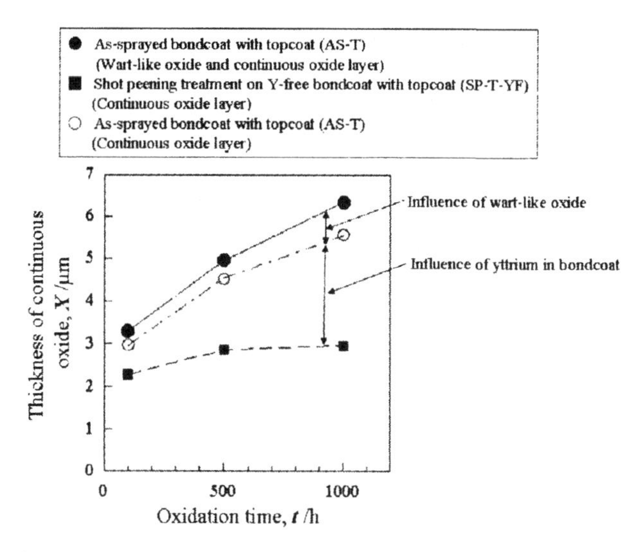


Figure 13 Factors of oxide growth acceleration in TBC

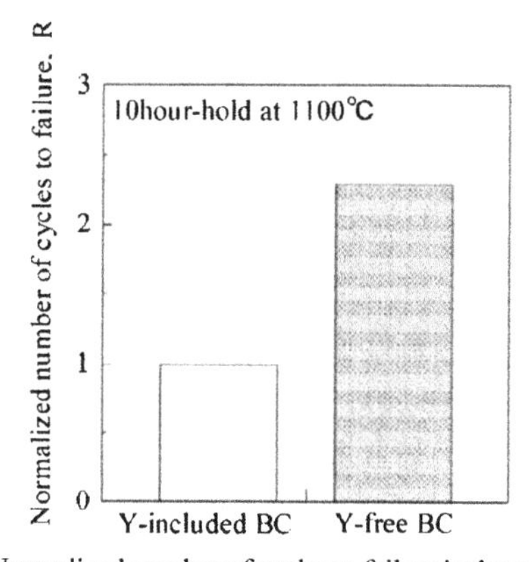


Figure 14 Normalized number of cycles to failure in thermal cycling test

can remove the unmelted particles and but can maintain the rough boundary.

On the other hand, yttrium in bond coat is included in the boundary oxide, and it accelerates its growth. The surface treatment and/or BC without yttrium can restrain the growth of the continuous oxide layer. Yttrium in CoNiCrAlY coating is added in order to improve the adhesion between the protective oxide layer of alumina and the coating[8]. It is, therefore, necessary to examine the influence of yttrium on the resistance to delamination.

Resistance to thermal cycling of TBC with Y-free bond coat

The thermal cycling test was performed by means of the specimens with Y-included BC and Y-free BC. Figure 14 shows the normalized number of cycles to failure of each specimen. The normalized number of heat cycles to failure of the specimen with Y-free BC was about twice larger. These results reveal that the CoNiCrAl (Y-free) bond coat is effective in order to improve the durability of TBC. One of the reasons is that it restrains the growth of continuous oxide layer, but the further analysis is necessary.

## SUMMARY

The TBC specimen with shot-peening treatment on bond coat surface was prepared in order to clarify the influence of the surface treatment on the growth of the oxide layer at the boundary between bond coat (BC) and top coat (TC) in thermal barrier coating (TBC). Its high-temperature oxidation in air was compared with that of the specimen with as-sprayed BC. The high-temperature oxidation test, moreover, was carried out by means of the specimen with Y-free BC, and the influence of yttrium on the oxide layer growth was also examined. The results are as follows.

(1) The continuous oxide layer was formed at the BC/TC boundary of the specimen with as-sprayed BC, and the wart-like oxide, which resulted from unmelted particle of BC, was observed. In the continuous oxide layer, Y-rich granular microstructure was found.

(2) The shot-peening treatment on BC surface removed the unmelted particles, and the formation of the wart-like oxide was restrained. The content of yttrium, moreover, was decreased in the continuous oxide layer, and its growth was also restrained.

(3) In the specimen with Y-free BC, yttrium did not exist in the continuous oxide layer. And, whether or not the surface treatment was applied, its growth was restrained. There is a possibility that Y-free BC can restrain the growth of the continuous oxide layer.

## REFERENCES

[1]H. Arikawa and Y. Kojima, "Heat Resistant Coatings for Gas Turbine Materials (in Japanese)," J. the Surface Finishing Soc. Japan, **52**, 11-15 (2001).

[2]M.Narumi, Z.Yu, K.Taumi and T.Narita, "Oxidation Behavior of Plasma Sprayed CoNiCrAlY/YSZ Film at 1173K and 1273K in Air (in Japanese)," Zairyo-to-Kankyo, **50**, 466-471 (2001).

[3]A.Rabiei and A.G.Evans, "Failure Mechanism Associated with the Thermally Grown Oxide in Plasma-sprayed Thermal Barrier Coatings," Acta mater., **48**, 3963-3976 (2000).

[4]S.Takahashi, M.Yoshiba and Y.Harada, "Nano-Characterization of Ceramic Top-Coat/ Metallic Bond-Coat Interface for Thermal Barrier Coating Systems by Plasma Spraying," Materials Transactions, **44**, 1181-1189 (2003).

[5]L.Lelait, S.Alpérine and R.Mévre, "Alumina scale growth at zirconia-MCrAlY interface: a microstructural study," Journal of Materials Science, **27**, 5-12 (1992).

[6]T.Teratani, T.Suidzu, K. Tani and Y.Harada, "Formation of Alumina Protective Layer on MCrAlY Atmospheric Plasma Sprayed Coating by Chromate Processing (in Japanese)," J. High Temperature Soc. **29**, 247-252 (2003).

[7]J. Klöwer, "Factors affecting the oxidation behaviour of thin Fe-Cr-Al foils Part • •The effect of alloying elements: Overdoping," Materials and Corrosion. **51**, 373-385 (2000).

[8]K. Shimotori and T. Aisaka, "The Trend of MCrAlX Alloys for High-temperature-Protective Coatings –On the Effects of Alloy Compositions- (in Japanese)," Tetstu-to-Hagane, **69**, 1229-1241 (1983).

# CRACK GROWTH AND DELAMINATION OF AIR PLASMA-SPRAYED $Y_2O_3$-$ZrO_2$ TBC AFTER FORMATION OF TGO LAYER

Makoto Hasegawa[1], Yu-Fu Liu[1] and Yutaka Kagawa[1,2]
[1]Research Center for Advanced Science and Technology, The University of Tokyo,

4-6-1 Komaba, Meguro-ku,
Tokyo, 153-8904, Japan

[2]Center for Collaborative Research, The University of Tokyo,
4-6-1 Komaba, Meguro-ku,
Tokyo, 153-8904, Japan

## ABSTRACT

Failure behavior in an air plasma-sprayed $Y_2O_3$-$ZrO_2$ thermal barrier coating systems (TBCs) under tensile loading has been examined. The coating system is heat exposed in ambient air at 1423 K for 10, 50 and 100 h. Tensile testing of the system is conducted to observe cracking behavior in the TBC layer. Periodic cracking in the TBC layer, which is located perpendicular to loading direction, is observed after tensile loading. Density of the cracks increases with the increase in heat exposure time.

## INTRODUCTION

Thermal barrier coating systems (TBCs) are widely used to protect the hot section components in gas turbine blades and vanes from high temperature [1, 2]. They usually consist of outer ceramic TBC layer and inner intermetallic bond coat (BC) layer for protecting the nickel-base superalloy substrate from the high temperature and from the oxidation. Failure of TBCs certainly shortens the service life of components and hence understanding of the failure behavior of the TBCs is important for improving the performance of components.

Failure of TBCs under unexpected over-load under service leads damage in the TBC systems. Therefore, it is important to understand damage of TBC systems under tensile force. Studies on the damage under a tensile load have been reported [3, 4]. However, the reports do not reveal the effect of heat exposure on the failure behavior. The properties of constitutes and entire system of a TBC system strongly depends on a history of heat exposure, as reported elsewhere [5-8]. However, the cracking behavior of the TBC layer under in-plane tensile force is not well known. The present study has been focused on the experimental observation of the failure behavior in TBC layer under tensile loading.

## EXPERIMENTAL PROCEDURE

Thermal barrier coating layer (TBC) of an $Y_2O_3$ (7~8 mass%)-partially-stabilized-$ZrO_2$ was deposited on bond coat using an atmospheric plasma spray process. Bond coat (BC) material was a NiCoCrAlY alloy (Co22 Cr17 Al12.5 Y0.6 and balance Ni in mass%) that was coated on a nickel-base superalloy substrate (Inconel 738, Co8.5 Cr16 Al3.5 Ti3.5 W2.6 Mo1.8 Nb0.9 and balance Ni in mass%) by a low-pressure plasma spraying process. The thickness of TBC and BC layers, respectively, was ~200 and ~100 μm. The coated materials were heat exposed in ambient air at 1423 K for 10, 50 and 100 h. Microstructural characterizations on as-sprayed and heat-exposed TBC systems were performed on polished transverse plane of the TBC system using optical microscope (OM) and scanning electron microscope (SEM).

Specimens for tensile test were prepared from as-sprayed and heat-exposed TBC systems by diamond-saw cutting and polishing processes. **Figure 1** shows testing configurations of the specimen. Specimen size was 40 mm in length, $\ell$, 2 mm in width, w, and 3.4 mm in thickness, t, which included TBC and BC layers. Side surfaces of the specimen were polished up to 0.5 μm diamond paste finish to allow direct observation of the failure behavior.

Tensile tests were performed in ambient air atmosphere at 295 K by using a screw-driven type testing machine. Tests were done with a crosshead speed of 0.1 mm/min. Tensile strain was measured with a strain gage (effective gage area: 2 × 0.84 mm). The strain gage was attached to the center of the substrates bottom. After tests, fracture behavior of the TBC systems was observed by OM and SEM.

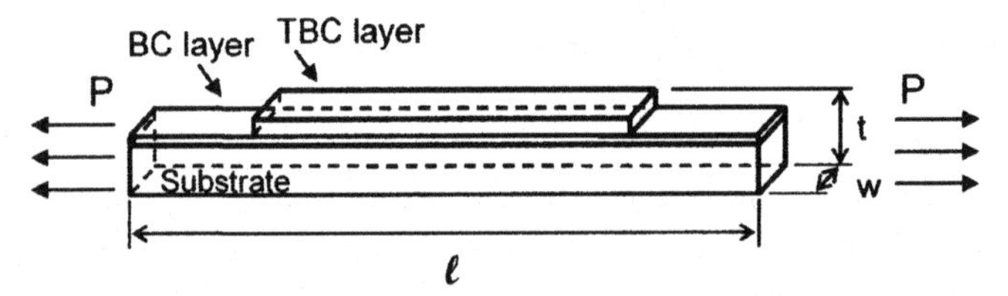


Fig. 1 Shape of tensile test specimen (P: applied load).

## RESULTS AND DISCUSSION

**Figure 2** shows entire view of the polished sections of the thermal barrier coating systems before and after heat exposure. Hereafter, respective layers are designated as thermal barrier coating (TBC) layer, bond coat (BC) layer and thermally grown oxide layer (TGO). Defects such as pores and inter-splat boundaries are observed in the polished cross section of the

TBC layer as black spots and black lines, respectively. Voids and inter-splat boundaries are observed in as-sprayed state (Fig. 2(a)) and the numbers of them seem to decrease with the increase in heat exposure time (Fig. 2 (b) ~ (d)). Such the change in microstructure is caused by

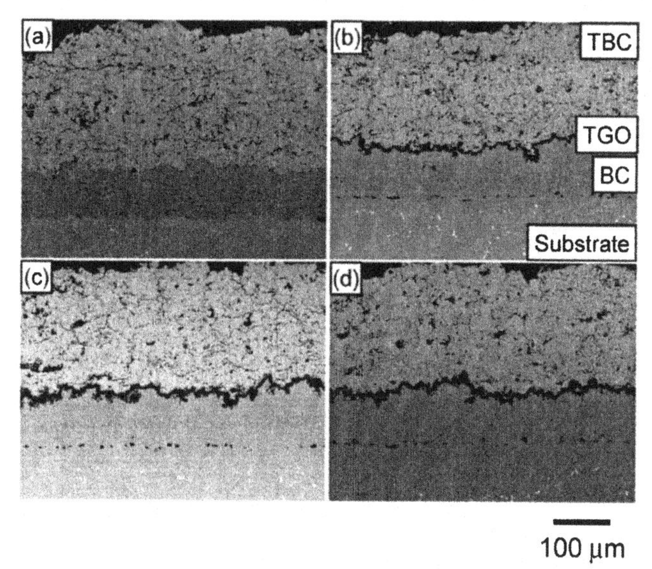


Fig. 2 SEM micrographs of polished transverse section of TBC system: (a) as-sprayed, (b), (c) and (d) after heat exposure for 10, 50 and 100 h, respectively.

shrinkage of TBC layer by progress of sintering. More detail sintering behaviors of the same material are reported elsewhere [5, 6].

The stress-strain curves obtained by tensile test of as-sprayed and heat-exposed TBC specimens are shown in **Fig. 3**. The specimens were loaded at a strain $\varepsilon = 0.01$. For all curves derived from TBC system specimens, starts from a linear regime and then shows nonlinear behavior. This behavior is observed independent of heat exposure time. In the as-sprayed specimen, a maximum tensile stress was ~900 MPa and the maximum stress tends to decrease with the increase in the heat exposure time. The range of the maximum stress in the heat-exposed specimen was 650~620 MPa. Transverse cracking of TBC is observed when the

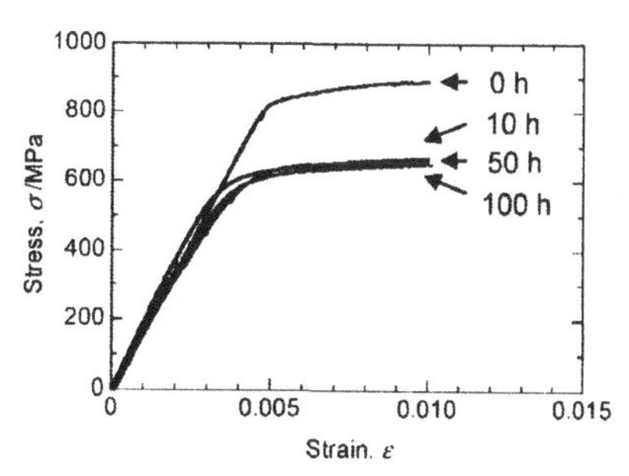


Fig. 3 Tensile stress-strain curves of as-sprayed specimen and heat-exposed specimens with heat exposure time.

stress reaches ~ 600 MPa and the cracks are formed with and equal spacing. **Figure 4** shows typical optical micrographs of the transverse cracks in the as-sprayed and heat-exposed (50 h) specimens. The transverse cracks are formed perpendicular to loading direction with an almost equal spacing. However, the crack spacing depends on heat exposure time. The average spacing of transverse cracks in as-sprayed state is the largest, and the spacing decreases with the increase in heat exposure time. This means that the density of transverse cracks under tensile loading in the present coating system tends to increase after heat exposure.

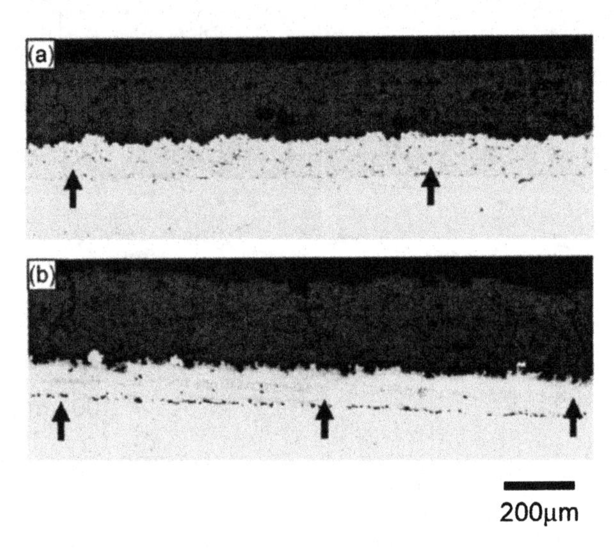


Fig. 4 Optical micrographs showing transverse cracks in (a) as-sprayed specimen and in the specimens subjected to heat exposure for (b) 50 h.

## CONCLUSIONS

Change in transverse cracking behavior in air plasma-sprayed TBC layer under tensile load has been observed. The transverse crack density is highly dependent on heat exposure time of TBC systems. The density increases with the increase in heat exposure time. Full understanding of this behavior needs more detail understanding of the change in constitute properties, however.

Acknowledgement

A part of this study was performed by Nanostructure Coating Project carried out by New Energy and Industrial Technology Development Organization (NEDO).

REFERENCES

[1] S. M. Meier and D. K. Gupta, " The Evolution of Thermal Barrier Coatings in Gas Turbine Engine Applications," *Trans. ASME J. Eng. Gas Turbines Power* **116**, 250-257 (1994).

[2] A. G. Evans, D. R. Mumm, J. W. Hutchinson, G. H. Meier and F. S. Pettit, " Mechanisms Controlling the Durability of Thermal Barrier Coatings," *Prog. Mater. Sci.* **46**, 505-553 (2001).

[3] Y. C. Zhou, T. Tonomori, A. Yoshida, L. Liu, G. Bignall and T. Hashida, " Fracture Characteristics of Thermal Barrier Coatings after Tensile and Bending Test," *Surf. Coat. Technol.* **157**, 118-127 (2002).

[4] L. Qian, S. Zhu, Y. Kagawa and T. Kubo, " Tensile Damage Evolution Behavior in Plasma-Sprayed Thermal Barrier Coating System," *Surf. Coat. Technol.* **173**. 178-184 (2003).

[5] S. Guo and Y. Kagawa, " Young's Moduli of Zirconia Top-Coat and Thermally Grown Oxide in a Plasma-Sprayed Thermal Barrier Coating System," *Script. Mater.* **50**, 1401-1406 (2004).

[6] A. Shinmi, M. Hasegawa, Y. Kagawa, M. Kawamura and T. Suemitsu, " Change in Microstructure of Plasma Sprayed Thermal Barrier Coating by High Temperature Isothermal Heat Exposure," *J. Jap. Inst. Metals* **69**, 67-72 (2005).

[7] M. Tanaka, M. Hasegawa , A.F. Dericioglu and Y. Kagawa, " Measurement of Residual Stress in Air Plasma-Sprayed $Y_2O_3$-$ZrO_2$ Thermal Barrier Coating System using Micro-Raman Spectroscopy," Accepted in *Mater. Sci. Eng. A*

[8] M. Hasegawa and Y. Kagawa, "Microstructural and Mechanical Properties Changes of A NiCoCrAlY Bond Coat with Heat Ecposure Time in Air Plasma-Sprayed $Y_2O_3$-$ZrO_2$ TBC Systems," Submitted in *Inter. J. Appl. Ceram. Technol.*

# LANTHANUM-LITHIUM HEXAALUMINATE – A NEW MATERIAL FOR THERMAL BARRIER COATINGS IN MAGNETOPLUMBITE STRUCTURE – MATERIAL AND PROCESS DEVELOPMENT

Gerhard Pracht, Robert Vaßen and Detlev Stöver
Forschungszentrum Jülich GmbH, Institute for Materials and Processes in Energy Systems 1, Jülich, Germany, D-52425 Jülich

## ABSTRACT

The first study on a new material composition for thermal barrier coating applications is presented. The new material has the chemical composition $LaLiAl_{11}O_{18.5}$ and belongs to the group of hexaaluminate compounds. It is comparable to $LaMgAl_{11}O_{19}$, which has been discussed for TBC application for some years. The preparation and characterization of the composition as well as first plasma spraying experiments will be described. In particular, the thermal expansion mismatch of the ceramic coating on a steel substrate is discussed which leads to strong segmentation cracks of the plasma-sprayed and heat-treated coating. The substitution of lithium in the structure supported the segmentation process of the coating.

## INTRODUCTION

Thermal barrier coatings are widely used in stationary and airborne gas turbines to thermally insulate air-cooled metallic components from the hot gases in the engine. Standard materials for TBC s are based on ceramic materials especially with yttria partially stabilized zirconia (YSZ). These coatings are typically applied by either plasma spraying or physical vapor deposition (EB-PVD).

In order to enhance the efficiency of gas turbines the temperature in the combustion chamber must be increased. However, this is not possible with YSZ coatings because the stability of this material is limited to a temperature of 1200 °C [1]. Additionally, sintering of the ceramic at operation temperature sets a limit to the lifetime of these coatings [2 - 4]. Besides the enhanced efficiency, a prolonged lifetime of the coating material is also desired.

To meet these requirements, new materials are under consideration [5, 6]. One very interesting candidate is $LaMgAl_{11}O_{19}$ – a hexaaluminate, which crystallizes in the magnetoplumbite structure [7]. Gadow et al. published first investigations on this material for TBC application as well as a patent [8 - 11].

On the basis of these investigations, compounds of the $ABAl_{11}O_{19}$ type were screened with respect to the influence of doping in order to study changes in their properties due to different A and B cations. The magnetoplumbite structural type was to be maintained and the compound should, if possible, be stable without secondary phases. $Bi^{3+}$, $Gd^{3+}$, und $Y^{3+}$ were also tested as alternatives for the A site and $Fe^{2+}$, $Ti^{4+}$, $Zr^{4+}$ und $Li^{+}$ for the B site. The substitution of titanium, zirconium and yttrium did not lead to the desired structural type as the main phase, which is why these experiments were discontinued. Of the other substitutions, only the $LaLiAl_{11}O_{18.5}$ obtained will be discussed in detail here. The other substitutions will be described at a later time. For example, Saruhan et al. [12] also performed investigations on the substitution of manganese on the B site.

Although the magnetoplumbite structure is more complex than corundum or zirconia, it has a high hexagonal symmetry. The complex structure is important for a low thermal conductivity and high symmetry for good high-temperature stability. The hexaaluminates are based on an alumina structure and exhibit an additional plane in the composition filled up with lanthanum and magnesium atoms. In opposite to zirconia $LaMgAl_{11}O_{19}$ exhibits no ionic conductivity at high temperature. Furthermore, a high melting point is assumed.

Table I: Key properties of standard YSZ for application in gas turbines.

| Property | Value / limits |
|---|---|
| Thermal expansion coefficient | $10 - 10.5\ 10^{-6}\ K^{-1}$ [13] |
| Thermal conductivity | $< 2.3\ Wm^{-1}\ K^{-1}$ (1000 °C) [13] |
| Sintering rate | $d(\Delta L/L_0)/dt \approx 0.1 - 0.35\ \%\ (100\ h)^{-1}$ |
| Temperature stability | Phase change above 1200°C |
| Young's modulus | E = 50 – 200 GPa [2, 13] |
| Porosity (APS coatings) | Standard 12 – 25% [13] |

New materials have to fulfill certain requirements such as higher temperature stability, lower thermal conductivity, adapted thermal expansion coefficient, lower sintering rate and a small Young´s modulus (s. table 1). From the industrial perspective, it is not necessarily a low thermal conductivity that is most important but rather a long lifetime. For a thermal barrier coating, apart from TGO formation and other properties, the lifetime depends in particular on the thermal mismatch.

In this paper, we present a new compound from the hexaaluminate family -$LaLiAl_{11}O_{18.5}$. This is not only a new material for thermal barrier applications, but it is also new with respect to the chemical composition itself. The properties obtained were therefore different as well as unexpected, when compared to standard YSZ, and similar to $LaMgAl_{11}O_{19}$.

EXPERIMENTAL

$LaLiAl_{11}O_{18.5}$ was prepared by a solid-state reaction of $\alpha$-$Al_2O_3$, $La_2O_3$ and $Li_2CO_3$ powders. All substances used here had a purity ≥ 99% ( $Li_2CO_3$ 99.0% Alfa Aesar, Karlsruhe; $Al_2O_3$ 99.8 % Martoxid MR70, Martinswerk, Bergheim, Germany; $La_2O_3$ 99,0% Projector GmbH, Duisburg, Germany). $La_2O_3$ was heat treated to eliminate any hydroxides. Than the powders were mixed and thoroughly milled in an ethyl-alcohol-based suspension to obtain a homogeneous reaction to a single phase. The powder was milled for longer than 12 hours to a particle size of less than $d_{50} < 1.5$ µm. The reaction temperature of the solid-state reaction was higher than 1450 °C [14] (different batches between 1450 and 1600°C). Lithium carbonate was decomposed during the reaction to lithium oxide.

All the prepared raw materials were analyzed by X-ray diffraction before being used for further investigations to guarantee high-quality powders. The phase analysis was conducted using an XRD (Model D5000, Siemens, Karlsruhe, Germany) in the 2θ range from 15° to 70° with $CuK_\alpha$ radiation (wavelength 1.5406 Å).

The physical properties of the new material were investigated on bulk samples in order to exclude the influence of the porosity and microstructure. The thermal expansion coefficient was obtained by dilatometric measurements and the thermal conductivity by the laser flash method.

A high-temperature dilatometer (Setaram, Caluire, France) was used to measure the thermal expansion coefficient. Prior to the dilatometer measurement, the required cylindrical specimens were cold-isostatic-pressed at a pressure of nearly 400 MPa and subsequently sintered at 1450 °C for 14 hours. Inside the dilatometer, the sample to be measured was positioned in an $Al_2O_3$ cylinder and the thermal expansion was recorded by monitoring the change in height of an $Al_2O_3$ rod placed on the sample. In order to subtract the thermal expansion of the holder and rod, for every dilatometry measurement a blank measurement without sample was also carried out.

The thermal diffusivity of the specimens was measured using the laser flash method (Theta Industries, Port Washington, NY) with a neodymium-glass laser (wavelength 1065 nm). The laser-flash method requires dense specimens, because the thermal diffusivity of the bulk material is the point of interest. The specimens used here were disks with a diameter of 10 mm and a thickness of 2 mm. These specimens were fabricated by uniaxial hot pressing at a temperature of 1450°C and a pressure of nearly 87 MPa in an argon atmosphere. A detailed description is given by Lehmann et al. [16].

The thermal conductivity ($\lambda$) was calculated from the values for thermal diffusivity ($\alpha$), specific heat capacity ($C_p$), and measured density ($\rho$), using the relation

$$\lambda = \alpha \cdot C_p \cdot \rho \qquad (1)$$

( $\alpha$ = thermal diffusivity and $\rho$ = theoretical density)

The phase stability up to high temperature and values for the specific heat capacity were obtained on powder from DSC measurements (differential scanning calorimetry - DSC Model 404c, Netzsch, Germany) at up to 1300 °C in air. Data were recorded during the period of rising temperature at 20 K $min^{-1}$ in platinum crucibles.

To use the material for plasma spraying, it was necessary to spray-dry the powder. Therefore a large batch of the initial powder was produced which contained only a very small amount of impurities (only perovskite phase). The chemical analysis is shown in table 2. A suspension for spray drying was prepared by mixing the milled powder ( 78 wt.%) with ethanol and 1.2 wt % of PEI ($[CH_2CH_2NH]_x$ m.w. 10,000, Polyscience).

The spray dryer was a Mobile Minor Ex Model H (Niro A/S, Denmark) with $N_2$ as the drying medium. A detailed description of the process is given by Cao et al. [15]. After spray drying, the powder was sintered at 1300 °C for 2 hours to remove the PEI binder. The tapped density of the spray-dried powder was 1.62 g $cm^{-3}$. For plasma spraying, a sieved fraction of between 36 and 125 µm of this spray-dried powder was used. The grain size distribution is characterized by $D_{10}$ = 11.4 ; $D_{50}$ = 23.4 ; $D_{90}$ = 70.0 µm. The BET surface of this powder is 0.69 $m^2$ $g^{-1}$. An SEM micrograph of a spray-dried powder particle is presented in fig. 1.

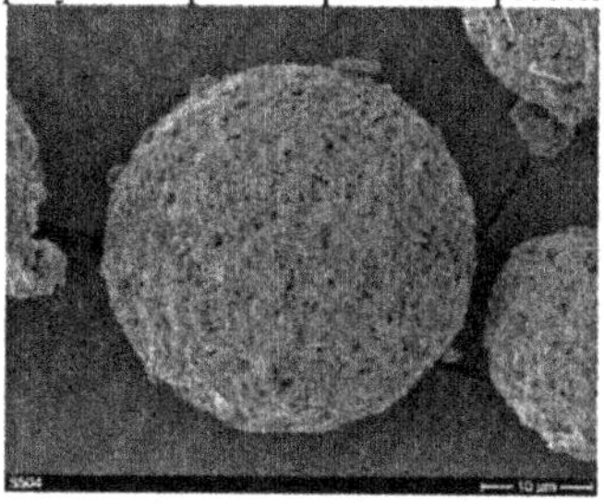

Fig. 1: SEM micrograph of spray-dried powder of $LaLiAl_{11}O_{18.5}$.

Plasma-sprayed coatings of $LaLiAl_{11}O_{18.5}$ were produced by atmospheric plasma spraying with a Sulzer Metco F4 gun on different substrates. Argon and hydrogen were used as plasma gases. Parameters and substrate pretreatment will be described below together with the results.

The coating porosity was determined by mercury porosimetry with a Pascal 140 and 440 supplied by CE instruments (Milan, Italy) operating in a pressure range between 0.008 and 400 MPa corresponding to pore diameters between 3.6 nm and 90 μm. These porosimetry measurements were performed on freestanding coatings and can only determine the open porosity.

## RESULTS AND DISCUSSION

### MATERIAL CHARACTERIZATION (POWDER AND BULK MATERIAL)

The X-ray pattern of the powder (s. fig 2) showed the same peaks as those of $LaMgAl_{11}O_{19}$. This new material therefore crystallized in the hexagonal magnetoplumbite structure similarly to $LaMgAl_{11}O_{19}$. The lattice constant of this hexagonal crystallizing composition was determined as a = 556.33 ± 0.55 pm and c = 2193.07 ± 2.86 pm. In fig. 2, an X-ray diffraction pattern of $LaLiAl_{11}O_{18.5}$ with a very small amount of the second phase of $LaAlO_3$ is shown. Additional phases often found during reaction to $LaLiAl_{11}O_{18.5}$ were $LaAlO_3$, $\alpha$-$Al_2O_3$ and spinel phases.

Lithium oxide was used in the structure to fill up the vacancies in the lattice. Using X-ray diffraction it was not possible to measure whether all the vacancies in the lattice were completely filled, or whether some lithium oxide was lost during reaction because the magnetoplumbite structure is also formed without filling up the vacancies. In the X-ray diffraction pattern, the difference between $LaAl_{11}O_{18}$ (structure with empty vacancies) and $LaLiAl_{11}O_{18.5}$ or $LaMgAl_{11}O_{19}$ (structure with filled vacancies) therefore only leads to a slightly different pattern, which cannot be distinguished when typical secondary phases are formed hiding these different patterns.

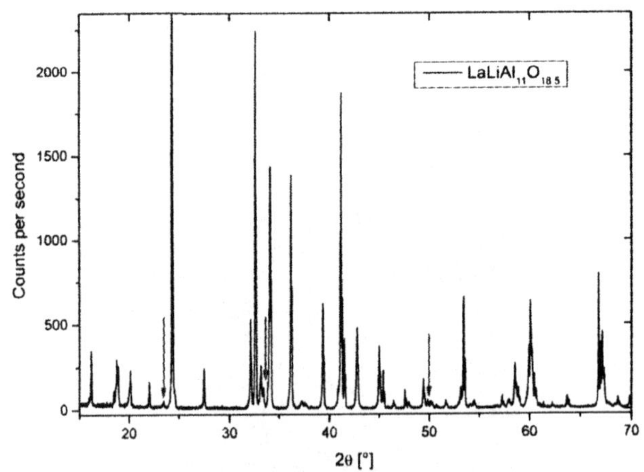


Fig. 2: X-ray diffraction pattern of $LaLiAl_{11}O_{18.5}$ fitting to magnetoplumbite structure. Gray arrows denote additional perovskite phases.

From chemical analyses (at ZCH, Forschungszentrum Jülich) it was confirmed that there was hardly any loss of lithium during powder preparation, spray drying and plasma spraying (tab. 2).

The thermal conductivity of this new material (investigations with bulk material) is 3.8 $Wm^{-1}K^{-1}$ at a temperature of 1000 °C. This is much higher than that of YSZ ( 2.2 $Wm^{-1}K^{-1}$ at 1000 °C [16]).

Table II: Chemical analyses by ICP-OES of two samples, one raw powder and the spray-dried powder, used for plasma-sprayed coatings. The amount of oxygen was analyzed by hot extraction in helium gas in combination with IR spectroscopy.

| Wt. % | Powder for plasma spraying | Raw powder | Theoretical values |
|---|---|---|---|
| Al | 40.1 | 40.1 | 40.2 |
| La | 18.2 | 18.3 | 18.8 |
| Li | 0.71* | 0.77* | 0.94 |
| O | 37.0 | 37.8 | 40.1 |
| Si | 0.044* | 0.061* | - |
| Ca | 0.026* | 0.029* | - |

* deviation of values ± 20%; all other values ± 3%

The measurements of the thermal expansion coefficient with a dilatometer led to unforeseen results. All the specimens showed shrinkage during heating although the samples were isostatically pressed and sintered for 14 hours at 1450 °C. The specimens were then heat treated again for 14 hours at 1450 °C, but the result did not change. The dilatometric analysis in figure 3 was obtained during heating, resulting in a thermal expansion coefficient between $6{\cdot}10^{-6}\ K^{-1}$ at 200 °C and $10.2{\cdot}10^{-6}\ K^{-1}$ at 1400 °C. Here it has to be mentioned that the sintering of the sample influenced the measurement significantly. The material could therefore display higher thermal expansion if sintering effects are excluded.

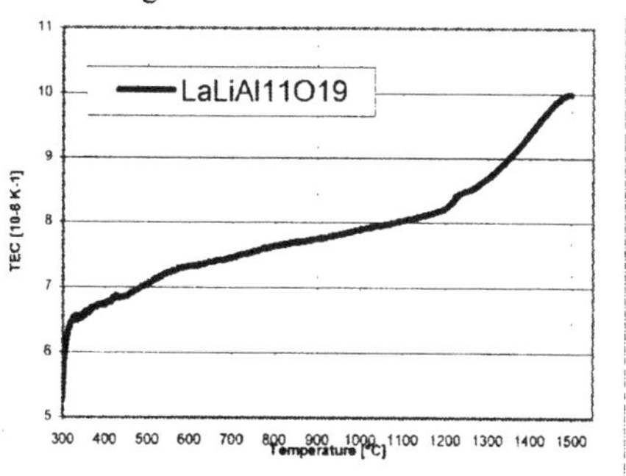


Fig. 3: Thermal expansion coefficient in a temperature range between 300 and 1500 °C.

The sintering was studied by keeping the sample at a high temperature (1500 °C) for 10 hours, as shown in figure 4. From this experiment it follows that a strong sintering process takes place during the whole period and was not completed even after 10 hours.

For plasma-sprayed coatings this sintering behavior is expected to be much stronger because of the porosity and cracks, which were dependent on deposition conditions. During deposition, the material is partly amorphous and these amorphous parts crystallize during heating. Shrinkage during crystallization is much stronger than pure sinter shrinkage. Further investigations to clarify the sintering mechanism are still in progress.

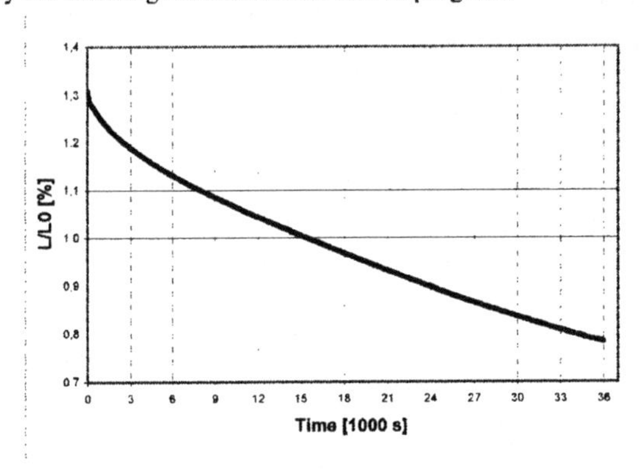


Fig. 4: Sintering of a dense sample of $LaLiAl_{11}O_{18.5}$ at 1500 °C. The sintering started at 1100 °C during heating and was not completed after 10 hours.

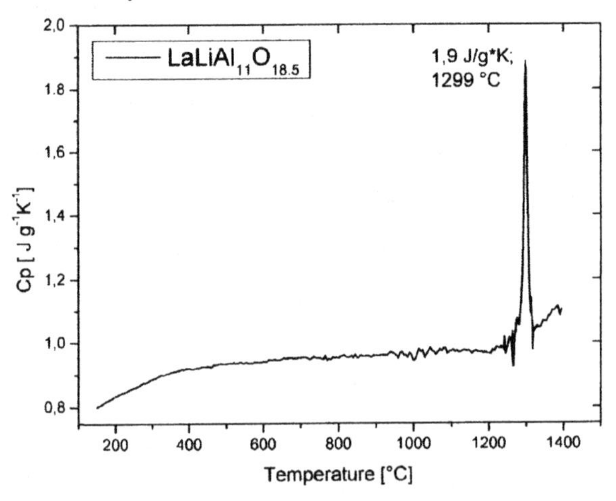


Fig. 5: DSC measurement of $LaLiAl_{11}O_{18.5}$ in the temperature range of 200 – 1400 °C. The material displays a significant endothermic peak at 1299 °C.

The DTA and DSC analyses of $LaLiAl_{11}O_{18.5}$ showed a reversible. endothermic peak at 1299 °C. In figure 5 a DSC diagram is presented. In the cooling curve the exothermic peak was at 1283 °C (not shown here). The powder showed no traces of melting . The reason for the peak at 1299 °C is unknown. In-depth investigations using high-temperature X-ray analysis on a platinum heating element in the area of 30 to 39 2θ at 1150, 1350 and 1400 °C found no phase transition. Only a small change in the content of main and secondary phases was observed. The amount of the perovskite phase decreased with higher temperature. During cooling this content increased again. So it is suggested from these high temperature X-ray results that the powder was not completely reacted to the composition of $LaLiAl_{11}O_{18,5}$.

## COATING DEPOSITION AND CHARACTERIZATION

The spray-dried powder was used to deposit $LaLiAl_{11}O_{18.5}$ by plasma spraying using a F4 torch. Some typical parameters are shown in table 3.

Table III: Plasma-spraying parameters with an F4 torch for $LaLiAl_{11}O_{18.5}$. The plasma gas was usually argon (41 slpm) and hydrogen (10 slpm).

| Coating no. | 287 | 288 | 289 |
|---|---|---|---|
| Distance [mm] | 110 | 100 | 100 |
| Power [kW] | 48.4 | 40.9 | 48.1 |
| Current [A] | 601 | 500 | 601 |
| Powder [g/ min] | 13 | 4.3 | 4.3 |
| Cycles | 38 | 60 | 60 |
| Translation speed [mm / s] | 500 | 500 | 500 |
| Porosity [%]* | 9.4 | 14.7 | 7.9 |
| Coat. thickness [µm] | 1758 | 825 | 940 |

*Porosity determined by mercury porosimetry.

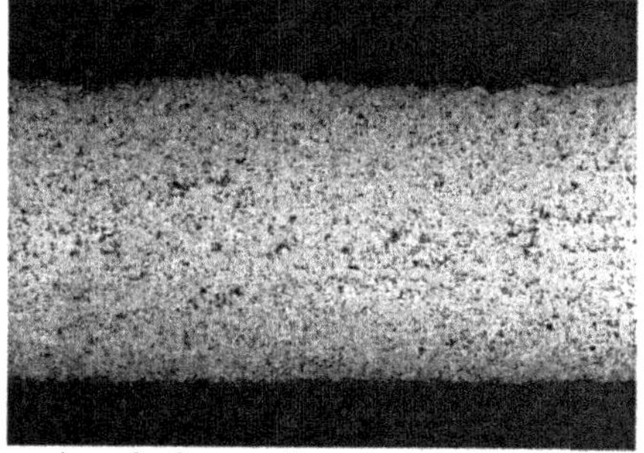

Fig.6: Micrograph of a cross section of a freestanding plasma-sprayed $LaLiAl_{11}O_{18.5}$ coating. No lamellar structure and no cracks were visible.

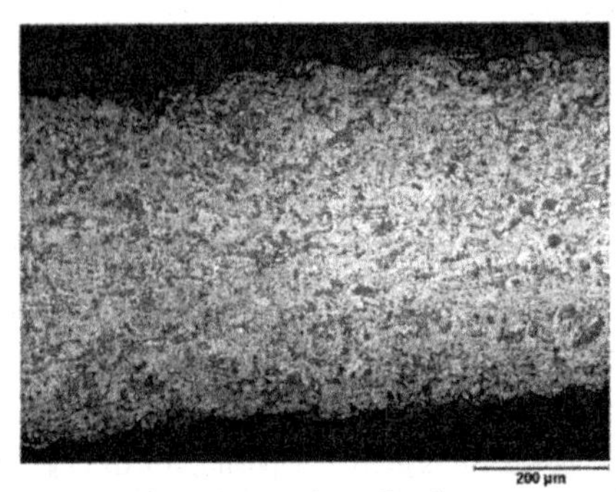


Fig. 7: For comparison, a micrograph of a cross section of a plasma-sprayed $LaMgAl_{11}O_{19}$ coating.

The plasma-sprayed coatings of $LaLiAl_{11}O_{18.5}$ and $LaMgAl_{11}O_{19}$ shown in fig. 6, 7 and 8 were homogeneous and the microstructure looked very different compared to standard zirconia coatings because there is no lamellar structure visible as is typical of YSZ (fig. 7). In the cross-section micrographs in fig. 6 and 7, a great deal of open porosity is visible, specially for $LaMgAl_{11}O_{19}$. In the SEM pictures, $LaLiAl_{11}O_{18.5}$ appeared to be more dense than in the micrographs. This was typical of all coatings of $LaLiAl_{11}O_{18.5}$ in these investigations.

In fig. 8b some areas with small particles (marked by arrows) were obtained. These areas could have been formed by overspray during plasma spraying. However, from the evidence of a large number of evaluated pictures of $LaMgAl_{11}O_{19}$ and $LaLiAl_{11}O_{18.5}$ these particle areas appeared to be very typical of both these hexaaluminate materials. Friedrich et al.[17] gave a corresponding description of this microstructure.

During plasma spraying of hexaaluminate materials only a small amount of the material is deposited in a crystalline form. Especially for $LaLiAl_{11}O_{18.5}$ a large amount of amorphous phase was found in X-ray investigations (see fig. 9). This amorphous phase is partly soluble in acids, e. g. hydrochloric acid. Hence some coatings were deposited on a substrate which had been previously coated with a rock salt in order to remove the upper coating without using any acid. This is an easy way to obtain freestanding coatings by plasma-spraying.

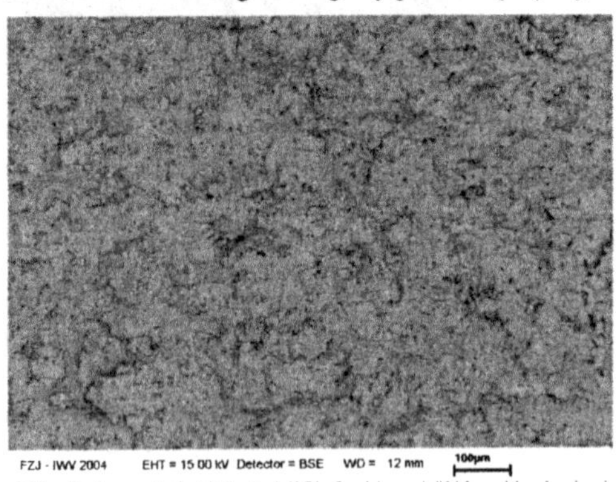

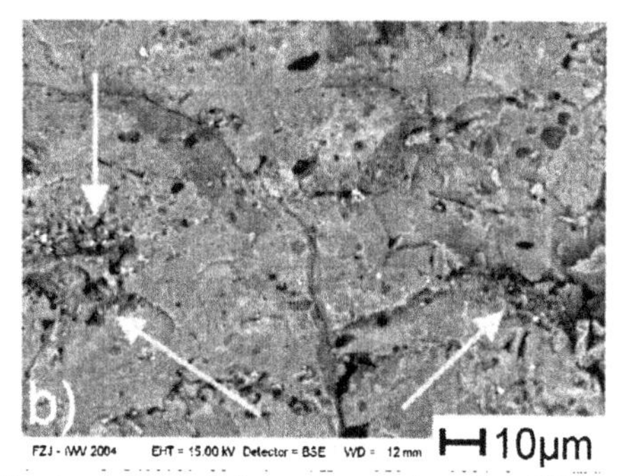


Fig 8 a+b: SEM pictures of fracture edges of plasma-sprayed coating (F4 torch; 287). The microstructure was significantly different to plasma-sprayed zirconia coatings. The arrows mark an area of small particles.

For a thermal barrier coating material, the properties at higher temperature are very important. For both hexaaluminates ($LaLiAl_{11}O_{18.5}$ and $LaMgAl_{11}O_{19}$) the partly amorphous phase in plasma-sprayed coatings have to crystallize during heat treatment. Additionally, for $LaLiAl_{11}O_{18.5}$ an intensive sintering was observed at higher temperature. Heat treatment of a plasma-sprayed coating of $LaLiAl_{11}O_{18.5}$ should therefore lead to a strong shrinkage due to sintering and recrystallization.

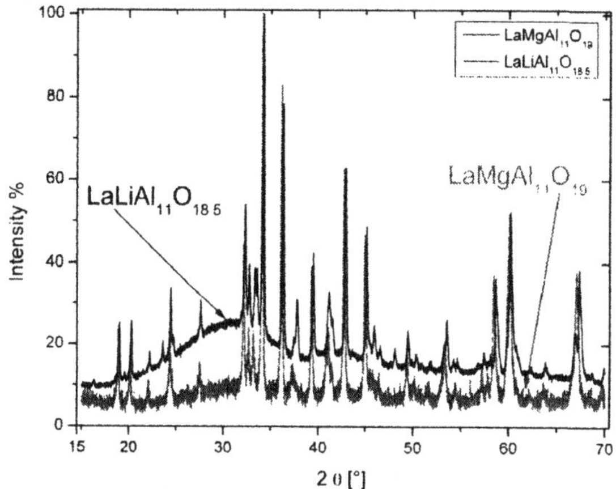


Fig.9: X-ray diffraction pattern of plasma-sprayed coatings of $LaMgAl_{11}O_{19}$ (gray) and $LaLiAl_{11}O_{18.5}$ (black). Both patterns display amorphous phase, but for $LaLiAl_{11}O_{18.5}$ the amount is significantly higher.

To check the shrinkage a simple test was performed on an steel substrate. Some test plates were coated by plasma spraying with YSZ, $LaMgAl_{11}O_{19}$, $LaFeAl_{11}O_{19}$ and $LaLiAl_{11}O_{18.5}$. The test substrates were stainless steel (type 1.4571) with a thermal expansion coefficient of 19.0 $10^{-6}$ $K^{-1}$ (between 20 and 500°C). The coating thickness was between 300 and

1000 μm, always a single layer without a bond coat. These samples were heated together to 1200 °C in a vacuum oven for 5 hours. It was expected in the case of poor adhesion that the coating would spall off from the substrate because of the thermal mismatch of metal and ceramic and the reasons mentioned above. With another material, where the coating had good adhesion on the plate, it was observed that the metal plate became strongly bent after this heat treatment.

This test has an unforeseen result (s. fig. 10 - 12). The metal plates for thick coatings of YSZ, $LaMgAl_{11}O_{19}$ and $LaLiAl_{11}O_{18.5}$ were only weakly bent, but on the surface of all hexa-aluminates segmentation cracks were obtained. Both YSZ coatings had no segmentation cracks and did not spall off. Only for the thick YSZ coating was a weak bending obtained and some spallation in the marginal zone . The coating of the $LaFeAl_{11}O_{19}$ spalled off completely, even though segmentation cracks arose, too. The $LaLiAl_{11}O_{18.5}$ coating had very large cracks while all other hexaaluminates showed only small segmentation cracks ( for details s. fig. 11).

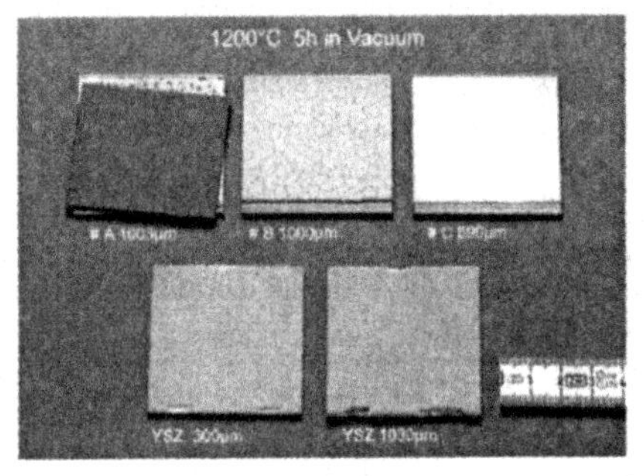


Fig. 10: Test plates with plasma-sprayed coatings after heat treatment (1200 °C 5 h in vacuum). A = $LaFeAl_{11}O_{19}$ ; B = $LaLiAl_{11}O_{18.5}$ ; C = $LaMgAl_{11}O_{19}$ ; YSZ = yttrium-stabilized zirconia. Different coating thicknesses between 300 and 1000 μm.

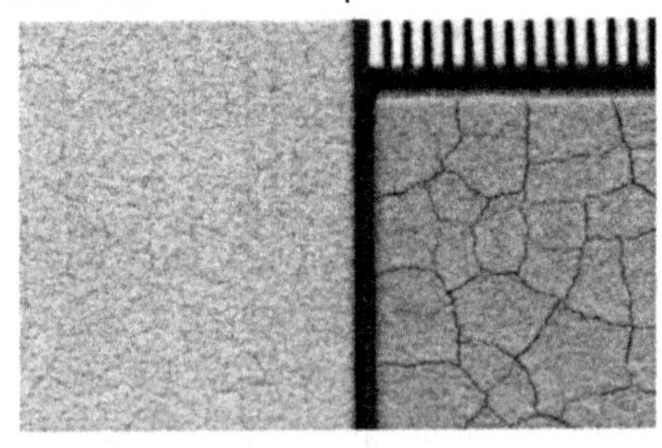

Fig. 11: Test plates with coatings of $LaMgAl_{11}O_{19}$ (Mg) and $LaLiAl_{11}O_{18.5}$ (Li) with higher magnification. The same plates as in fig. 10. On top right side, millimeter lines of a ruler.

The segmentation cracks had crack widths of 10 to 70 μm, which are visible in the micrograph in fig. 12.

The formation of segmentation cracks in the plasma-sprayed ceramic layers is rather unusual. However, in this test the high thermal expansion coefficient of the stainless steel substrate and the large layer thickness play an important part. A standard turbine alloy, such as IN738, has a thermal expansion coefficient of about $16 \cdot 10^{-6}$ $K^{-1}$ and the usual layer thickness for an APS coating is only 300 µm. Nevertheless, the tendency to crack during high-temperature treatment has not been previously observed for materials such as zirconia or aluminum oxide in comparable coatings.

200 µm

Fig. 12: Micrograph of a cross section of the plasma sprayed and heat treated $LaLiAl_{11}O_{18.5}$ coating (different specimen than in fig.10; thinner coating). The segmentation cracks run from surface to substrate. Crack widths of 10 up to 50 µm can be observed. Coating thickness is 700 µm.

If one assumes that in the course of further heat action the cracks can no longer be closed by sintering then these cracks represent a special expansion tolerance for the hexaaluminates which is independent of thermal mismatch. This property should have a very positive effect on the lifetime of the thermal barrier coatings.

This property is particularly pronounced for $LaLiAl_{11}O_{18.5}$ especially since the adhesion to the substrate is sufficiently good, in contrast to $LaFeAl_{11}O_{19}$.

## CONCLUSION

A new material was presented for application as a thermal barrier coating in gas turbines. This new material has a significantly higher thermal conductivity than comparable materials and demonstrates phase transition in the high-temperature range.

Nevertheless, the material has the unusual and probably very beneficial property that it can form segmentation cracks during heat treatment, which should lead to a considerable expansion tolerance. The extent to which the lifetime can be prolonged in applications as thermal barrier coating material still remains to be investigated.

## ACKNOWLEDGEMENTS

The authors thank Dr. José Luis Marques for the dilatometric measurements and Dr. W. Fischer for the high-temperature X-ray investigations. Likewise thanks are due to P. Lersch for the X-ray measurements and K.-H. Rauwald and M. Kappertz for preparing the plasma-sprayed samples. The authors are also grateful to D. Pitzer (FZJ B-NZ) for laser flash and DSC investigations and to D. Sebold for the SEM measurements.

REFERENCES

[1]Jaeyun Moon, Hanshin Choi, Hyungjun Kim, Changhee Lee, "The effects of heat treatment on the phase transformation behavior of plasma-sprayed stabilized $ZrO_2$ coatings", *Surface and Coatings Technology* **155**, 1 – 10 (2002).

[2] B. Siebert, C. Funke, R. Vaßen, D. Stöver, "Changes in porosity and Young's modulus due to sintering of plasma sprayed thermal barrier coatings", *J. Materials Processing Technology* **92 – 93**, 217 – 223 (1999).

[3] J. A. Thompson, W. Ji, T. Klocker, T. W. Clyne, "Sintering of the top coat in thermal spray TBC systems under service conditions", *Superalloys 2000*; T. M. Pollock et al. (eds) Seven Springs, USA TMS (The Minerals, Metals & Materials Society), 685 – 692 (2000).

[4] R. Vaßen, M. Ahrens, A. F. Waheed, D. Stöver, "The influence of the microstructure of thermal barrier coating systems on sintering and other properties", *International Thermal Spray Conference*, E. Lugscheider and C. C. Berndt (Eds.), Pub. DVS Deutscher Verband für Schweißen, Germany, 879 – 883 (2002).

[5] R. Vaßen, D. Stöver, "Conventional and New Materials for Thermal Barrier Coatings", *NATO Science Series II: Mathematics, Physics and Chemistry;* Kluwer Academic Publishers, Dordrecht, NL, 16, 199 – 216 (2001).

[6] D. Stöver, G. Pracht, H. Lehmann, M. Dietrich, J.-E. Döring und R. Vaßen, "New material concepts for the next generation of plasma-sprayed thermal barrier coatings"; *Thermal Spray 2003:* Advancing the Science and Applying the Technology; Proceedings of the 2003 International Thermal Spraying Conference, Vol. 2, 3rd International Conference on Thermal Spray Technology, Orlando Fl, USA, 1455-1462 (2003).

[7] Like Xie, A. N. Cormack, "Defect solid state chemistry of magnetoplumbite-structured ceramic oxides II: Defect energetics in $LaMgAl_{11}O_{19}$", *J. Solid State Chem.* **88**, 543 - 554 (1990).

[8] R. Gadow, G. W. Schäfer, "Thermal insulating material and method for producing same", *Patent* WO 99/42630, (1998).

[9] C. J. Friedrich, R. Gadow, T. Schirmer, "Lanthane Aluminate – A new material for atmospheric plasma spraying of advanced thermal barrier coatings", *Proceedings ITSC 2000*, Thermal Spray: Surface Engineering via Applied Research, C. C. Berndt (Ed.), Pub. ASM International, Materials Park, OH-USA, 1219 – 1226 (2000).

[10] S.R. Choi, N.P. Bansal and D. Zhu, Mechanical and Thermal Properties of Advanced Oxide Materials for High-Temperature Coatings Applications, Ceram. Eng. Sci. Proc., **26** [3] 11-19 (2005).

[11] R. Gadow, M. Lischka; "Lanthanum hexaaluminate – novel thermal barrier coatings for gas turbine applications – materials and process development"; *Surface and Coatings Technology* **151 – 152**, 392 – 399 (2002).

[12] B. Saruhan-Brings, U. Schulz, C.-J. Kröder, Wärmedämmmaterial mit im wesentlichen magnetoplumbitischer Kristallstruktur, "Thermal barrier coatings with mainly magnetoplumbite crystal structure", *European Patent EP* 1256 636 A2 (2002).

[13] Nitin P. Padture, Maurice Gell, Eric H. Jordan, "Thermal barrier coatings for gas-turbine engine applications", *Science*, 296, 280 – 284 (2002).

[14] R. C. Ropp and G. G. Libowitz, "The nature of the alumina-rich phase in the system $La_2O_3$-$Al_2O_3$", *J. Am. Ceram. Soc.* **61** [11 – 12], 473 – 475 (1978).

[15] X. Q. Cao, R. Vassen, S. Schwartz, W. Jungen, F. Tietz, D. Stöver, "Spray-drying of ceramics for plasma-spray coatings", *J. Europ. Ceram. Soc.* **20**, 2433 – 2439 (2000).

[16] H. Lehmann, D. Pitzer, G. Pracht, R. Vaßen, D. Stöver, "Thermal conductivity and thermal expansion coefficients of the lanthanum-rare earth element-zirconate system", *J. Am. Ceram. Soc.*, **86** [8],. 1338 – 44 (2003).
[17] C. J. Friedrich, R. Gadow and M. H. Lischka, "Lanthanum Hexaaluminate Thermal Barrier Coatings", *Proceedings Cocoa Beach* 2001, USA, 375 – 382 (2001).

* Corresponding author: Gerhard Pracht e-mail address: gerdpracht@gmx.de

# Modeling and Life Prediction of Thermal Barrier Coatings

# SIMULATION OF STRESS DEVELOPMENT AND CRACK FORMATION IN APS-TBCS FOR CYCLIC OXIDATION LOADING AND COMPARISON WITH EXPERIMENTAL OBSERVATIONS

R. Herzog, P. Bednarz, E. Trunova, V. Shemet, R.W. Steinbrech, F. Schubert, and L. Singheiser

Institute of Materials and Processes in Energy Systems 2,
Research Centre Juelich, 52425 Juelich, Germany

## ABSTRACT

Oxidation induced spallation of plasma-sprayed thermal barrier coatings (APS-TBCs) is regarded as one major failure mode of ceramic coated gas turbine components. A failure crack path, which is located partly in the thermally grown oxide (TGO) and partly in the TBC is typical for this kind of failure (grey failure). Recent investigations have shown that the related damage evolution starts within the first 10% of life by the formation of micro cracks at the TGO and by opening of pre-existing micro cracks in the TBC. Crack growth and linking of these cracks along the interface lead to final spallation. However, parameters, which govern the kinetics and thus the life-time are not sufficiently known. Finite element simulations of the stress response near the TGO at micrometer scale were conducted corresponding to cyclic furnace tests with identical loading. The load cycle consisted of thermal cycling between 20°C and 1050°C and a dwell-time of 2 h at 1050°C. Continued TGO growth was considered (thickness increase and lateral growth). To include realistic material data, the deformation properties of both the actual NiCoCrAlY bond coat and the plasma-sprayed TBC ($ZrO2$ with 7-8 wt. % $Y2O3$) as well as the oxidation kinetics have been experimentally determined and implemented in the FE code. The stress calculations showed two distinct features: (i) a fast development of high tensile stresses in the bond coat with a maximum value directly at the interface bond coat / TGO below a roughness peak, which occurred during the cooling stage and which were maximum at the lowest cycle temperature, and (ii) a development of a lateral region of larger tensile stresses alongside the roughness peak over roughness valleys. The simulation of crack formation at the interface bond coat / TGO using cohesive elements resulted in an early formation of a micro crack at the roughness peak .

## INTRODUCTION

Thermal barrier coatings (TBCs) offer a considerable potential for a further increase of the turbine inlet temperature (TIT) and thus for a further increase of the efficiency of gas turbines and combined cycle power plants. Even so TBCs are still applied predominantly as a wrapping against unexpected overheating and for an increase of component life without increasing the TIT, because basic aspects of damage and spalling of thermal barrier coatings have not been understood as yet, and the time or number-of-cycles to failure cannot be reliably predicted.

Finite element analysis (FEA) can contribute to understand the degradation and damage processes by providing a tool for analysing the stress response in thermal barrier systems and for simulating the evolution of stresses and the formation and growth of cracks during cyclic and high temperature loading. By doing so, finite element analysis principally allows to separate the parameters, which affect the stress response and to assess their respective impact.

A large number of research groups have reported results from numerical simulations of the stress response at the metal/ceramic interface of TBCs ([1,2,3,4,5,6,7,8,9.10,11,12]). Different approaches have been chosen in the past years. Some workings comprise parametrical studies with respect to the thickness of the thermally grown oxide (TGO) by implementing various but constant thickness values. In these cases TGO growth stresses were neglected, because bond coat oxidation was not considered and implemented as a continuous process at high temperature. Some include oxidation as a continuous process, but did not consider creep and high temperature stress relaxation. Others again comprise plastic and partly creep and stress relaxation properties of the coatings, but generally those properties were taken from third publications and it was not possible to conclude whether the material data were representative for a real TBC composite or not. Thus, the demand for more realistic FE simulations for thermal barrier coatings based on more realistic material properties for the coatings increased in the last years. The present work aims at a further step towards FE simulations for APS TBCs with improved significance. The simulations were conducted with material data, which have been determined predominantly on actual coatings used in corresponding life-time experiments. The data comprise plastic deformation and creep/stress relaxation at high temperature. The numerical calculations include furthermore the simulation of crack formation at the bond coat / TBC interface. Both, simulations of the stress response and crack formation are compared with experimental observations.

## EXPERIMENTAL OBSERVATIONS

The thermal barrier composite, which was experimentally investigated, consisted of the single-crystalline Ni-alloy CMSX-4, a NiCoCrAlY bond coat made by vacuum plasma-spraying and an APS TBC of zirconia doped with about 7-8 wt% yttria. Micrographs of the materials are displayed in Fig. 1.

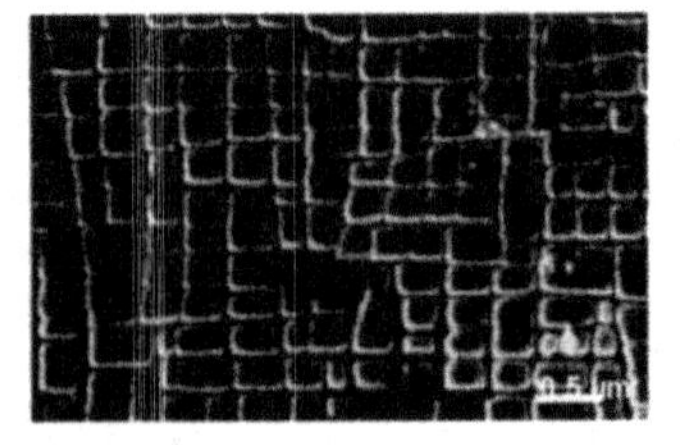
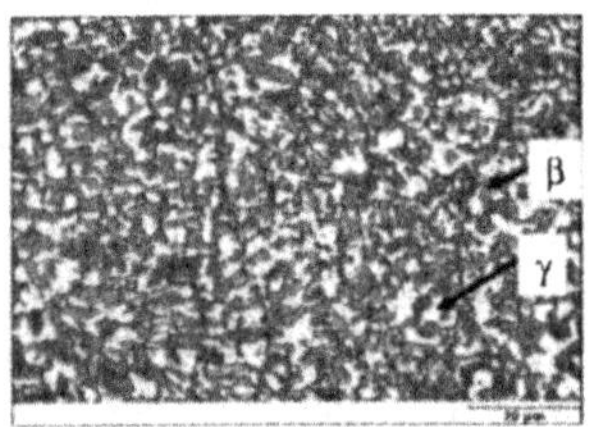

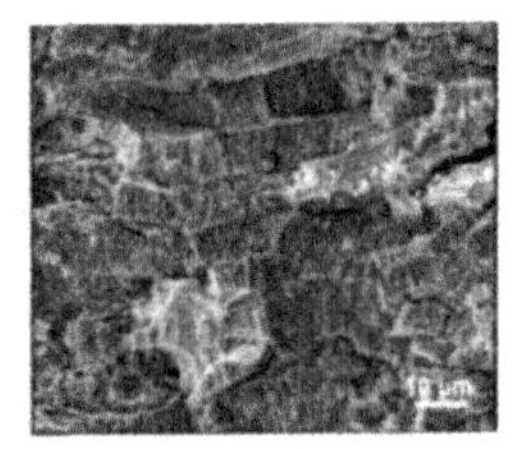

Fig. 1: Left: base material CMSX-4, center: NiCoCrAlY bond coat; right: APS TBC of zirconia doped with about 7-8 wt% yttria.

The thermal barrier composite was cyclically exposed between 60°C (approx.) and 1050°C with a dwell-time of 2 hours at 1050°C. The material was exposed until macroscopic spallation of the TBC was observed. The resulting failure mode is represented in Fig. 2. The failure crack path was located partly in the TBC and partly in the TGO. The mean TGO thickness was >10μm after failure.

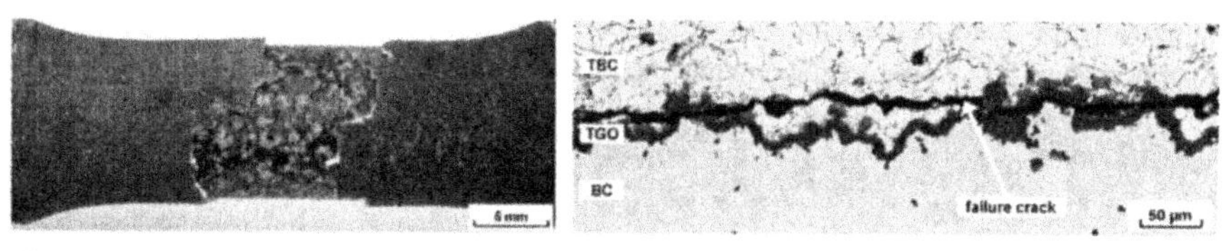


Fig. 2: left: observed failure of the plasma-sprayed TBC after cyclic thermal tests; right: metallographically prepared cross section directly next to the spalled area revealing a failure crack path which is located partly in the TBC and partly in the TGO.

Additional experiments were conducted up to selected fractions of life. The specimens were metallographically prepared to document the damage state. Fig. 3 shows examples of crack pattern observed at early stages of exposure (up to about 30% of time-to-failure). Frequently, micro cracks were observed at or near the interface bond coat / TGO at roughness peaks. Some were directly located at and along the interface downwards both sides of a roughness peak. They partly crossed the TGO towards the TBC and the crack tips were sometimes located within the TBC. Other micro cracks showed a similar shape, but were located partly in the TGO with some distance to the bond coat / TGO interface. Those types followed also the shape of the roughness peak, crossed the TGO with their tips partly penetrating the TBC above roughness valleys.

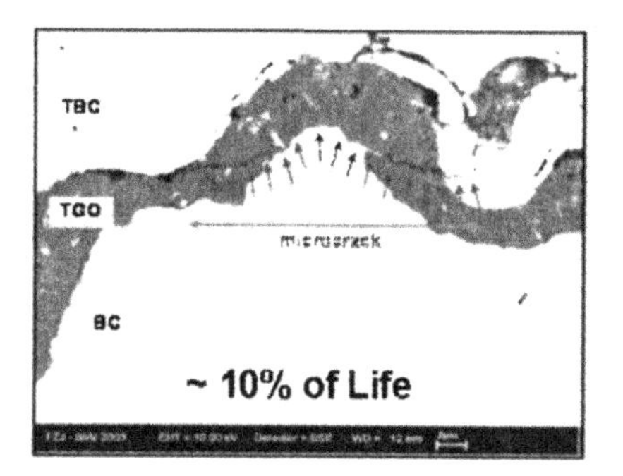


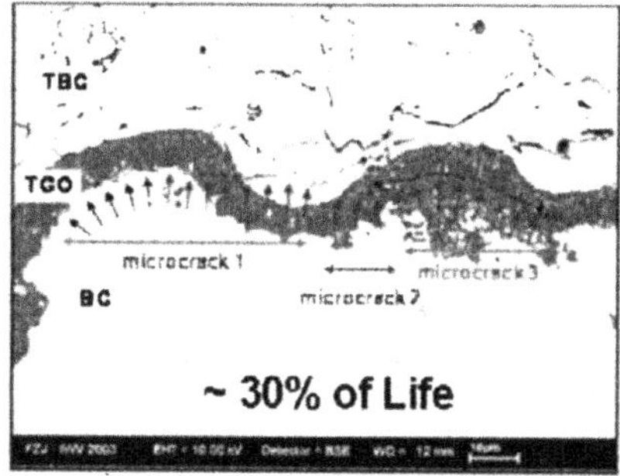


Fig. 3: Frequently observed crack pattern at early stages of exposure indicating weak points for crack formation and initial crack growth.

It has to be assumed that crack pattern which were frequently observed at early stages of exposure indicate weak points for crack formation and initial crack growth. The present results suggest that for the material investigated and the type of loading the interface bond coat / TGO at roughness peaks or the area directly above the bond coat / TGO interface within the TGO have to be considered as weak points for crack formation. Further weak points, which are indicated by the observed crack pattern, are the TGO at roughness flanks, the interface TGO / TBC at roughness valleys and the area above roughness valleys within the TBC. It has to be further taken into account that single micrographs taken from selected fractions of life are snapshots of a continuous damage evolution. Thus, cracks which on one micrograph are located in the TGO above the bond coat / TGO interface at roughness peaks might have been formed directly at the interface bond coat / TGO at an earlier stage of exposure. Prolonged bond coat oxidation and growth of the TGO would let them appear as if being formed and located within the TGO. See also [7] for a similar and more detailed discussion.

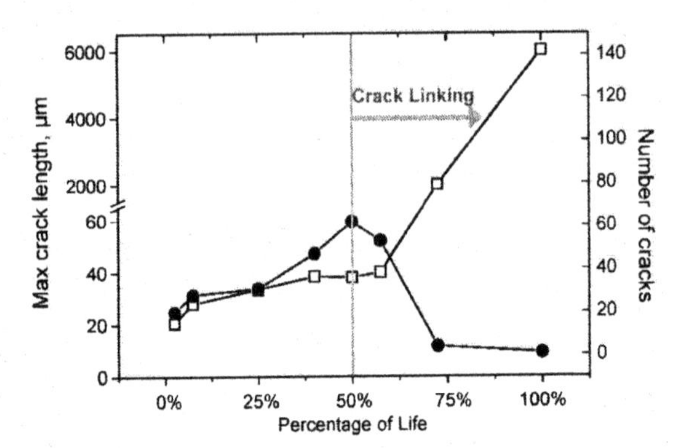


Fig. 4: Maximum crack length and number (density) of micro cracks, which have been observed in or near the TGO, for selected exposure times, here plotted against the fraction of life. The data covers the range from early stages of exposure to macroscopic spalling. At about 50% of life, micro cracks started to form links resulting in accelerated crack growth.

Similar micrographs were taken from each sample, which had been exposed to a certain fraction of life. To analyse and quantify the damage state, an area of the cross section, constant for all samples, was defined. All micro cracks from this area associated with the TGO were counted and characterized with respect to their length and location. From these crack data, the maximum crack length and the number (density) of micro cracks were determined for each exposure time and plotted against the fraction of life. The results are displayed in Fig. 4. Both curves reflect the evolution of damage at the interface of the investigated thermal barrier composite with respect to the applied load type.

The number density of micro cracks increased within the first 50% of life indicating that more and more micro cracks were formed along the interface. The maximum crack length increased also within the first 50% of life, but less steeply. It is worth mentioning that the shape of the crack growth curve indicates decelerated crack growth, that is the crack increment per time or number of cycles was decreasing within the first 50% of life. The maximum crack length was restricted to values below 50µm. Above 50% of life both curves reflect a different behaviour. The number density of cracks was significantly reduced indicating that individual micro cracks started to form links. At the same time the evolution of the maximum crack length was significantly accelerated until final macroscopic failure occurred.

The crack growth curve as well as the number density of micro cracks vs. exposure time in Fig. 4 highlight spalling failure of TBCs as a consequence of a continued evolution and accumulation of damage. A damage process, which starts below 10% of life with the formation of micro cracks along the bond coat / TBC interface and ends with macroscopic spalling of the TBC. Thus, failure and life time of this type of plasma-sprayed thermal barrier composite under cyclic furnace testing appears to be determined by the kinetics of the whole and quite slow damage process. More detailed results about the evolution of damage in APS TBCs were presented for instance by Echsler and Trunova [13,14]. The Finite Element Analysis (FEA), which is described in the next section, aims at simulating the local stress response, the evolution of local stresses and the crack formation near the TGO during the initial stage of exposure to obtain more information about the kinetics of the observed damage evolution.

## FINITE ELEMENT SIMULATION

### *Mesh and boundary conditions*

A cylindrical geometry had been chosen for the FE model, which corresponds to the specimen geometry used in the experiment. It consists of four layers or material constituents: base material, bond coat, thermally grown oxide and TBC. The base material had an outer diameter of 10 mm. The thickness of the bond coat was 150 μm and of the TBC 300 μm. The initial thickness of the TGO was 0.5 μm. The undulations (or roughness profile) at the bond coat / TBC interface have been approximated as sinusoidal. The sinus function has been parameterized by an amplitude of 15 μm and a wavelength of 60 μm. The mesh consists of 4 node Generalized Plane Strain elements with reduced Gauss integration (CPEG4R, HKS/ABAQUS[1] FEA). Geometry and mesh are shown in Fig. 5. Regarding a cylindrical co-ordinate system, the nodes which are lying on the edges of the segment (Fig. 5: right) has been constrained normal to the edges.

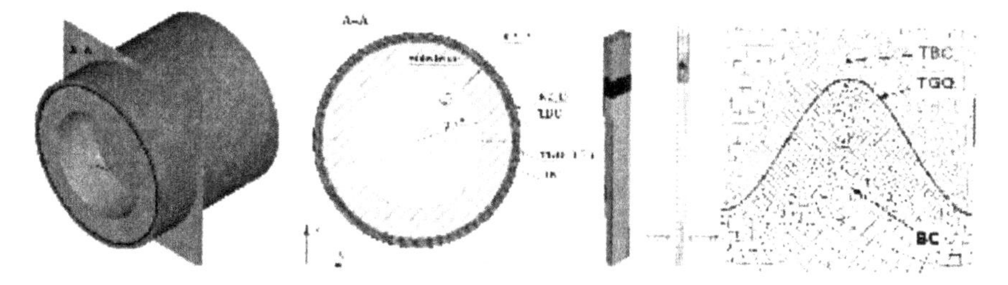


Fig. 5: Geometry and mesh of the FE-model

### *Material data and bond coat oxidation*

The base material (CMSX-4) was treated as an entirely elastic material, because of the low stresses which occur during pure thermal cycling without additional creep or fatigue loading. The bond coat was considered as elastic-visco-plastic and the TGO as elasto-plastic. Elastic and creep data has been considered for TBC deformation. All material properties were temperature dependent. Fig. 6 shows the data for the coefficient of thermal expansion (CTE) and the Young's modulus. Fig. 7 displays creep data for the TBC determined by compression creep tests with stand-alone coatings [15]. Creep properties of the bond coat were determined by compression creep tests with stand-alone coatings and shear deformation experiments on TBC composites [16]. Creep of bond coat and TBC was generally considered for T ≥ 750°C, whereas bond coat creep was substantial at 750°C, but significant primary creep rates of the TBC occur only above

[1] distributed by Habbit, Karlsson & Sorensen

950°C. Primary and secondary creep stages were taken into account. The data were implemented using Eq. (1):

$$\dot{\varepsilon} = A' \cdot \sigma^{n'} \cdot e^{-\frac{\varepsilon}{\varepsilon'}} + A'' \cdot \sigma^{n''} \cdot e^{-\frac{\varepsilon}{\varepsilon''}} + A \cdot \sigma^{n} \qquad (1)$$

where $\dot{\varepsilon}$ is the deformation rate, A, A', A'' are pre-factors, n, n', n'' are stress exponents, ε is the strain, ε' and ε'' are model parameters for the primary creep stage and σ is the stress. The right hand term covers steady state creep, the first two cover primary creep. The model parameters are temperature dependent.

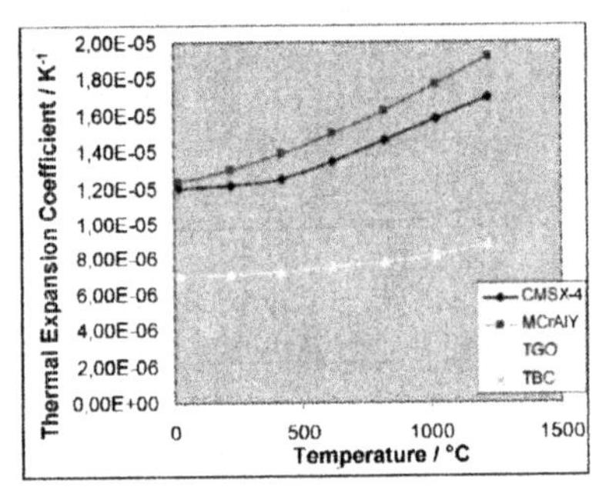


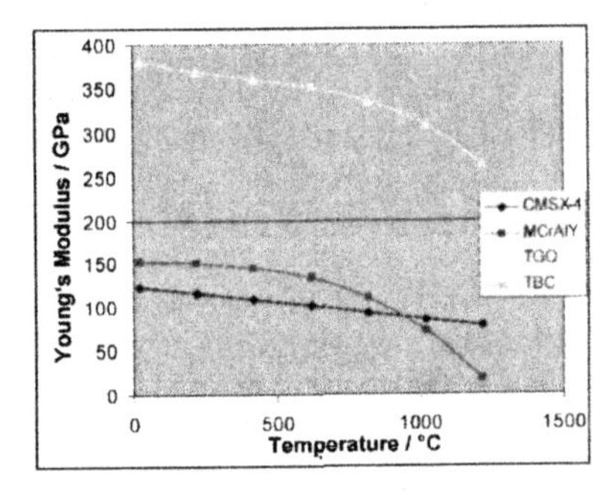


Fig. 6: Data for CTE and Young's modulus, which have been used for the simulations.

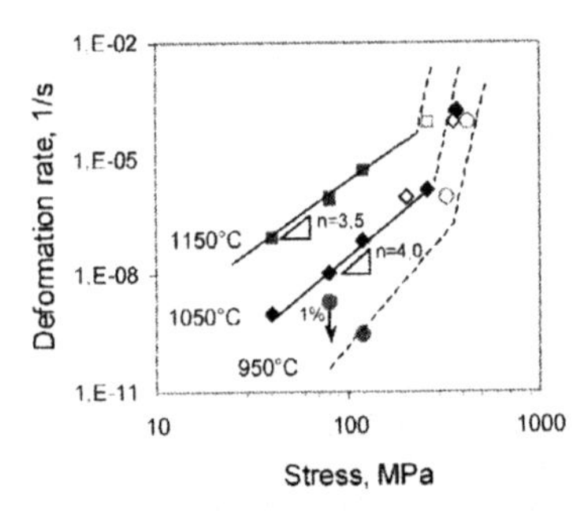


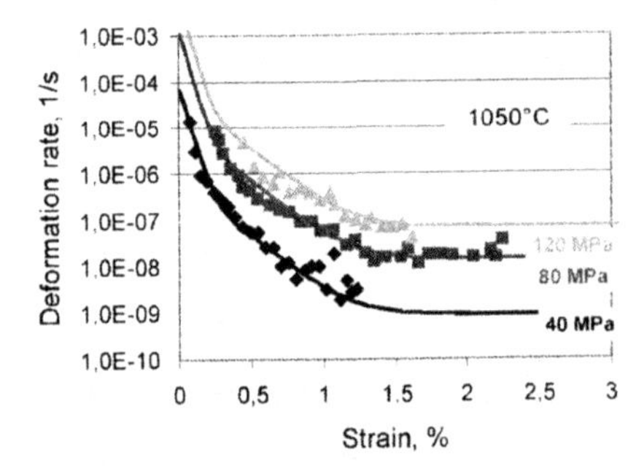


Fig. 7: Creep data for the APS TBC; left: Norton plot (minimum deformation rate vs. stress); right: deformation rate vs. strain and model curves with primary and secondary creep stages as an example for 1050°C at 3 different stress levels.

Growing of the alumina scale at high temperature due to bond coat oxidation was simulated as a continued process using the swelling option in ABAQUS. It was modeled as an orthotropic swelling strain of the TGO, whereby lateral TGO growth (length increase) was considered as a constant amount of thickness growth (generally 5%). The initial thickness of

TGO was defined as 0.5 µm. The oxidation kinetics of this thermal barrier composite has been experimentally determined for 3 different temperatures (950°C, 1000°C, 1050°C) [13] and were implemented for this temperature range using a parabolic time law.

*Load parameters*

The simulated load cycle consists of thermal cycling and high temperature exposure corresponding to the experiments. It consists of four steps (Fig. 3): (i) heating from 20°C to 1050°C in 103 s (10°C/s), (ii) dwell-time at 1050°C for 2 hours, (iii) cooling from 1050°C to 20°C in 103 s (10°C/s), (iv) dwell-time at low temperature (20°C) for 15 min. A temperature of 200°C was selected at which the TBC composite was initially stress free. It matches approximately with the material temperature during the air-plasma-spraying process. The simulations comprised generally 160 load cycles, what amounted to about 30% of life. The TGO thickness was about 5.7 µm after the last cycle.

*Results*

All presented results from the numerical calculations comprise the stress response near the TGO at room temperature after the $160^{th}$ load cycle. The displayed stresses are radial stresses.

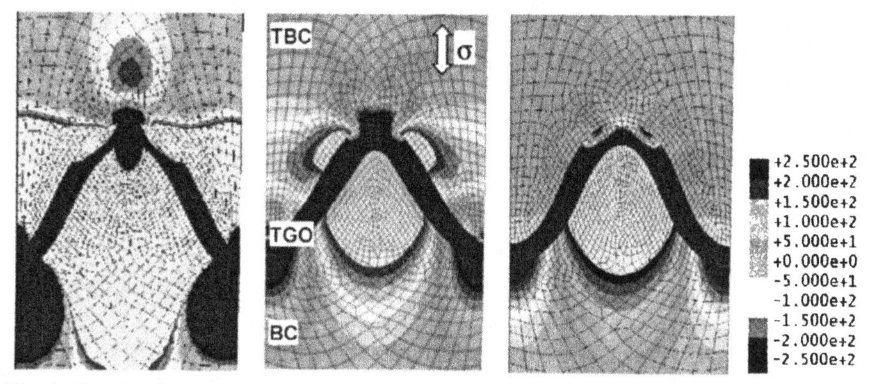


Fig. 8: Simulated stress response near the TGO after 160 load cycles at 20°C with continued TGO growth (here: TGO thickness = 5.7 µm); 1. case (left): all materials elastic; 2. case (center): like 1. case, but BC and TGO additionally with plastic properties; 3. case (right): like 2. case, but BC and TBC additionally with creep properties; light grey regions = tensile > 250 MPa, dark grey regions = compression < 250 MPa

Fig. 8 represents three different stress distributions, whereby the applied material properties were different for each case. The stress response at the left hand side (1. case) is a result from an entirely elastic calculation but with continued TGO growth (thickness increase and 5% lateral increase) to get information about the undisturbed effect of growth stresses in combination with thermal cycling. Noticeable are rather high and localized compressive and tensile stresses. High tensile stresses (light grey regions) were present in the TBC and in the bond coat as well as in some smaller regions in the TGO (off-peak). Compressive stresses were developed primarily in the TGO and in the BC below roughness valleys, and also within smaller regions in the BC directly below the peak and in the TBC directly above the valley. The absolute stress values were quite high (> 10 GPa and <-10 GPa) and probably not realistic.

The second stress distribution (Fig. 8, center) results from a calculation for which the plastic deformation properties of bond coat and TGO have been additionally taken into account in contrast to the first case. The main effects were an overall stress decrease and some redistributions of local stresses. The largest tensile stress was 1040 MPa and was located directly at the interface bond coat / TGO at the roughness peak. At this position, the stresses were changed from compression to tension compared to the first case. The maximum compressive stress occurred directly above the largest tensile stress in the TGO with appr. 2400 MPa. The third case (Fig. 8, right) comprised additionally creep in bond coat and TBC and thus the possibility of stress relaxation. By comparing Fig. 8 (center) and Fig. 8 (right), at first, a decrease of the high tensile stresses alongside the roughness peak in the TBC becomes apparent, and secondly a shape change of the interface. The curvature at the peak became less sharp. This effect was directly due to stress relaxation, which relaxed the entire structure. In contrast, the tensile stresses at the bond coat / TGO interface were decreased only slightly by less than 10%. However, this region showed the largest tensile stresses.

The material properties used in the last case were taken as a reference parameter set. It includes plastic and creep properties corresponding to the experimentally investigated MCrAlY bond coat, plastic properties of the TGO and creep properties of the experimentally investigated plasma-sprayed TBC. One of the first questions was how the stress response is developing with increasing number of load cycles. In particular the stress distribution at 20°C was of interest, because the largest stresses appeared at the lowest cycle temperature. Fig. 9 displays the stress response after selected load cycles at 20°C. Two remarkable features characterize the simulated stress response. At first, high tensile stresses occurred at the bond coat / TGO interface even after the first load cycle. Thus, early crack formation at this site appears quite likely depending of course on actual interface shape, material properties (deformation properties as well as resistance against crack formation at the interface) and load parameters. The corresponding cyclic furnace tests revealed crack formation at the bond coat / TGO interface within the first 10% of life (about 50 cycles). For comparison see also Fig. 3.

Secondly, a coherent lateral region of tensile stresses was developing in the TBC at both sides of the roughness peak indicating higher loaded regions. According to the cyclic tests, the regions in between roughness peaks and over valleys showed cracking. However, crack formation and propagation in the TBC would be critically affected by pre-existing splat boundaries, micro cracks and pores. Thus, the numerical simulations provide here only a tool for merely rough estimations of the load situation directly in the TBC.

An additional result is indicated by Fig. 9. Up to the $10^{th}$ cycle the tensile stresses alongside the roughness peak in the TBC increased due to the initially fast oxidation and thus large growth stresses. Afterwards the tensile stresses decreased in this region, because the oxidation rate decreased (parabolic behavior) and the stress relaxation got relatively more influence on the stress response. Then again, a narrow zone of large tensile stresses was emerging directly from the TGO and was growing in lateral direction. This particular result indicates the fairly complex interaction of growth stresses, thermal stresses and stress redistributions due to plastic deformation and even more due to stress relaxation.

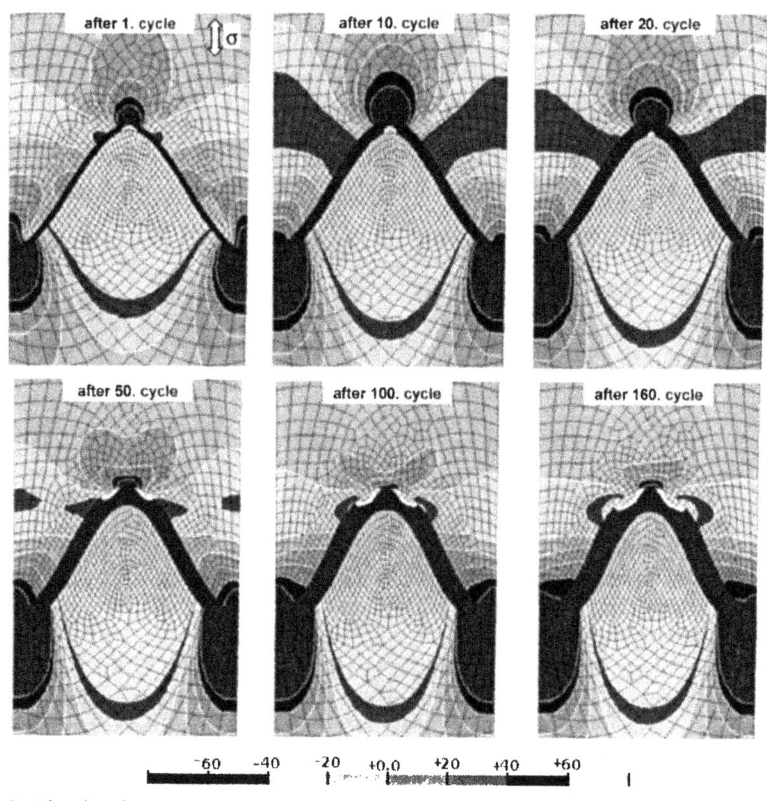


Fig. 9: Simulated stress response near the TGO after selected load cycles at 20°C with continued TGO growth using the reference parameter set, which is described in the text.

Starting from the reference simulation a couple of systematic variations of certain material parameters, such as CTE, Young's modulus and creep rate of the materials were conducted to analyze their influence. Here, only the influence of the stiffness variation of the thermal barrier coating should be described exemplarily. Fig. 10 displays the stress response associated with the reference parameter set (Fig. 10, center) as well as simulation results with a 50% higher (Fig. 10, left) and lower (Fig. 10, right) Young's modulus of the TBC (reference value: 17.5 GPa). The images show the stress distribution at 20°C after the $160^{th}$ cycle.

The stress plots indicate that an increase of the stiffness in the TBC increases predominantly the tensile stresses in the TBC except the small regions of compressive stresses directly above the roughness peak and directly above the roughness valley. Particularly the lateral region of tensile stresses alongside the roughness peak in the TBC increased by the stiffness increase of the TBC. The stresses in the lateral tensile zone directly at the boundary of the unit cell are increased approximately linearly, the stresses directly at the interface TGO /

TBC in the TBC are affected more than linearly. Beyond that, the tensile stresses at the bond coat / TGO interface were increased. On the other hand, the tensile stresses in the TBC were generally decreased by decreasing the stiffness and above the lateral region of tensile stresses they were even changed into small compressive stresses.

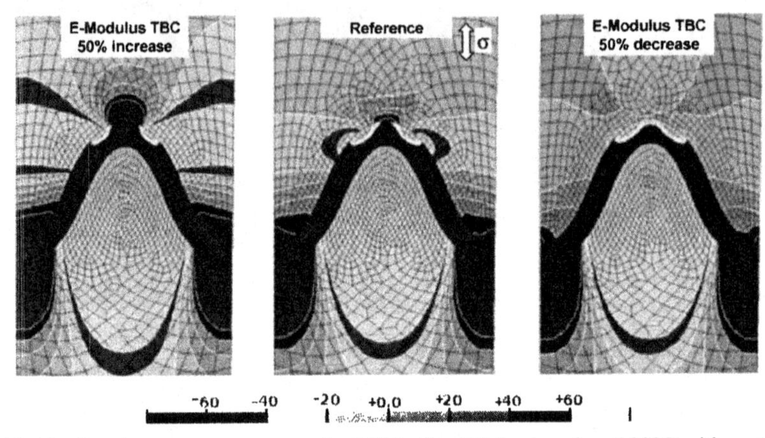


Fig. 10: Simulated stress response near the TGO after 160 load cycles at 20°C with continued TGO growth; the Young's modulus of the TBC was modified by ± 50% with respect to the reference parameter set.

One could use the calculation with the higher stiffness value also as a rough impression of how the stress response would change for the case high temperature exposure causes a time dependent stiffness increase of the TBC.

In addition to the simulation of the stress response, simulations of crack formation and initial crack growth have been conducted using cohesive elements [17]. Due to the fact that the highest local tensile stresses were developing at the interface bond coat / TGO, cohesive elements were implemented directly at the interface. Critical stress values of 600 MPa (normal) and 1200 MPa (shear) have been applied for crack formation. For crack opening a critical strain energy release rate of 20 N/m has been taken into account for normal and shear loading. Fig. 11 represents the stress state after two selected cycles taking into account the reference parameter set. After the $19^{th}$ load cycle the maximum tensile stress exceeded the critical stress value of 600 MPa at the roughness peak and a crack was formed during cooling (Fig. 11, left).

The formation of the micro crack at the interface affected the stress field substantially. The tensile stress region directly below the crack was relaxed. In contrast, small regions with high tensile stresses occurred in the TGO close to the crack tips at both sides. After 160 cycles the crack was elongated downwards the roughness profile along the interface at both sides of the peak. However, between the $19^{th}$ and the $160^{th}$ cycle the crack was not growing steadily. As a result of the interaction between thermal stresses, growth stresses and stress relaxation the crack was after certain cycles partly closed and opened again. No tendency was found for the crack to propagate further downwards the roughness profile up to the $160^{th}$ cycle. In contrast, the crack tip saw high tensile stresses in the adjacent TGO indicating a potentially bending and a penetration of the crack into the TGO. This behavior would correspond to the frequently observed crack pattern, which are exemplarily shown in Fig. 3.

Further development of the simulations are planned including the prediction of crack growth direction as well as dynamical re-meshing.

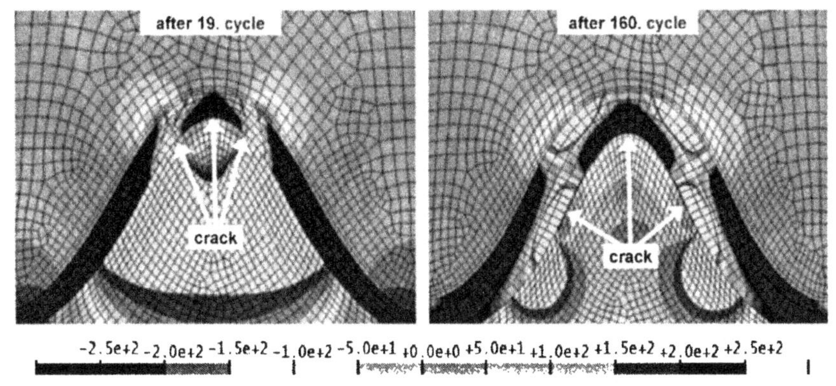


Fig. 11: Simulated stress response near the TGO and formation of a micro crack at the bond coat / TGO interface after 19 cycles and initial growth after 160 cycles at 20°C. It shall be noticed, that the scale of the stress distribution is different with respect to the other stress plots.

## CONCLUSIONS

Simulations of the response in a plasma-sprayed thermal barrier system were presented and compared with corresponding cyclic thermal tests including a dwell time of 2 hours at high temperature (cyclic furnace tests). The material properties of the MCrAlY bond coat and the ceramic thermal barrier coating were determined in order to obtain more realistic deformation properties, particular creep and thus stress relaxation properties as input data for the FE simulations. The simulation results showed that considering a continued oxidation process by simulating a continued TGO growth at high temperatures was required to cover the influence of growth stresses on the stress response. In general, the stress response in and near the TGO at the rough bond coat / TBC interface and its evolution during prolonged cyclic loading, was the consequence of the complex interaction of thermally induced stresses, oxidation induced growth stresses and redistribution of stresses due to plastic deformation and even more due to stress relaxation as a result of local creep processes. Pure stress calculations showed two distinct features: (i) the fast development of high tensile stresses in the bond coat with a maximum value directly at the interface bond coat / TGO below a roughness peak, which occurred during the cooling stage and which were maximum at the lowest cycle temperature. and (ii) the development of a lateral region of larger tensile stresses alongside the roughness peak over roughness valleys. The simulation of crack formation at the interface bond coat / TGO using cohesive elements resulted in an early formation of a micro crack at the roughness peak (after the $19^{th}$ cycle). Up to 160 cycles the crack was elongated downwards the roughness profile, but was also partly closed and opened again and no tendency was observed for the crack to propagate further downwards. Instead, high tensile stresses in the adjacent TGO would suggest a penetration of the crack into the TGO in the upper half of the sinusoidal interface, but this process was not covered by the simulations, at this time. The results of the numerical simulations were in agreement with the experimental observations of crack pattern obtained within the first 30% of life from the corresponding cyclic furnace tests.

## ACKNOWLEDGMENTS

The German Science Foundation (DFG) is greatly acknowledged for the financial support of the work within the Sonderforschungsbereich 370. Furthermore, many thanks are due to Priv.-Doz. Dr. R. Vaßen for manufacturing the coatings at the Institute of Materials and Processes in Energy Systems (IWV 1) of Research Centre Juelich.

## REFERENCES

[1] G.C. Chang, W. Phucharoen, R.A. Miller: *Surf. Coat. Technol.*, 1987, *30*, 13.
[2] R. Vaßen, G. Kerkhoff, D. Stöver: *Mater. Sci. Eng. A*, 2001, *303*, 100.
[3] G. Kerkhoff, R. Vaßen, C. Funke, D. Stöver: *Proceedings of the 6th Liege Conference on Materials for Advanced Power Enineering 1998*, 1998, 1669.
[4] M. Ahrens, R. Vaßen, D. Stöver: *Surf. Coat. Technol.*, 2002, *161*, 26.
[5] A. M. Freborg, B.L. Ferguson, W.J. Brindley, G.J. Petrus: *Mater. Sci. Eng. A*, 1998, *245*, 182.
[6] A.G. Evans, D.R. Mumm, J.W. Hutchinson, G.H. Meier, F.S. Pettit, in *Prog. Mater. Sci.*, 2001, *46*, 505.
[7] J. Cheng, E.H. Jordan, B. Barber, M. Gell: *Acta Mater.*, 1998, *46*, 5839.
[8] J. Rösler, M. Bäker, M. Volgmann: *Acta Mater.*, 2001, *49*, 3659.
[9] J. Rösler, M. Bäker, K. Aufzug: *Acta Mater.*, 2004, *52*, 4809.
[10] E.P. Busso, J. Lin and S. Sakurai: *Acta Mater.*, 2001, *49*, 1529.
[11] K. Sfar, J. Aktaa, D. Munz: *Mater. Sci. Eng. A*, 2002, *333*, 351.
[12] P. Bednarz, R. Herzog, E. Trunova, R.W. Steinbrech, L. Singheiser: *Proceedings of the 29th International Conference on Advanced Ceramics and Composites 2005,* Ceramic Engineering and Science Proceedings, 2005, *26, 3,* 55.
[13] H. Echsler: *PhD thesis, RWTH Aachen University*, 2003, ISBN 3-8322-1895-5.
[14] E. Trunova: *PhD thesis, submitted to RWTH Aachen University*, Berichte des Forschungszentrums Jülich, Reihe Energietechnik/Energy Technology, to be published in 2006.
[15] R. Herzog, E. Trunova, R.W. Steinbrech, E. Wessel, R. Vaßen, F. Schubert, L. Singheiser: Proceed. of the International Conference "Creep and Fracture in High Temperature Components – Design & Life Assessment Issues", 12-14 September 2005, Institution of Mechanical Engineers, Central London, UK
[16] P. Majerus, R.W. Steinbrech, R. Herzog, F. Schubert: Proceedings of the 7th Liege Conference 2002, 30. September - 2. October 2002, Liege, Belgium, Materials for Advanced Power Engineering 2002, ISBN 3-89336-213-2
[17] M. Cliez, J.-L. Chaboche, F. Feyel, S. Kruch: *Acta Mater.*, 2003, 52, 1133-1141

# NUMERICAL SIMULATION OF CRACK GROWTH MECHANISMS OCCURRING NEAR THE BONDCOAT SURFACE IN AIR PLASMA SPRAYED THERMAL BARRIER COATINGS

A. Casu, J.-L. Marqués, R. Vaßen, D. Stöver
Institut für Werkstoffe und Verfahren der Energietechnik (IWV1), Forschungszentrum Jülich GmbH, Jülich/Germany

ABSTRACT

Under thermal cycling, the failure of an air plasma sprayed thermal barrier coating (TBC) on a metallic bondcoat (BC) usually occurs near the interface between both coatings. The local curvature of such an interface is responsible for the stress components which lead to the growth of micro-cracks already produced during the plasma spraying. The growth of oxide scales (TGO) between BC and TBC at high temperatures determines the stress level near the TGO-TBC interface during cooling, where the main stress source is the mismatch in thermal expansion.

A failure mechanism based on finite-element calculations of thermal stress within the TBC is presented, which models the TGO-TBC interface as a sinusoidal profile. Assuming the coating system has completely relaxed its stresses during the thermal cycling hot phase, sub-critical crack growth after cooling to room temperature is calculated for horizontal cracks starting at every hill of the curved TGO-TBC interface profile. Failure is assumed when the growing cracks cover one whole profile wavelength. In a second step, an extension of the presented model is discussed where crack growth follows the path where the energy release rate becomes maximum. Finally, the crack path is implemented directly in the finite-element mesh. The conclusions drawn from the numerical calculations are compared to crack configurations near the TGO-TBC interface, taken from micrographs of thermally cycled samples.

## INTRODUCTION

The increasing temperatures inside a gas turbine require the metallic components to be protected against the high thermal load. This is carried out by depositing a thermally insulating ceramic coating, the thermal barrier coating (TBC), on top of the metallic component. Due to the different thermal expansion coefficient of both materials, high stress levels develop in the ceramic coating during temperature changes, which lead to crack growth of previously existing micro-cracks and, eventually, to the failure of the ceramic coating. A reliable prediction method for the lifetime of the metallic-ceramic system is, nevertheless, still to be achieved.

The most used material as TBC is yttria partially stabilized zirconia (YSZ), and one of the most extended techniques for depositing it on a metallic component is using atmospheric plasma spraying (APS). In this technique the ceramic TBC material, injected into the plasma jet as particles of size 10-100μm, is molten and the fluid particles are flattened on the metallic component to build up the thermal barrier coating. In order to improve the adhesion between the metallic substrate and the ceramic layer, a metallic bondcoat (BC) is sprayed first on the metallic component before depositing of the TBC. The roughness of the resulting BC, sprayed under very low gas pressure to avoid its oxidation, ensures the interlocking of the BC with the ceramic TBC deposited on it. Additionally, the aluminium content within the BC material, usually MCrAlY (M=Ni, Co), eventually forms an alumina scale, impeding the oxidation of the metallic component below the BC. This oxide scale produced during the exposure of the system to high temperatures is called the thermally grown oxide (TGO). The whole system consisting of a metallic component, the metallic bondcoat and the ceramic thermal barrier coating is denoted TBC system. The main aim of the present work is to develop a better understanding of the crack growth mechanism for such a system under thermal cycling, in order to improve the lifetime estimation of the TBC system.

The paper is organized as follows: the next section presents a new technique to calculate the most relevant length scales for the BC-TBC interface profile. These length scales determine the out-of-plane components of the thermal stress, responsible for the crack growth within the TBC near the BC-TBC interface. In the third section a simplified lifetime model will be discussed, which estimates the horizontal growth of TBC cracks based on the thermal stress distribution calculated by means of the Finite Element method (FEM), but without including explicitly the crack into the FEM mesh. This approach is improved in a second step by determining at every growth step the crack path for the maximum opening stress. The results collected from these two steps are used in the final section to develop a full FEM lifetime model by tracking explicitly within the FEM mesh the crack propagation through the TBC for the maximum energy release rate. The paper is concluded with the discussion of the results obtained.

## CHARACTERIZATION OF THE BC-TBC INTERFACE PROFILE

The failure of the TBC system under thermal cycling, with a high temperature above 1000 °C and below 1300°C, usually takes place near the interface between the bondcoat and the thermal barrier coating. This is due to the growth of delamination cracks which run approximately parallel to the BC-TBC profile.[1] For a perfectly flat BC, the compression/tension stress developed within the TBC during the thermal cycling has only in-plane components, parallel to the BC-TBC interface, which therefore cannot lead to the just mentioned delamination crack growth. In order to produce stress components which would eventually open those cracks, a local curvature in the BC-TBC profile is required. Such condition fully applies to plasma sprayed BCs (and TBCs), where their inherent disordered character and fragmentation of the sprayed molten particles lead to an irregular, and thus curved, BC surface.

In a first approximation for the description of TBC failure, the BC-TBC interface will be considered periodic, with a certain characteristic wavelength $\lambda_0$ still to be determined. The idea behind this assumption is to consider that the failure crack growth is a local process, not influenced by other cracks outside the length scale $\lambda_0$, and taking place more or less simultaneously at similar places along the whole BC-TBC interface. The segment of the BC-TBC profile where this representative single crack growth occurs will be taken as a cosine function

$$y = A_0 \cos\left(\frac{2\pi}{\lambda_0} x\right) \tag{1}$$

The amplitude $A_0$ is determined through the measured average roughness of the BC profile, either the mean roughness $R_a$ or the root mean square roughness $R_q$

$$R_a = \frac{1}{L}\int_0^L |h(x)| dx \quad \text{or} \quad R_q = \sqrt{\frac{1}{L}\int_0^L (h(x))^2 dx} \tag{2}$$

where $L$ is the sampling length along the BC profile and $h(x)$ denotes the profile height (or deviation) with respect to the line of averaged height. Assuming that the periodic function in equation (1), with $L=\lambda_0$, describes effectively the BC surface profile whose experimental roughness is given by $R_a$ or $R_q$, the amplitude $A_0$ becomes

$$A_0 = \frac{\pi}{2} R_a \qquad \text{or} \qquad A_0 = \sqrt{2} R_q \tag{3}$$

For a typical 150μm thick MCrAlY bondcoat sprayed under low pressure conditions and injecting particles with diameter about 40μm, the resulting roughness is $R_a \approx 6\mu m$ and $R_q \approx 7\mu m$, which in both cases leads to an amplitude $A_0 \approx 10\mu m$.

The other characteristic length scale necessary to describe locally the BC-TBC profile is $\lambda_0$. The plasma sprayed BC (as well as the TBC) is formed through the flattening of impinging molten particles, called splats, and the subsequent disordered stacking of these splats into lamellae. The surface profile resulting from such process is of course non-periodic and thus a simple Fourier analysis of the BC profile cannot lead to a reliable estimation of $\lambda_0$ as the effective longitudinal scale of the profile. On the contrary, one could consider the sprayed profile to be completely random. In such case the BC surface would not contain any kind of preferred length scale and would result to be self-similar. A double logarithmic representation of the Fourier components as a function of the corresponding wavelengths would yield one straight line, with the slope related to the fractal dimension of the BC profile.

The surface of plasma sprayed coatings contains many irregularities with size scales ranging from 1μm to 1mm and is therefore rough. Nevertheless, the Fourier components for the profile of such coatings, when represented double logarithmically, are not arranged along one single straight line as it would correspond to a pure random profile. For plasma sprayed coatings, an inflection point characteristically separates two branches of different slope, one of higher inclination for small length scales (small wavelength $\lambda$ or equivalently large wave vector $k=2\pi/\lambda$) and second one of lower inclination for the large length scales corresponding to near macroscopic undulations. The length scale for this inflection denotes the smallest size at which the Fourier components deviate from a pure random signal. Hence the BC profile, although apparently rough and random, does contain a characteristic longitudinal scale. Further, the length value of this inflection point is well correlated to the size of the particle injected into the plasma jet to produce the coating.[2]

The existence of a length scale in the BC profile, manifested as the just mentioned inflection, should actually be expected from the very nature of the deposition technique: it is the footprint of the size of the flattened particles (splats) building the plasma sprayed coating. This interpretation is further supported from the possibility to derive the existence of the inflection for the Fourier spectrum and the observed values for the inclination of both branches from a simple stochastic model. Such model describes the splat flattening as a diffusion process under the additional influence of a spatially correlated noise corresponding effectively to processes at very small length scales (splat fragmentation, local flattening hindrance).[2]

Figure 1 shows a typical example for the Fourier transformation of a MCrAlY bondcoat surface profile, where the diameter of the sprayed particles ranges from 25μm to 55μm. A perthometer Mahr M2 rasters the height profile of the BC surface along a sampling length of $L$=4.1mm with a diamond head of radius 2μm, and this profile is Fourier transformed. The double logarithmic representation of the squared Fourier transformation components $\tilde{h}(k)$, $\tilde{D}(k) = 2\pi|\tilde{h}(k)|^2$, averaged over five different measured profile lines of the same sample, is represented as a function of the length scale $\lambda$ described by the wave vector $k=2\pi/\lambda$; the wavelength is measured in micrometer. The Fourier components are grouped into two different branches: the red line corresponding to the linear fitting of the range of large length scales (small wave vector $k$) and the green line to those of very small length. The inflection point in the Fourier spectrum for this bondcoat sample corresponds to a length scale of 70±10μm, which is equated to the wavelength $\lambda_0$ for the effective periodic function (1) describing locally the BC-TBC interface. This is the smallest length scale incorporated in the BC profile not being pure random noise and thus characteristic for the BC surface. It is also the length yielding the strongest curvature (the curvature of a cosine is a decreasing function of the wavelength) and therefore the highest stress level leading to the delamination of the ceramic TBC deposited on the BC. Hence, such $\lambda_0$ is the length scale determining the shortest lifetime of the TBC system.

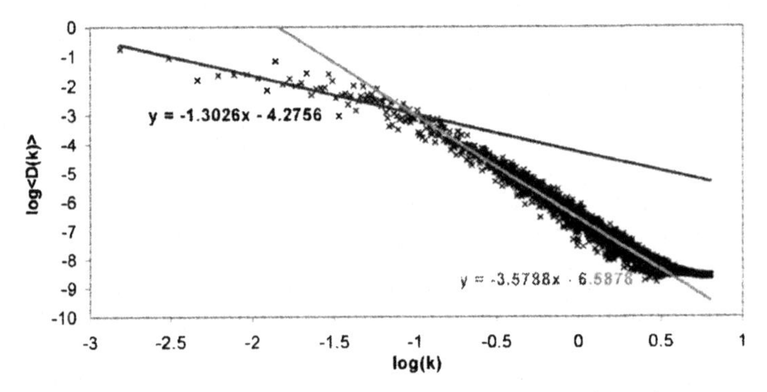


Figure 1. Double logarithmic representation of the averaged squared Fourier components for a plasma sprayed bondcoat, displaying the characteristic inflection point.

## SIMPLIFIED LIFETIME MODEL FOR HORIZONTAL AND INCLINED CRACKS

For the following discussion, the TBC system to be investigated consists of a flat metallic substrate 3mm thick made of IN738, a 150μm thick vacuum plasma sprayed (VPS) MCrAlY bondcoat with a resulting surface roughness of $R_a \approx 6$μm, and a 300μm thick atmospheric plasma sprayed (APS) thermal barrier coating made of YSZ.[3] The thermal cycling consists, firstly, of heating up the TBC surface for 5min by means of a gas burner and cooling simultaneously the back side of the metallic substrate, in order to achieve a controlled BC-TBC interface temperature about 1050°C; and secondly, removing the burner and, by means of compressed air, cooling rapidly the TBC surface down to a temperature slightly above 30°C. The sample is maintained at the latter temperature for 2min, before repeating the whole process.[4] During the thermal cycling, the TBC surface temperature is scanned with a pyrometer, and the temperature in the metallic substrate with a thermocouple. Under thermal cycling, the main stress sources within the TBC system are the thermal stress at heating up or cooling down, as well as the stress created by the growth of the TGO scale at the BC-TBC interface. Nevertheless, since both the thermal stress during the rapid heating and the growth stress occur at a high temperature, the TBC system, particularly the sprayed materials at the highest temperatures, is able to relax these stresses in a short time, without initializing crack growth. Only during the cooling down the temperature near to the BC-TBC interface goes down very rapidly and no relaxation process can be efficiently activated to reduce the resulting thermal stress. Hence crack growth, resulting in the TBC system failure, takes place only during the cooling down. This is confirmed by the measurement of acoustic emission (AE), which is directly correlated to crack growth, during the thermal cycling. As shown in Fig. 2, AE signals arriving at sensors located at three different positions on the back side of the TBC system mainly occur during the whole cooling down phase, not during the heating up or the hot phase.

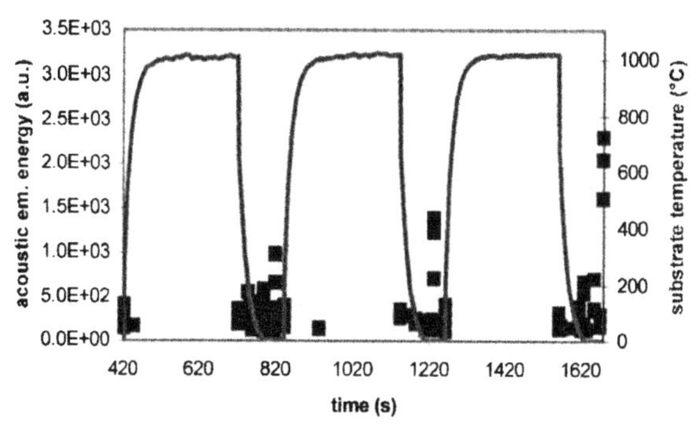


Figure 2. Acoustic emission energy during 3 cycles, with the corresponding temperature in the metallic substrate, after 1173 cycles.

Table 1. Thermo-mechanical parameters for the TBC system materials at room temperature.[7]

| | metallic substrate, IN378 | metallic BC, MCrAlY | oxide TGO, $Al_2O_3$ | ceramic TBC, YSZ |
|---|---|---|---|---|
| thermal expansion coeff. $\alpha$ | $15.8 \times 10^{-6} K^{-1}$ | $17.5 \times 10^{-6} K^{-1}$ | $8.0 \times 10^{-6} K^{-1}$ | $10.7 \times 10^{-6} K^{-1}$ |
| elastic modulus $E$ | 191GPa | 140GPa | 360GPa | 25GPa |
| Poisson number $\nu$ | 0.3 | 0.3 | 0.22 | 0.22 |

Let us discuss qualitatively the evolution of the thermal stress, firstly for the initial TBC system state, without an oxide scale separating the TBC from the BC. The thermal expansion coefficient of the ceramic TBC is lower than that of the underlying metallic layer (see Table 1). A local convex curvature (or "hill") along the BC-TBC interface can be approximated as a small BC lump surrounded by a quite extended TBC. A simple way to estimate the thermal stress in the neighborhood of the BC hill is to consider the surrounding TBC as an inert background and to subtract its thermal expansion coefficient from that of the materials locally in contact with the TBC. To the BC hill, thus, a positive effective thermal expansion coefficient $\alpha_{BC}-\alpha_{TBC}>0$ is assigned. During the cooling down, for a negative temperature change about $\Delta T$=-1000K near the BC-TBC interface, the top of the BC hill tries to contract but it is held by the inert TBC: the TBC above the hill wants to be pulled apart in radial direction. This TBC part is put under radial tension and horizontal TBC cracks situated near such a hill can grow due to the tensile stress. On the other hand, if the oxide scale separating the BC from the TBC has grown to a thickness high enough to screen the BC from the TBC, the previous reasoning is reversed. Now the small hill is made of a material with a lower $\alpha$ than the surrounding extended TBC and can be taken as behaving with a negative thermal expansion coefficient $\alpha_{TGO}$-$\alpha_{TBC}$<0 against an inert TBC. During the cooling down, the TGO hill tries now to expand but it is hindered by the TBC: the TBC region above the hill is now put under radial compression and cracks located there will become closed. The previous discussion applies analogously for a concave curvature (a "valley") in the BC-TBC interface, by interchanging the role of tensile and compressive stresses.

Summarizing, at the start of thermal cycling, those cracks running parallel to the BC-TBC interface which are placed near to convex curvatures (hill) are able to grow in the cooling phase. But only until they penetrate in a TBC region of concave curvature (valley) where they become temporarily halted. With the increasing oxidation of the BC during thermal cycling, the radial compressive stress around the BC-TBC valley gets progressively converted into tensile: the previously stopped delamination can continue growing. Assuming that such a process occurs simultaneously at every hill-valley, the failure of the TBC system takes place when the single crack vertex reaches a location above the valley middle point.[3]

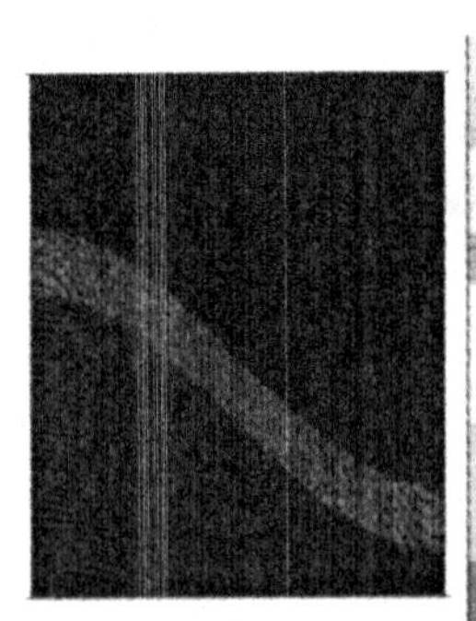

Figure 3. Left, FEM mesh of the upper BC part, a 5μm TGO and the lower TBC part, for a cosine profile with $A_0$=10μm and $\lambda_0$=70μm. Right, distribution of the thermal stress vertical component (range –250MPa to +250MPa) at the BC-TBC interface without oxide scale (middle) and at the TGO-TBC interface for a 5μm TGO (right).

As the next step, this qualitative failure mechanism is numerically implemented by means of the Finite Element method (FEM) using the software ANSYS (Ansys Inc., Canonburgh, Pittsburgh, PA, USA). With the already mentioned dimensions for the TBC system, and assuming the hill-valley structure of the BC-TBC profile to be the equation (1) with $A_0$=10μm and $\lambda_0$=70μm, half wavelength of such profile is generated with a fine meshing of approximately square cells of 0.5μm side for the region neighboring the BC-TBC interface. A plane strain state is considered. At one vertical flank a periodic boundary condition is fixed and at the opposite side a free displacement in horizontal direction for the whole flank is chosen. As shown in Figure 3, within the TBC and when no TGO is present, the vertical stress component $\sigma_{yy}$ is positive (tensile) above the hill, able to open horizontal cracks, but negative (compressive) above the valley. For an intermediate state in the TGO growth, the tensile region above the hill has disappeared, the compressive region above the valley has been reduced and the positive vertical stress has moved to the lateral flank of TGO-TBC profile. As already discussed, only the thermal stress during the cooling down will be considered here. Actually, a free standing plasma sprayed TBC is also able to display creep even at room temperature,[5] although such effect, regarding the crack growth at low temperatures in the TBC, will be not considered. The reason is that TBC creep is mainly a collective effect of all the sprayed lamellae when they slightly slide over each other. For a thermal cycling high temperature phase below 1300°C, the crack growth leading to the TBC delamination occurs only within the first two deposited lamellae, which are quite good adhered to each other and thus partially impeded to participate in the collective creep relaxation. Furthermore, the brittle TBC coating is assumed to behave elastically. The effect of plastic stress relaxation very near to the crack vertex is included effectively in the TBC fracture toughness value, being the calculation pure elastic for thermal stress at distances above 0.5μm from vertex.

Crack growth inside the TBC is sub-critical and proceeds slowly. Let $a$ denote the current crack length: the sub-critical crack growth under mode I (crack opening mode) is described by the Paris-Erdogan law as function of the stress intensity factor $K_I(\sigma,a)$ for mode I

$$\frac{da}{dt} = v_0 \left( \frac{K_I(\sigma,a)}{K_{I,crit}} \right)^m \quad (4)$$

with $K_{I,crit}$ the fracture toughness for mode I, approximately equal to 1MPa $m^{1/2}$ for plasma sprayed YSZ;[6] $v_0$ is the crack velocity at critical conditions, equal to $7.6x10^{-5}$m/s for YSZ, and $m$=18.[7] For a simplified formulation of the lifetime model, the thermal stress is obtained from the numerical FEM calculations (as in Fig. 3), without including explicitly the crack in the FEM mesh. And the stress intensity factor is approximated by that of a crack in an infinite plane under uniform stress, $K_I(\sigma,a) \approx \sigma\sqrt{\pi a/2}$, being now $\sigma$ the thermal stress component perpendicular to the flank of a fictive crack (which grows without modifying the stress distribution) evaluated at the vertex location of such crack. It is clear that this can only lead to an estimation of the lifetime prediction, since the crack does locally modify the stress distribution. However it will be used as a first approximation to discuss later the most rapid form of crack growth.

The TGO growth determines the sign of the thermal stress and thus whether a crack is able to increase its length or not. Additionally, another time dependent effect strongly contributes to accelerate the crack growth: the sintering of the plasma sprayed TBC due to its relative high porosity (about 10%). The crack growth evolution is therefore controlled by the current TGO thickness and sintering of the TBC elastic modulus, both of them described by a diffusion law $\sim t^{1/2}$ with a temperature dependent activation factor

$$d_{TGO} = A_{TGO} e^{-E_{TGO}/k_B T} \sqrt{t} \quad \text{and} \quad E_{TBC}(t) = E_{TBC}(t=0) + A_{sint} e^{-E_{sint}/k_B T} \sqrt{t} \quad (5)$$

with $A_{TGO}=3.715x10^{-3} m/s^{1/2}$ and $A_{sint}=2.49x10^{19} Pa/s^{1/2}$, and activation energies $E_{TGO}$=1.435eV and $E_{sint}$=3.10eV;[7] $T$ is to the temperature at the BC-TBC interface during the hot temperature phase. The elastic modulus for YSZ in as-sprayed state is taken as $E_{TBC}(t=0)=25x10^9$Pa.

Now the lifetime of the TBC system is estimated as follows: the stress distribution is numerically calculated for a homogeneous temperature change of $\Delta T$=-1000K and for different

TGO thickness between 0μm and 30μm. An initial horizontal crack of length 20μm, assumed to be already produced during the solidification of the sprayed molten lamellae, is situated (fictively, without being included in the FEM mesh) and centered over the BC-TBC hill. At every thermal cycle, during the 5min of the hot phase the two processes described by equation (5) take place. They determine the thermal stress which arises during the subsequent cooling down phase of 2min duration at which the crack grows according to equation (4). The thermal stress distribution is obtained by interpolation between the two corresponding FEM simulations nearest to the current TGO thickness. The process is repeated until the crack has covered a complete wavelength of the BC-TBC profile. For a crack running horizontally, opened only by a positive vertical stress component $\sigma_{yy}$, the evolution of crack length as a function of the increasing TGO is represented in Fig. 4 (black line, left diagram).

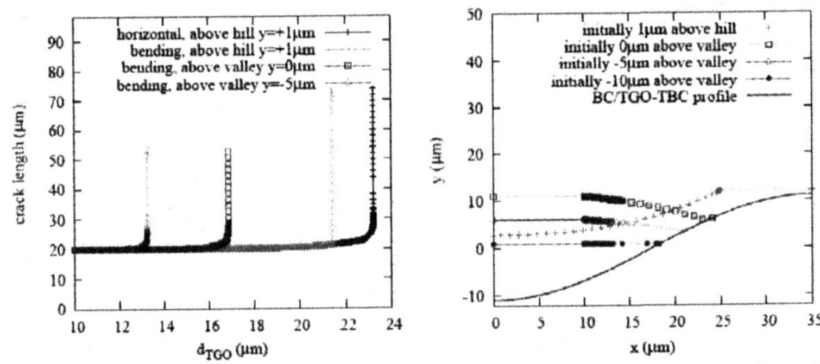


Figure 4. Crack length as function of TGO thickness and crack path, for a horizontal crack (black) and bending cracks, starting centered from the BC-TBC hill (green) or from the valley (red). The initial crack length is 20μm. Growth for valley cracks stopped when reaching TGO.

An improvement to the previous lifetime model is achieved if the crack is now allowed to bend and choose at each growth step the direction at which the crack vertex is "seeing" the maximum opening stress. Hence a faster crack growth results. Let $\sigma_{xx}$, $\sigma_{xy}$ and $\sigma_{yy}$ be the local stress distribution at the point where the fictive crack vertex is placed, then the azimuthal or hoop tension and the shear component along an inclination $\theta$ with the horizontal axis are

$$\sigma_{\theta\theta} = \sigma_{xx} \sin^2\theta - \sigma_{xy}\sin(2\theta) + \sigma_{yy}\cos^2\theta, \quad \sigma_{r\theta} = -0.5(\sigma_{xx} - \sigma_{yy})\sin(2\theta) + \sigma_{xy}\cos(2\theta) \quad (6)$$

The inclination $\theta = \beta$ at which the hoop stress is maximized is given by

$$\tan(2\beta) = 2\sigma_{xy} / (\sigma_{xx} - \sigma_{yy}) \quad (7)$$

together with the additional condition $\sigma_{xx} < \sigma_{yy}$ for ensuring $\left.\frac{d^2\sigma_{\theta\theta}}{d\theta^2}\right|_{\theta=\beta} < 0$ at $\beta \leq 45°$. Such inclination also corresponds to one local principal stress direction due to relation $\frac{d\sigma_{\theta\theta}}{d\theta} = -2\sigma_{r\theta}$.

From the simulated local distribution of thermal stress for the current location of the (fictive) crack tip, the bending angle $\beta$ is determined from equation (7), and the corresponding maximum stress component which makes the crack grow under mode I from

$$\sigma_{open} = \sigma_{\theta\theta}(\theta = \beta) = \frac{1}{2}\left[(\sigma_{xx} + \sigma_{yy}) + \sqrt{(\sigma_{xx} - \sigma_{yy})^2 + 4\sigma_{xy}^2}\right] \tag{8}$$

The stress intensity factor for the crack growth equation (4) is taken, in analogy to circularly curved cracks in an infinite plane under homogeneous stress[8], as $K_I \approx \sigma\sqrt{\pi a'/2}$, being $a'$ the crack length projected onto the horizontal direction.

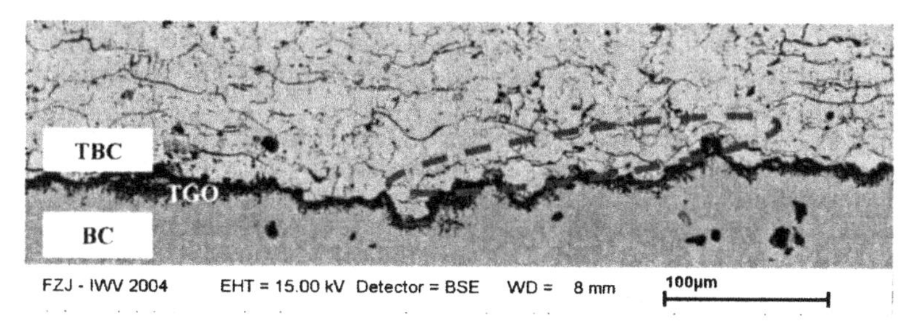


Figure 5. Micrograph of a TBC system after 1500 cycles with a temperature at the TBC surface of 1250°C and of 1020°C in the metallic substrate. The intermediate growth state of a delamination crack parallel to the TGO-TBC interface, and covering several hill-valley groups, is visible. The TGO is about 6µm thick..

The crack growth evolution thus achieved is shown in Fig. 4. Cracks starting from the BC-TBC hill, when the TGO reaches a thickness high enough, bend downwards toward the valley, where for the growing TGO the opening stress becomes increasingly larger. Such a crack, however, does not penetrate deep into the valley since a very large TGO thickness would be required to yield a positive opening stress. The crack growth now proceeds more rapidly than when the growth direction was fixed to be horizontal. The downward trajectory is also observed experimentally (see Fig. 5). Further, and according to the simplified lifetime model, cracks starting centered over the BC-TBC valley grow even faster than when starting over the hill. Here, the crack is initially not able to grow since it is located in the valley region of vertical compressive stress; only after a thick enough TGO has pulled down the compression region, the crack starts to grow rapidly, even for a TGO thickness lower than for hill cracks. The growth direction is slightly downwards, pointing to the interior of the TGO flank. In the latter case, failure is assumed when the crack reaches the TGO. For an inclined crack there exists in the TGO flank a stress level high enough to drive crack growth. This growth is additionally enhanced through the TGO elastic modulus, higher by one order of magnitude than that of the plasma sprayed YSZ, for quite similar fracture toughness values in both layers. Hence crack growth in the TGO can be assumed in a first approximation to be near critical and thus very fast.

However, the result on the starting point for fastest crack growth can only be provisional as discussed above. For the crack bending and growth only the stress distribution at the current crack tip location is used for deriving the stress intensity factor, without considering whether the remaining crack part is still open or closed. Hence, for a crack centered over the valley, the vertex can soon "feel" the concentration of opening stress being developed on the TGO-TBC flank (see Fig. 3, right), different from the crack starting over the hill, where its vertex is more distant from such a flank. The latter, nevertheless, extends for most of its length within a non-compression region, whereas a large part of the valley crack is under compression.

## EXTENDED LIFETIME MODEL: CRACK PATH IMPLEMENTED IN THE FEM MESH

In the previous simplified model, although the crack growth direction has been correctly reproduced, the necessary TGO thickness at failure is unrealistic high (above 13μm), when compared to actual failed TBC systems. Further, neglecting the extension and presence of the crack itself in the FEM calculation leads to a conclusion about which initial crack location is more relevant for the lifetime which can be wrong. Hence the model will be extended by explicitly tracking the crack growth in the TBC.

The accurate calculation of the stress intensity factor in equation (4) has to be carried out through a path integration around the crack tip, which requires locally a very fine mesh and, moreover, to re-adapt the mesh every time the crack grows. This path integral can be avoided by considering the change in elastic potential energy of the TBC system when the crack grows. Herein, the crack will be modeled by giving very soft properties (elastic modulus of 100Pa, negligible compared to that of the other materials, vanishing thermal expansion coefficient) to the mesh cells occupied by the crack. This can be easily implemented in the FEM simulation. At every crack growth step, the elastic energy $U$ is calculated for the current crack state and 9 further extensions: case 1, crack extended one cell ahead; case 2, one cell up; case 3, one cell down; cases 4/5, one cell up/down, one cell ahead; cases 6/7, one cell up/down, two cells ahead; cases 8/9, one cell up/down, three cells ahead (see Fig. 6, left). Then the energy release rate for each of the 9 cases is calculated, defined as the relative energy change between the current crack state of length $a$ and each crack extension, $G(\sigma,a)=\dfrac{U(\sigma,a)-U(\sigma,a+\Delta a)}{\Delta a}$, and the maximum value is selected. Since $G$ is equal, up to a material dependent parameter, to the squared stress intensity factor, the sub-critical growth equation (4) can be re-formulated as

$$G_{I(,crit)}(\sigma,a)=\frac{1-\nu_{TBC}^2}{E_{TBC}}\left(K_{I(,crit)}(\sigma,a)\right)^2 \;\Rightarrow\; \frac{da}{dt}=v_0\left(\frac{G_I(\sigma,a)}{G_{I,crit}}\right)^{m/2} \tag{9}$$

Nevertheless, one point should be considered. The energy release rate is always a positive scalar and thus the softening of a cell ahead of the crack vertex always reduces the energy, even if that cell was under compression and thus the crack would not be able to grow there. Therefore, the maximum energy release rate for crack growth has to be found out but only among the cases where the opening stress on the crack tip is positive. Hence at every growth step and for the current crack state, the stress distribution averaged over the cell ahead of the crack vertex is read (cell average to avoid the stress singularity just at the crack tip). As next step, angle $\beta$ is determined according to equation (7) and then the maximum energy release rate for mode I, $G_I$, is selected for only those two extension cases corresponding to the two directions nearest to the just calculated $\beta$. Such $G_I$ and crack extension direction are used to calculate the current growth velocity (eq. (9)). Since angle $\beta$ is referred to the horizontal, whereas the 9 crack extensions are defined on the local mesh (Fig. 6, left), angle $\varphi$ has to be used to convert $\beta$ into the corresponding crack extension case.

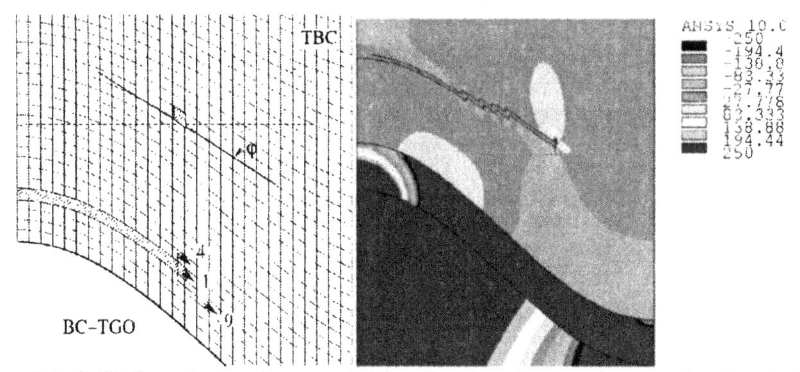


Figure 6. Left, TBC crack extension cases 1-4-9 and angle $\varphi$ between horizontal and crack tip cell. Right, vertical stress $\sigma_{yy}$ (range −250MPa to +250MPa) at an intermediate growth step for a TGO thickness (middle layer) of 5.2μm. Initial 20μm crack centered 10μm above hill.

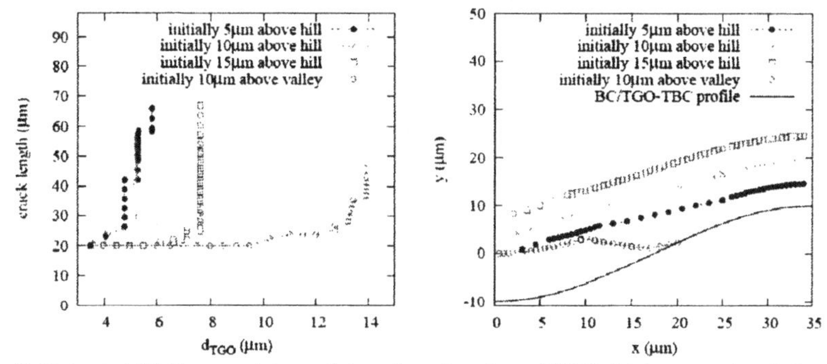


Figure 7. Extended lifetime model: crack length as function of TGO thickness and path for a crack starting centered from the BC-TBC hill (green and blue) or from the valley (red). The initial crack length is 20μm. Growth for valley crack stopped when reaching TGO.

Now the growth for initial 20μm cracks is simulated, both starting from the BC-TBC hill and valley. Since the typical thickness of the first sprayed TBC lamella is about 10μm, this is the reference height above the BC-TBC interface for the initial crack. In Figure 6 (right) the vertical stress distribution at an intermediate growth state for a hill crack is shown when the TGO has reached a thickness of 5.2μm. The crack path is similar to that in the simplified model of the previous section (Fig. 4, right) but now a quite lower TGO is required for the crack to cover a whole wavelength of BC-TBC hill-valley. Figure 7 displays the crack length and growth path for different initial crack locations. The shortest growth time corresponds to cracks starting from the hill (not from the valley as in the simplified model), particularly that initially placed 10μm above, which requires a 5.7μm TGO at failure (actually below 5.3μm for most of the crack growth). This value lies well within the values measured experimentally for delamination cracks (see Fig. 5). Also the crack path, following approximately parallel the hill-valley structure, is the correct one. A crack closer to the BC-TBC interface (Fig. 7, blue line) starts earlier but then grows more slowly across the valley.

It should be noted that cracks initially centered over the valley need a much thicker TGO to be able to start growing. The path followed then bends towards, and eventually penetrates into, the TGO. It might be that if the cooling down phase would start after a very long hot phase, such that the TGO has already grown above, let's say, 12μm, then valley cracks, being already in a positive tension region closer to the TGO-TBC interface and "feeling" thus a higher stress, would grow faster than hill cracks. This could yield an explanation why the failure for long isothermal oxidation occurs rather inside the TGO, in contrast to the thermal cycling where failure characteristically takes place within the first sprayed TBC lamellae.

## CONCLUSIONS

A lifetime model has been developed, which tracks directly in the FEM mesh the sub-critical growth of delamination cracks near the curved BC-TBC interface for an increasing TGO growth during thermal cycling. The force driving the crack growth is the thermal stress during the cooling down phase. The crack growth velocity is determined through a combination of trajectory for maximal opening stress and maximum energy release rate, and the resulting TGO thickness at failure well corresponds to experimental values.

## ACKNOWLEDGMENTS

The authors thank Mr K.-H. Rauwald and Mr R. Laufs for the sprayed samples, Mr M. Kappertz for the cross-section preparation and Dr D. Sebold for the careful micrographs.

## REFERENCES

[1]R. Vaβen, F. Traeger, D. Stöver, "Correlation Between Spraying Conditions and Microcrack Density and Their Influence on Thermal Cycling Life of Thermal Barrier Coatings," *J. Thermal Spray Technol.* **13**, 396-404 (2003).

[2]S. Giesen, "Characterization of plasma sprayed coatings by means of Fourier analysis and stochastic equations," diploma thesis, Fachhochschule Aachen/Jülich (2005), in German.

[3]R. Vaβen, G. Kerhoff, D. Stöver, "Development of a micromechanical life prediction model for plasma sprayed thermal barrier coatings," *Mater. Sci. Eng. A* **303**, 100-109 (2001).

[4]F. Traeger, R. Vaβen, K.-H. Rauwald, D. Stöver, "Thermal Cycling Setup for Testing Thermal Barrier Coatings," *Adv. Eng. Mater.* **5**, 429-432 (2003).

[5]M. Ahrens, S. Lampenshcerf, R. Vaβen, D. Stöver, "Sintering and Creep Processes in Plasma-Sprayed Thermal Barrier Coatings," *J. Thermal Spray Technol.* **13**, 432-442 (2003).

[6]S.R. Choi, D. Zhu, R. Miller, "Mechanical Properties/Database of Plasma-Sprayed $ZrO_2$-8wt% $Y_2O_3$ Thermal Barrier Coatings," *Int. J. Appl. Ceram. Tehcnol.* **1**, 330-342 (2004).

[7]F. Traeger, M. Ahrens, R. Vaβen, D. Stöver, "A life time model for ceramic thermal barrier coatings," *Mater. Sci. Eng. A* **358**, 255-265 (2003).

[8]H. Tada, P.C. Paris, G.R. Irwin, "The Stress Analysis of Cracks Handbook," Del Research Co., Pennsylvania, section 21 (1973).

# COMPARISON OF THE RADIATIVE TWO-FLUX AND DIFFUSION APPROXIMATIONS

Charles M. Spuckler
NASA Glenn Research Center
21000 Brookpark Rd.
Cleveland Ohio 44145

## ABSTRACT

Approximate solutions are sometimes used to determine the heat transfer and temperatures in a semitransparent material in which conduction and thermal radiation are acting. A comparison of the Milne-Eddington two-flux approximation and the diffusion approximation for combined conduction and radiation heat transfer in a ceramic material was preformed to determine the accuracy of the diffusion solution. A plane gray semitransparent layer without a substrate and a non-gray semitransparent plane layer on an opaque substrate were considered. For the plane gray layer the material is semitransparent for all wavelengths and the scattering and absorption coefficients do not vary with wavelength. For the non-gray plane layer the material is semitransparent with constant absorption and scattering coefficients up to a specified wavelength. At higher wavelengths the non-gray plane layer is assumed to be opaque. The layers are heated on one side and cooled on the other by diffuse radiation and convection. The scattering and absorption coefficients were varied. The error in the diffusion approximation compared to the Milne-Eddington two flux approximation was obtained as a function of scattering coefficient and absorption coefficient. The percent difference in interface temperatures and heat flux through the layer obtained using the Milne-Eddington two-flux and diffusion approximations are presented as a function of scattering coefficient and absorption coefficient. The largest errors occur for high scattering and low absorption except for the back surface temperature of the plane gray layer where the error is also larger at low scattering and low absorption. It is shown that the accuracy of the diffusion approximation can be improved for some scattering and absorption conditions if a reflectance obtained from a Kubelka-Munk type two flux theory is used instead of a reflection obtained from the Fresnel equation. The Kubelka-Munk reflectance accounts for surface reflection and radiation scattered back by internal scattering sites while the Fresnel reflection only accounts for surface reflections.

## INTRODUCTION

Thermal barrier coatings (TBCs) are being developed for high temperature applications in gas turbine engines. Some of the ceramic materials being considered for TBCs are semitransparent in the wavelength range where thermal radiation can be important. For example, zirconia can be transparent up to about 5 μm (refs 1 and 2). In a semitransparent material, combined conduction and radiation determine the temperature inside and the heat transferred through the material. The radiative heat transfer in a semitransparent material is determined by the absorption, emission, scattering, and refractive index. The reflection at an interface is determined by the refractive index of the materials on each side of the interface. For diffuse thermal radiation going from a material with a higher refractive index to a lower refractive index the surface reflection is increased by the total reflection of the incident radiation at angles greater than the critical angle. The thermal radiation emitted internally and by an opaque material into a semitransparent material depends on the refractive index squared. The internal thermal radiation transmitted through the interface of a semitransparent layer is reduced by the internal surface reflection, which may include total internal reflection, so the energy emitted will not exceed that of a blackbody. Therefore, the refractive index of a semitransparent material can have a considerable effect on the temperatures in a material.

The amount of thermal energy absorbed, emitted, and scattered by a material is determined by the absorption and scattering coefficients. The coefficients have units of reciprocal length. The reciprocal of these coefficients can be considered to be the mean distance traveled before absorption or scattering occur if the coefficients don't vary along their path (ref. 3 page 424). Because the temperature changes as a result of absorption or emission, these processes have a direct effect on the temperature. Scattered thermal radiation does not affect the temperature unless it is absorbed. Scattering in some cases can act as additional absorption in determining the temperature profiles in a material ref. 4. Part of the scattered radiation that is not absorbed will be scattered back out of the layer increasing its reflectivity.

The setting up and the solution of the exact spectral radiative transfer equation that include absorption, emission, and scattering is complex. Approximate solutions, such as the two-flux and diffusion methods, which are easer to solve, have been developed. The diffusion approximation is the simplest approximation with radiation treated as a diffusion process and absorption, emission, and reflection of thermal radiation occurring at the surfaces of the material. The two-flux approximation, which includes absorption emission and scattering, is more complicated and requires a computer solution. In the two–flux approximation, it is assumed that there is a radiative flux traveling in the positive and negative x-directions with radiation absorbed emitted and scattered inside the material and reflections occurring at the internal and external surfaces of the layer. The Milne-Eddington two-flux, diffusion, and exact solutions for an absorbing, emitting, and scattering plane layer were compared for a plane layer in ref. 5. The two-flux method was in good agreement with the exact solution for the conditions considered. The diffusion method was found to give good predictions for large optical thicknesses [optical thickness = (absorption coefficient + scattering coefficient) x thickness]. The diffusion approximation and the discrete ordinate method for a plane non-scattering glass layer on an opaque substrate and a two dimensional non-scattering glass in an opaque container were compared in refs. 6 and 7. The results from the diffusion approximation were reasonable for thick glass layers and greatly under predicted the temperature and heat flux for thin layers or layers with small opacity. The discrete ordinate and the diffusion method were used to predict the heat transfer in a cylindrical partially stabilized zirconia piece under going laser assisted machining ref. 8. The temperatures predicted by the diffusion solution are 130 K higher than those predicted by the discrete ordinate method, while the measured quasi-steady temperature approached an intermediate value.

In this paper diffusion and Milne-Eddington two flux solutions are compared for an emitting, absorbing, and scattering plane gray layer and a non-gray plane layer on a substrate. An absorption coefficient of $a = 0.1346\ cm^{-1}$ and a scattering coefficient of $\sigma_s = 94.38\ cm^{-1}$ were used as a base line. These coefficients are in the range of those of zirconia in the wavelengths where it is semitransparent ref. 2. To determine how scattering and absorption affect the accuracy of the diffusion solution compared to the Milne-Eddington two flux solution the absorption and scattering coefficients are increased and decreased from the base line. To try to get better agreement between the results of the diffusion and Milne-Eddington two-flux solutions, the reflectance used for the radiative heat input to the layer for the diffusion solution was changed from one that only has surface reflections to one that also included reflections from internal scattering and/or a substrate.

## MODEL

The models used are a semi-infinite plane gray semitransparent layer figure 1a and a

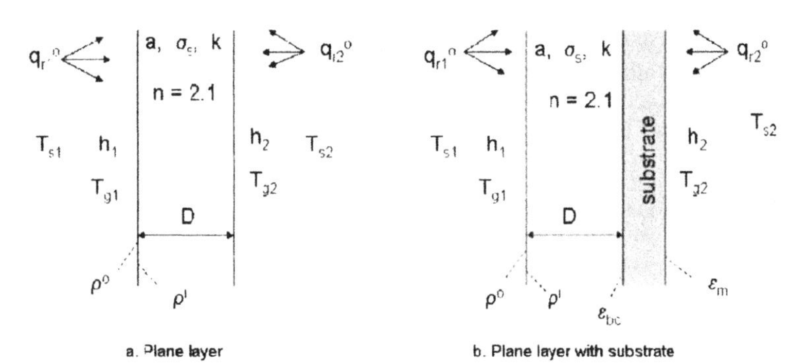


Figure 1 Heat transfer models

semi-infinite non-gray plane layer on a substrate figure 1b. The non-gray plane layer is opaque for wavelengths greater than 5 μm and semitransparent for wave lengths less than 5 μm. There is diffuse radiative and convective heat transfer on each side. The external radiative heating is $q_{r1}^{o}$ and $q_{r2}^{o}$. The hot side gas and surrounding temperatures, $T_{s1}$ and $T_{g1}$, are 2000K and the cold side temperatures $T_{s2}$ and $T_{g2}$ are 800K. The heat transfer coefficients are $h_1 = 250$ w/m$^2$K on the hot side and $h_2 = 110$ w/m$^2$K on the cold side. The plane layer and the layer on the substrate are 1 mm thick and have a thermal conductivity k = 0.8 w/mK. The substrate is 0.794 mm thick and has a thermal conductivity of 33 w/mK. The emissivity of the back side of the metal substrate, $\varepsilon_m$, is 0.6. The conditions on the layer with a substrate were used by Siegel (ref. 9) to determine internal radiation effects in a zirconia based TBC on a combustor liner. An infinitely thin bond coat is assumed between the semitransparent layer and the substrate and its emissivity is assumed to be 0.7 or 0.3. The refractive index, n, of the semitransparent layer is assumed to 2.1 which is in the expected range for zirconia ref. 1. The refractive index of the surrounding gas is assumed to be one. The external interface reflection, $\rho^o$, was obtained using the Fresnel equation for a non-absorbing layer (ref. 3 page 87). The non-absorption assumption should be good for the absorption coefficients used here (refs. 10 and ref. 3 page 88). The internal interface reflection which includes total internal reflections was obtained using eq. 35 in ref. 11. The Milne-Eddington two flux equations used are in ref. 5 and 12 and a method of solution is in reference 5. The Kubelka-Munk type two flux equations are in reference 1 and reference 13 page 193. The equation for the reflectance obtained from solving the Kubelka-Munk two flux equations is in the appendix. The diffusion equations were solved for both Fresnel reflectance and a Kubelka-Munk reflectance. The Fresnel reflectance only takes into account the surface reflection while the Kubelka-Munk reflectance takes into account surface reflection, radiation internally scattered back by the layer and substrate reflection if a substrate is present.

PLANE LAYER WITHOUT A SUBSTRATE

For the plane layer without a substrate, the Milne-Eddington two-flux solution was compared to the solution of the exact radiative transfer equations for an absorbing emitting and scattering layer for scattering coefficients from 0.944 $cm^{-1}$ to 943.81 $cm^{-1}$ and absorption coefficients from 0.0013 $cm^{-1}$ to 13.46 $cm^{-1}$. The exact radiative transfer equations along with a solution method are in ref. 4. The percent difference in the interface temperatures [100 x ($T_{2\text{-flux}}$– $T_{exact}$)/$T_{exact}$] was between - 0.075 and 0.28% for front gas-layer interface, between -0.44 and 0.21% for the back gas-layer interface and between 0.14 and 3.6% for the heat flux. Calculations indicate that if the number of spatial increments used in the solution of the exact equations is increased the percent difference in the heat flux would decrease. This shows that for the plane layer the results of the two flux solution are in good agreement with the exact solution. In the remainder of the paper the Milne-Eddington two-flux and diffusion solutions will be compared to determine the accuracy of the diffusion solution.

The percent difference in the temperature between the diffusion and Milne-Eddington two-flux solutions [100 x ($T_{diffusion}$ – $T_{2\text{-flux}}$)/ $T_{2\text{-flux}}$] for the front gas-layer interface of a gray

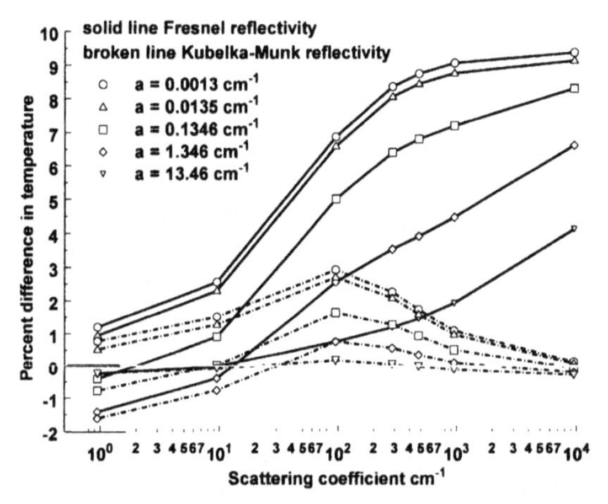


**Figure 2 Percent difference between diffusion and 2-flux solutions for front surface of plane layer**

plane layer is shown in figure 2. The symbols indicate points for which calculations were made. The percent difference in temperature calculated using Fresnel surface reflectivity and the Kubelka-Munk layer reflectivity for the radiative input in the diffusion solution are shown. The absolute value of the percent difference is less than 10% for all scattering and absorption coefficients considered. Using the Fresnel reflectivity, the percent difference in temperature for moderate to high scattering, increases with increased scattering and decreased absorption. For high scattering and low absorption the percent difference is leveling off at 9 to 10%. When the Kubelka-Munk reflectance was used, the maximum percent difference in temperature was less than 3% and that occurred for a scattering coefficient of 94.38 $cm^{-1}$ and the lowest absorption. For any absorption combined with high scattering the percent difference is near zero using the Kubelka-Munk reflectivity. For high absorption and all scattering considered and for high scattering and all absorption considered the diffusion approximation with a Kubelka-Munk reflectance gives good results, because radiation scattered back is taken into account.

The percent difference in the temperature of the back surface of a plane gray layer for the diffusion and Milne-Eddington two flux solutions is shown in figure 3. The absolute value of the

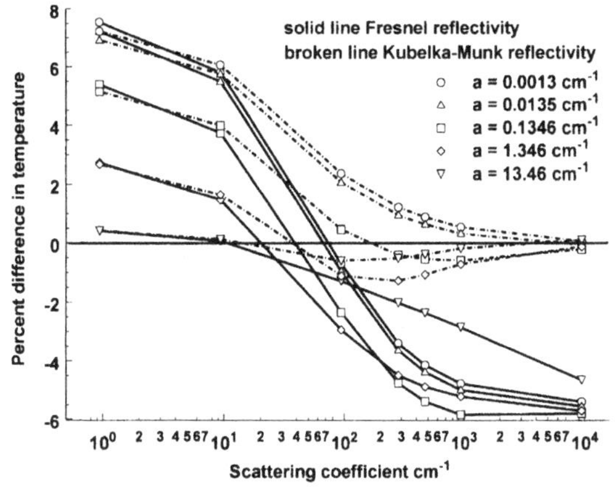


**Figure 3 Percent difference between diffusion and 2-flux solutions for back surface of plane layer**

percent difference is less than 8%. When the Fresnel reflectivity is used the percent difference goes from a positive value at low scattering to a negative value at high scattering. For low scattering the diffusion solution yields a higher temperature than the Milne-Eddington two-flux solution and the opposite occurs for high scattering. When the Fresnel reflectivity is used, the percent difference in temperature approaches a value of -5 to -6% for high scattering for all except the highest absorption considered where the percent difference is lower. For low scattering the percent difference in temperature is about the same whether the Fresnel or the Kubelka-Munk reflectivity are used. As the scattering is increased, the percent differences in temperature for the two different reflectivities diverge. When the Kubelka-Munk reflectance is used, the percent difference in temperature between the Milne-Eddington two-flux and diffusion solution is small for highest absorption used and all scattering considered. Also, for the Kubelka-Munk reflectance the percent difference approaches zero for all absorption considered and high scattering.

The percent difference in the heat flux calculated using the Milne-Eddington two-flux and the diffusion solutions, is shown in figure 4. The percent difference in heat flux using the Fresnel reflectivity for the diffusion solution is less than -8% for all absorption considered and a scattering coefficient of 94.38 $cm^{-1}$ or less. For a scattering coefficient greater than 94.38 $cm^{-1}$ the percent difference increases rapidly reaching over 200% for low to moderate absorption. The reason for this high a difference is that radiation scattered back is not accounted for when the Fresnel reflectivity is used. When the Kubelka-Munk reflectance is used the percent error is relatively constant for a scattering coefficient less than around 94.38 $cm^{-1}$ and is nearly -47% for the lowest absorption. The percent difference decreases as the absorption increases. Above scattering coefficient 94.38 $cm^{-1}$, the percent difference decreases with increased scattering reaching less than -10% for high scattering. For the highest absorption coefficient used the percent difference between the two solutions is less than 10%. For low to moderate scattering,

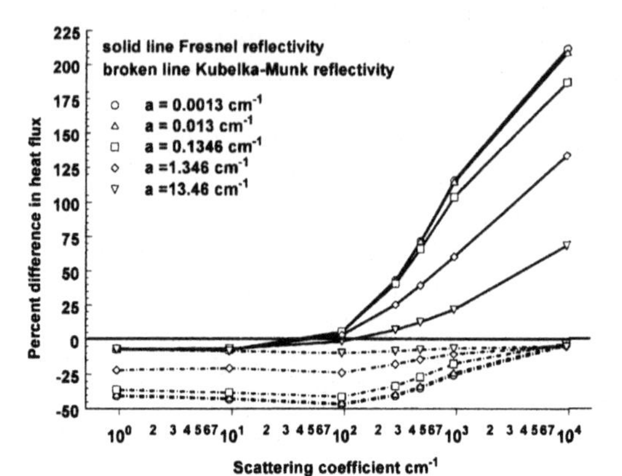


**Figure 4 Percent difference in heat flux between diffusion and 2-flux solutions for plane layer**

using the Fresnel reflectivity gives the best agreement between the diffusion and two flux solutions. When the Kubelka-Munk reflectivity is used the best agreement occurs for highest absorption and all scattering considered and for the highest scattering and all absorption used.

PLANE LAYER WITH SUBSTRATE

For the non-gray plane layer on a substrate, the layer was assumed to be semitransparent up to 5 μm wavelength. Above 5μm the material was considered to be opaque. In the Milne-Eddington two-flux solution, the Fresnel reflectance was used for all wavelengths. In the

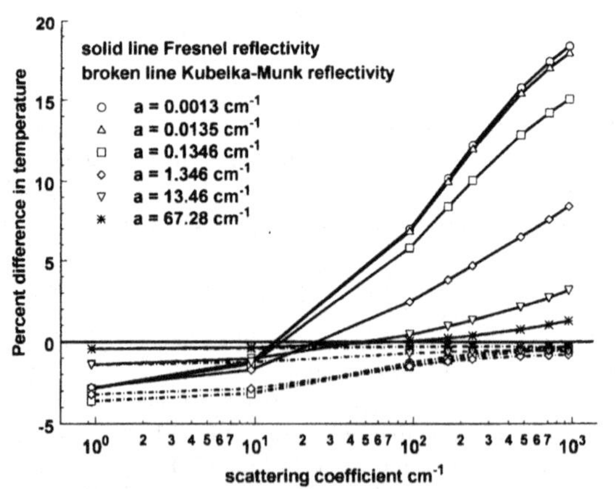


**Figure 5 Percent difference between diffusion and 2-flux solutions for front surface of layer with substrate reflectivity changes at 5μm**

diffusion solution when the term Fersnel reflectivity is used, means that the Fresnel reflectance was used for wavelengths above and below 5 μm. But, in the wavelength range less than 5 μm the material is semitransparent and internal scattering increases the reflectance of the layer. To try to account for this increase in reflectance the diffusion solution was also solved using the Kubelka-Munk reflectance at wavelengths less than 5 μm and the Fresnel reflectance for wavelengths greater than 5 μm; this is termed the Kubelka-Munk reflectivity.

The percent difference in temperature obtained using the Milne-Eddington two-flux and the diffusion solution for the front gas-layer interface of a layer on a substrate with a bond coat emissivity of 0.7 is in figure 5. Using the Fresnel reflection for all wavelengths, the percent difference in temperature goes from about a -3% at low scattering for all except higher absorptions to almost 20% for high scattering and low absorption. When The Kubelka-Munk reflectance was used for wavelengths less than 5 μm, the absolute value of the percent difference in temperature was less than -4% and decreased with scattering except for absorption coefficients of 13.46 $cm^{-1}$ and higher where there was a slight increase in the absolute percent difference for higher scattering. For high scattering coefficients the percent difference in temperature is less than -1%.

For the interface between the semitransparent layer and the substrate, the percent difference in the temperature using the diffusion and Milne-Eddington two-flux solutions for a bond coat emissivity of 0.7 is shown in figure 6. When the Fresnel reflectivity was used, the

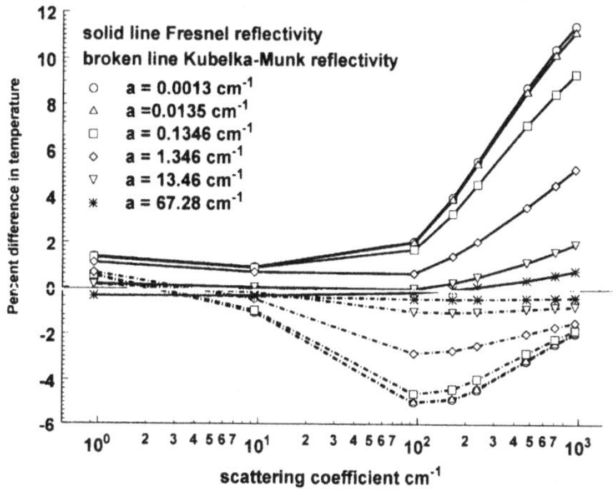


**Figure 6 Percent difference in temperature between diffusion and 2-flux solutions for front of substrate reflectivity changes at 5 μm**

percent difference in temperature is less than 1.4% for the lowest scattering. The percent difference at first decreases slightly with increasing scattering before increasing. The highest percent error occurs for high scattering and low absorption where the difference reaches about 12%. When the Kubelka-Munk reflectivity is used at the lower wavelengths, the percent difference in temperature is less than 0.7% for the lowest scattering. It decreases with scattering becoming negative and reaching a maximum negative value for a 94.38 $cm^{-1}$ scattering coefficient. The maximum negative percent difference occurs for the lowest absorption. For high absorption and all scattering the diffusion and Milne-Eddington two-flux solution are in good agreement. The percent difference in temperature for the back surface of the substrate is quite similar to the percent difference in temperature for the front surface of the substrate.

The percent difference in the heat flux using the diffusion and Milne-Eddington two-flux solutions for a layer on a substrate with a bond emissivity of 0.7 is shown in figure 7. The

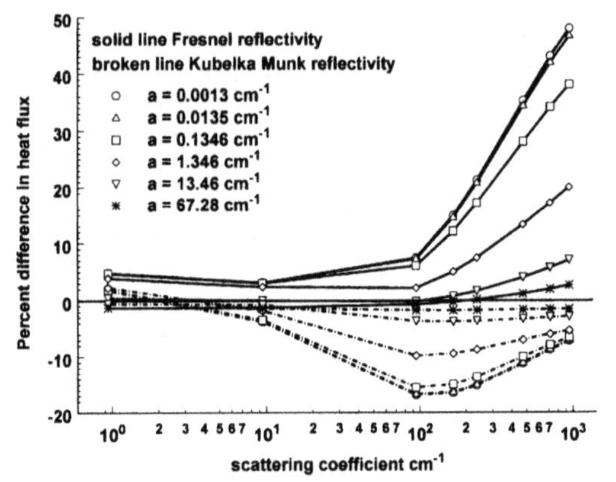


**Figure 7 Percent difference in heat flux between diffusion and 2-flux solutions for layer with substrate reflecitivity changes at 5μm**

shapes of the curves are similar to those in figure 6 for the percent difference in temperature of the substrate, but the percent difference in heat flux is 3 to 4 times higher than the percent difference in temperature. When the Fresnel reflectivity is used for the diffusion solution, the absolute value of the percent difference in heat flux is as high as 6% for the lowest scattering. For a 94.38 $cm^{-1}$ scattering coefficient, the percent difference in heat flux begins to increase rapidly for lower absorption. The percent difference reaches nearly 50% for the lowest absorption used. When the Kubelka-Munk reflectivity is used for the diffusion solution, the absolute value of the percent difference in heat flux is less than about 3% for the lowest scattering. The maximum percent difference which occurs at a 94.38 $cm^{-1}$ scattering coefficient is around -17%. For higher scattering the percent difference decreases with scattering. For absorption coefficients greater than 13.46 $cm^{-1}$ and all scattering the percent difference in heat flux was less than -4%.

When the emissivity of the bond coat was change from 0.7 to 0.3 and the Fresnel reflectivity was used, the change in the absolute value of the percent difference between the diffusion and Milne-Eddington two-flux solution was less than 3% for the surfaces temperatures and was 11.3% for the heat flux. When the Kubelka-Munk reflectivity was used the absolute value of the percent difference between the diffusion and Milne-Eddington two-flux solution was less than 2% for the surfaces temperatures and was 4.0% for the heat flux. This indicates that the Kubelka-Munk reflectivity can handle a change in bond coat emissivity better than the Fresnel reflectivity.

## CONCLUSIONS

A study was performed to determine the accuracy of the diffusion solution compared to the Milne-Eddington two-flux solution. A one dimensional model was used and a gray plane layer and a non-gray plane layer on a substrate were considered. The layer was 1 mm thick and the substrate was 0.794 mm thick. There is convective and diffuse radiative heat transfer on each side of the layer. For the diffusion solution Fresnel reflectivity, which only takes surface reflection into account, and Kubelka-Munk type reflectivity which take surface reflection and radiation scattered back into account were used. When the non-gray layer on a substrate was considered, Fresnel reflectivity or Kubelka-Munk type reflectivity was used for the wavelengths where the material was semitransparent and Fresnel reflectivity was used for the opaque regions for the diffusion solution. For the plane layer the Milne-Eddington two-flux solution was compared to the exact solution and found to be in good agreement with a percent difference less than 0.45% for the temperature and less than 3.6% for the heat flux. For the plane gray layer, the largest percent difference between the Milne-Eddington two-flux and the diffusion solutions occurred when Fresnel reflectivity was used and there was high scattering and low absorption, except for the back surface of the plane layer were the difference was higher for low scattering and low absorption. The percent difference in interface temperatures for the diffusion and Milne-Eddington two-flux solution for the plane layer were small for high absorption and all scattering considered, and for high scattering and all absorption considered if the Kubelka-Munk type reflectivity was used. The percent difference in heat fluxes was also small for high scattering if the Kubelka-Munk type reflectivity was used. For the non-gray layer on the substrate, the percent difference in temperature between the Milne-Eddington two-flux and the diffusion solutions were less than 5% for low scattering when either the Fresnel or Kubelka-Munk reflectivities were used. When the Kubelka-Munk reflectivity was used the percent difference in temperature was less than 6% for all scattering and absorption used, while the percent difference was nearly 12% for high scattering and low absorption when the Fresnel reflectivity was used. The percent difference in heat flux was nearly 50% for high scattering and low absorption when the Fresnel reflectivity was used. When the Kubelka-Munk type reflectivity was used the percent difference in heat flux was less than 17%. The diffusion solution can be improved under some conditions if a Kubelka-Munk type reflectivity is used. Also using the Kubelka-Munk type reflectivity seems to account for a change in substrate emissivity better than using the Fresnel reflection.

For the heating conditions of combined external radiative heat load and convection the absorption thickness (absorption coefficient x layer thickness) not the optical thickness [(absorption coefficient + scattering coefficient) x layer thickness] is the crucial factor in determining whether the use of the diffusion approximation is appropriate. The most inaccurate results from the diffusion approximation are generally for high scattering thickness (scattering coefficient x layer thickness) with no absorption or low absorption thickness (absorption coefficient x layer thickness) with only surface reflections accounted for. Using a Kubelka-Munk reflectivity, which includes radiation scattered back, can increase the accuracy of the diffusion solution approximation for large scattering thicknesses.

REFERENCES

[1]Wahiduzzaman, S and Morel,T., Effect of Translucence of Engineering Ceramics on Heat Transfer in Diesel Engines, ORNL/Sub/88-22042/2, April 1992.

[2]Makino, T., Kunitomo, T., Sakai, I., and Kinoshita, H., Thermal Radiation Properties of Ceramic Materials, *Heat Transfer-Japanese Research,* **13,** [4] 33-50 (1984).

[3]Siegel, R. and Howell, J. R. *Thermal Radiation Heat Transfer,* $4^{th}$ ed. Taylor & Frances, New York, 2002.

[4]Spuckler, C. M. and Siegel, R., "Refractive Index and Scattering Effects on Radiative Behavior of a Semitransparent Layer," *Journal of Thermophysics and Heat Transfer*, **7**[2], 302-10 (1993).

[5]Siegel, R. and Spuckler, C. M., "Approximate Solution Methods for Spectral Radiative Transfer in High Refractive Index Layers," *International Journal of Heat and Mass Transfer*, **37** [Suppl. 1] 403-13 (1994).

[6]Lee, K.H and Viskanta, R. "Comparison of the diffusion approximation and the discrete ordinates method for investigation of heat transfer in glass," *Glastech. Ber. Glass Technol.* 72[8], 254-266 (1999).

[7]Lee, K.H and Viskanta, "Two-dimensional combined conduction and radiation heat transfer: comparison of the discrete ordinates method and the diffusion approximation methods," *Numerical Heat Transfer, Part A*, 39, 205-225 (2001).

[8]Pfefferkorn, F. E., Incropera, F. P., and Shin, Y. C. "Heat Transfer model of semi-transparent ceramics undergoing laser-assisted machining," *International Journal of Heat and Mass Transfer*, 48, 1999-2012 (2005).

[9]Siegel, R. "Internal Radiation Effects in Zirconia Thermal Barrier Coatings," *Journal of Thermophysics and Heat Transfer*, **10**[4], 707-9 (1996).

[10]Cox, R. L., "Fundamentals of Thermal Radiation in Ceramic Materials,"; pp. 83-101 in Symposium on Thermal Radiation of Solids, edited by S. Katzoff, NASA SP-55, 1965.

[11]Richmond, J. C., "Relation of Emittance to Other Optical Properties," *Journal of Research of the National Bureau of Standards-C. Engineering and Instrumentation*, **67C** [3], 217-26 (1963).

[12]Siddall, R. G. "Flux methods for the analysis of radiant heat transfer," *Proceedings of the Fourth Symposium on Flames and Industry*, Paper 16, pp. 169-179, The Institute of Fuel (1972).

[13]Ishimaru A., *Wave Propagation and Scattering in Random Media Vol. 1 Single Scattering and Transport Theory*, Academic Press, New York, 1978.

APPENDIX EQUATIONS

The equation for the reflectance obtained from solving the Kubelka-Munk type two-flux equations is

$$R = \rho^{o} + \frac{(1-\rho^{i})}{2}\left\{A\left[\frac{\sqrt{1-\omega^{2}}}{(1-\omega)} - 1\right] - B\left[\frac{\sqrt{1-\omega^{2}}}{(1-\omega)} + 1\right]\right\} \quad (1)$$

Where A and B are

$$A = \frac{\dfrac{2.0(1-\rho^{o})(1-\omega)}{(1-\rho^{i})\sqrt{1-\omega^{2}} + (1+\rho^{i})(1-\omega)}}{1.0 - \left[\dfrac{(1+\rho_{bc})(1-\omega) - (1-\rho_{bc})\sqrt{1-\omega^{2}}}{(1+\rho_{bc})(1-\omega) + (1-\rho_{bc})\sqrt{1-\omega^{2}}}\right]\cdot\left[\dfrac{(1+\rho^{i})(1-\omega) - (1-\rho^{i})\sqrt{1-\omega^{2}}}{(1+\rho^{i})(1-\omega) + (1-\rho^{i})\sqrt{1-\omega^{2}}}\right]\cdot e^{-2\sqrt{1-\omega^{2}}\cdot\tau_{L}}} \quad (2)$$

$$B = -\frac{\dfrac{2.0\left(1-\rho^{o}\right)\left(1-\omega\right)}{\left(1-\rho^{i}\right)\sqrt{1-\omega^{2}}+\left(1+\rho^{i}\right)\left(1-\omega\right)}e^{-2\sqrt{1-\omega^{2}}\,\tau_{L}}}{\left[\dfrac{\left(1+\rho_{bc}\right)\left(1-\omega\right)+\left(1-\rho_{bc}\right)\sqrt{1-\omega^{2}}}{\left(1+\rho_{bc}\right)\left(1-\omega\right)-\left(1-\rho_{bc}\right)\sqrt{1-\omega^{2}}}\right]-\left[\dfrac{\left(1+\rho^{i}\right)\left(1-\omega\right)-\left(1-\rho^{i}\right)\sqrt{1-\omega^{2}}}{\left(1+\rho^{i}\right)\left(1-\omega\right)+\left(1-\rho^{i}\right)\sqrt{1-\omega^{2}}}\right]\cdot e^{-2\sqrt{1-\omega^{2}}\cdot\tau_{L}}} \tag{3}$$

$\rho^{o}$ = external interface reflection
$\rho_{bc}$ = reflectivity of the bond coat for a layer with a substrate
$\rho_{bc}$ = $\rho^{i}$ for a plane layer without a substrate
$\rho^{i}$ = internal interface reflection
$\tau_{L} = (2a+\sigma_{s})\cdot D$

$$\omega = \frac{\sigma_{s}}{\sigma_{s}+2a}$$

a = absorption coefficient
D = thickness of the semitransparent layer
$\sigma_{s}$ = scattering coefficient

## DAMAGE PREDICTION OF THERMAL BARRIER COATING

Y. Ohtake
Ishikawajima-Harima Heavy Industries Co., Ltd.
1, Shin-Nakahara-Cho, Isogo-ku,
Yokohama-shi, Kanagawa 235-8501, Japan

### ABSTRACT

Thermal barrier coatings are applied to many high temperature airplane engine and gas turbine hot section parts. The durability of the thermal barrier coating determines the life of the parts. Thus, the development of a thermal fatigue life prediction method of thermal barrier coating is paramount to the design of the parts. The fracture mechanism of thermal barrier coating by thermal cycle test and furnace heating test are examined here. All specimens fractured in delamination of the thermal barrier coating. The delamination driving forces were calculated using the finite element method. The stress in thermal barrier coating was calculated using two models; a full scale 3D model and a subscale 2D model to capture the waviness of the BC/TC interface. It was found that the delamination was due to the growth of thermal growth oxidation layer, thermal stress, the shape of the part, the configuration of the interface and the mismatch strain at the interface. Thus, this paper proposed a simplified method to predict the thermal fatigue life of thermal barrier coating based on the present results. A damage factor was proposed to evaluate the damage of the thermal barrier coating. The factor was constructed using three parameters: a) thickness of thermal growth oxidation layer, b) stress, and c) mismatch strain at the interface. These parameters are determined from a combination of the experiment and analytical results. The damage prediction method is very simple to apply and hence it is effective in the design of the high temperature coated parts.

### INTRODUCTION

Thermal barrier coating (top coat ; TC) is applied to the parts of the airplane engine and gas turbine. Top coat is used for a coating system together environment barrier coating (bond coat ; BC). Fracture of the top coat occurs by either delamination or the crack in the normal direction to the thickness of the top coat when the coating system is heated. The delamination of top coat occurs near the BC/TC interface by the growth of thermal growth oxidation (TGO) layer at the interface and other causes [1]-[3]. The vertical crack in top coat is due to the difference of thermal expansion of the materials. We had examined the damage of a plate specimen with a typical coating system by burner rig test [1]. The damage was the delamination of top coat at the interface [1]. The delamination of thermal barrier coating had been investigated by a lot of researchers until now [4]-[10]. This paper proposed a simplified method of thermal fatigue life to predict the delamination of top coat. The method should be used to predict thermal fatigue life of thermal barrier coating in the design and it may be used for the decision of the shape of application part and the compositions (thickness, material, layer number) of top coat in the design for the application of coating system to the part.

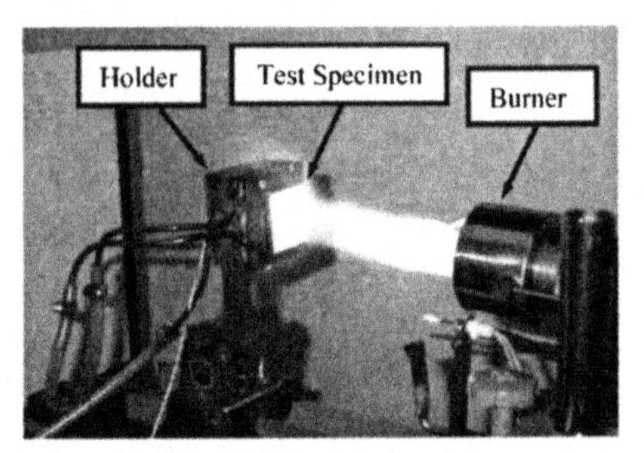


Fig.1 Appearance of thermal cycle test

EXPERIMENTAL PROCEDURE

The specimen was made of a coating system that was applied to single crystal CMSX-2 substrate of nickel base superalloys. The coating system consisted of bond coat and top coat. The bond coat was CoNiCrAlY that was deposited by low pressure plasma spray (LPPS). The top coat was 8 wt. percent yttria stabilized zirconia (YSZ) that was deposited by air plasma spray (APS). Thermal fatigue damage of top coat was examined in thermal cycle test by burner rig [1]. A plate specimen with corner edges was used for the test. The dimension of substrate of the specimen is length 500mm, width 500mm and thickness 3mm. Thickness of top coat and bond coat of the specimen are 0.5mm and 0.125mm, respectively. The total cycle time is 5 min with 3 min of holding time at maximum temperature for a total of 1000 cycles. Figure 1 shows a photograph of the burner rig test. The burner rig test heats on the surface of top coat of specimen by heating gas and cools on the back surface by air, and then the specimen has a temperature distribution in the direction of the thickness in heating. The distributions produce stress in the specimens because the thermal expansions of top and bond coats are different from one of substrate. The specimen was removed to examine the damage of top coat after the thermal cycling test. A second test was performed to examine the damage of top coat and the growth of TGO layer [2][3]. The test specimen consisted of small circular plate used in the furnace test. The dimension of substrate of the specimen is diameter 20mm and thickness 3mm. Two coating thickness were used 0.3mm and 0.5mm, with the same bond coat thickness of 0.125mm. The specimens were tested at 800°C, 900°C and 1200°C [2][3]. The heating intervals were 100h, 200h and 500h. The specimen was examined for the damage of top coat and to measure the thickness of the TGO layer. An experimental relation was deduced based on the heating time and the thickness of TGO layer to determine the TGO growth rates at temperatures.

ANALYTICAL PROCEDURE

The stresses in the thermal fatigue specimens were calculated using the finite element method [1]. Figure 2 shows the mesh of the analytical model (Specimen model). The model consists of 8-nodes three-dimensional solid elements. The effect of coating waviness was examined using a two-dimensional model to examine the stress in top coat near the interface between bond coat and top coat after the growth of TGO layer. Figurer 3 shows the mesh of the analytical model. The model consists of 4-nodes generalized plane strain elements. A TGO layer is inserted in the model and it grows with heating time. The BC/TC interface shape is modeled as cosine wave with three different amplitudes 5μm, 10μm and 20μm. The furnace temperature is set at 900°C uniformly. The TGO growth rate is 1.5μm/s and the total heating time is 6000 second. The results of burner rig test [1] were used to determine the relationship between the temperature and the growth rate of TGO layer. Table 1 denotes material properties of the coatings that were used to make the FEM analysis. Those properties of substrate are taken from [2] [11]. The material property of TGO layer was adopted from [12]. The bond coat was supposed for elastic-ideal plastic material [12] and is added creep property in combination with the Norton law [13]. The ABAQUS program [14] was used for the finite element calculations.

**Table.1 Material property**

| | Temperature K | Bond coat CoNiCrAlY | Top coat APS YSZ | Top coat EB-PVD YSZ |
|---|---|---|---|---|
| Thermal conductivity (W/mK) | 373 | 10.10 | 0.95 | 1.80 |
| | 773 | 19.30 | 0.96 | 1.80 |
| | 1273 | 25.32 | 0.97 | 1.84 |
| | 1473 | 29.87 | 0.98 | 1.84 |
| Specific heat (J/kg_K) | 373 | 429 | 608 | 498 |
| | 773 | 430 | 610 | 505 |
| | 1273 | 410 | 600 | 500 |
| | 1473 | 415 | 605 | 510 |
| Density ($kg/m^3$) | 298 | 7078 | 5723 | 3780 |
| Young's modulus (GPa) | 298 | 177.5 | 38.0 | 44.0 |
| | 873 | 151.3 | | – |
| Poission's ratio (–) | 298 | 0.3 | 0.3 | 0.3 |
| Thermal expansion coefficient ($\times 10^{-5}$/K) | 1273 | 15.70 | 11.10 | 8.12 |

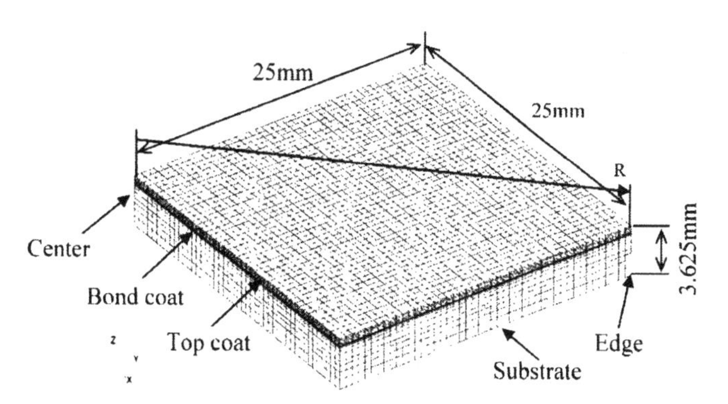


Fig.2 Modeling and division of mesh element for FEM analysis

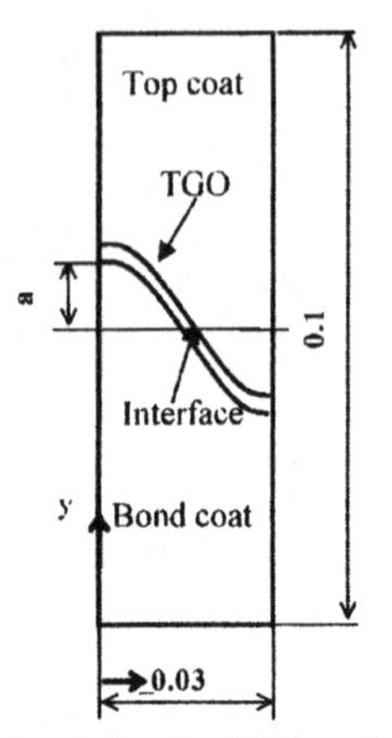


Fig.3 Modeling for FEM analysis

EXPERIMENTAL RESULTS

Figure 4 shows two photographs of the microstructure of the plate specimen after thermal cycle test after 1000 cycles. The observed damage as seen in Fig.4 was delamination of the top coat near the BC/TC interface. The delamination was detected at two locations at the center and the edges of the specimen. Many small cracks were observed in top coat near the interface. The cracks initiated from the pores in top coat. Thus, it is found that the delamination of top coating occurred by linking of the micro-cracks in the top coat near the interface. Moreover, a TGO layer was observed at BC/TC interface. The delamination of top coat near the interface may be affected by the growth of the TGO layer. Figure 5 shows the relationship between heating time and the thickness of TGO layer at the interface.

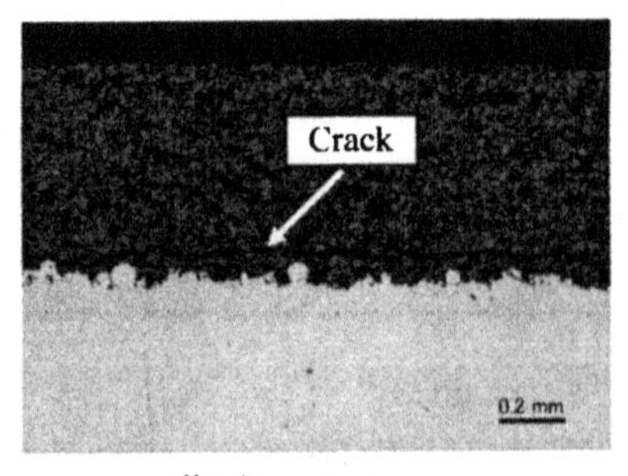


(i) At edge area (ii) At center area

Fig. 4 Microstructure of specimen at 1000 cycles

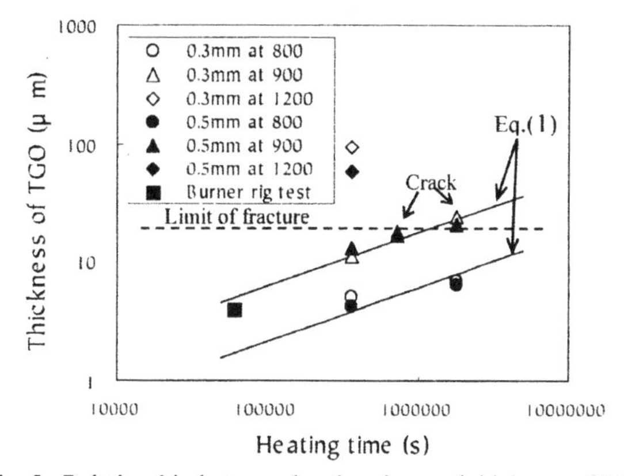


Fig. 5 Relationship between heating time and thickness of TGO layer

The results of heating test and thermal cycle test were denoted in Fig.5. It was found that the thickness of TGO layer increased with test temperature and heating time. The growth of TGO layer was almost independent of the initial thickness of the top coat. The growth of TGO layer can be easily modeled as a relationship in terms of thickness of TGO layer w. heating time t and two constants k and n.

$$w = kt^{n} \tag{1}$$

The line at 900°C in Fig.5 was k=0.035 and n=0.45 and the line at 800 was k=0.013 and n=0.45. The results of burner rig test could be predicted by using the 900°C curve of Fig. 5. Small cracks are detected in top coat of those specimens after 200h and 500h at 900°C. The cracks propagated parallel to the interface direction of top coating surface from pore near the interface. The delamination of top coat of the specimen occurred after 100h at 1200°C. Thus, it was found that TGO layer was grown as increasing test temperature and heating time, so that the failure occurred by delamination of top coat near the BC/TC interface.

## ANALYTICAL RESULTS

Figure 6 shows the calculation results of specimen at 20s after heating start in burner rig tests. Figure 6(i) shows the relationships between distance R and the stresses in full scale model in Fig.2. Those stresses are normal stress $\sigma_z$ and shear stress $\tau_{zx}$ in mesh of top coat at BC/TC interface. The maximum values of those stresses were generated at the corner edge of the model by the effects of both the shape of the specimen and thermal stress. It resulted the delamination of top coat at the edge area of the specimen in Fig.4(i). Figure 6(ii) shows the relationship between distance R in Fig.2 and mismatch strain. The mismatch strain had been proposed by Miller [4]-[6]. The maximum value of the mismatch strain occurred at the center of the full scale model. It was found that the mismatch strain caused the delamination of top coat at the center of the specimen in Fig.4(ii).

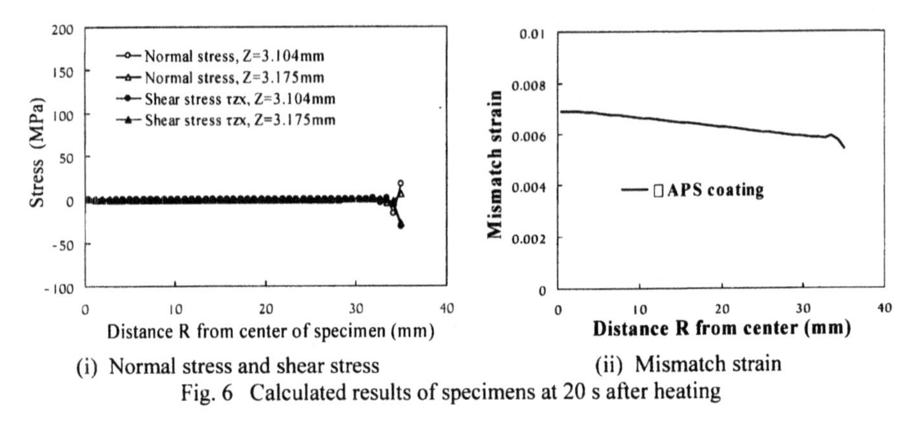


(i) Normal stress and shear stress (ii) Mismatch strain

Fig. 6 Calculated results of specimens at 20 s after heating

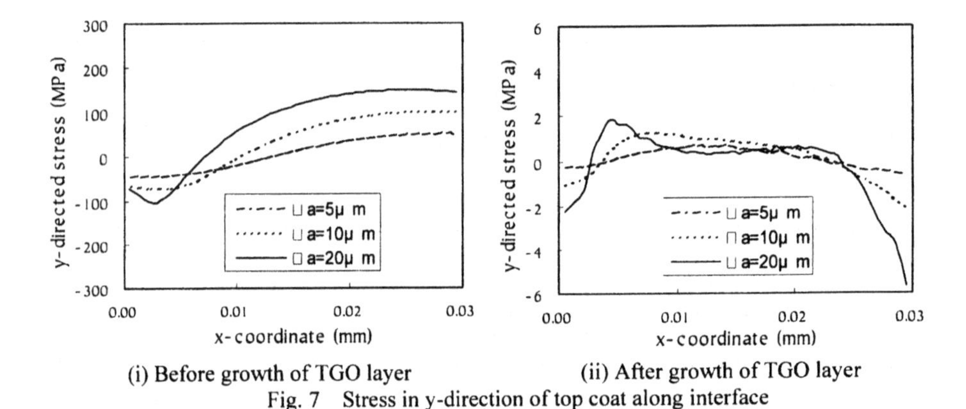


(i) Before growth of TGO layer (ii) After growth of TGO layer

Fig. 7 Stress in y-direction of top coat along interface

Figure 7 shows normal stress $\sigma_y$ in the y-direction in the mesh of top coat along BC/TC interface for x-coordinate of a subscale 2D model in Fig.3. The maximum values of the normal stress $\sigma_y$ in all models were increased as TGO layer glows with heating. The location of the maximum value moved to the convexity (x=0mm) from the concavity (x=0.3mm) of the BC/TC interface of the model in Fig.3. The simulation results coincided with the locations of the occurrence of the delamination near the convexity of the interface.

## THERMAL FATIGUE LIFE PROCEDURE

Thermal cycle test was performed by using the plate specimen with a coating system. The damage was the delamination of top coat near BC/TC interface at those locations of the center and the edge of the plate specimen after 1000 cycles. It was found that the delamination of top coat was due to the growth of TGO layer, thermal stress, the shape of the parts that applied top coat, the configuration of the interface and mismatch strain at the interface. Those fracture mechanisms were cleared from thermal cycle test, heating test and thermal stress analysis. Thus, this paper was investigated a simplified method to predict thermal fatigue life of top coat from those examinations. Equation (2) was proposed to predict the thermal fatigue life of top coat. The delamination of top coat was caused when the damage factor D in Eq.(2) was reached to one. The factor D was composed of three kinds of parameters $\varepsilon_o,\ \varepsilon_m\ \varepsilon_c$ in Eq.(2).

$$D = \frac{\varepsilon_o}{\varepsilon_{oc}} + \frac{\varepsilon_m}{\varepsilon_{mc}} + \frac{\varepsilon_c}{\varepsilon_{cc}} \tag{2}$$

The strain $\varepsilon_o$ is determined from the calculation of finite element analysis of the growth of TGO layer. The mismatch strain $\varepsilon_m$ and the strain $\varepsilon_c$ are decided by the calculation of the analysis model of the part, where $\varepsilon_c$ implies the strain that is caused by the effect of the shape of part. Three kinds of parameters $\varepsilon_{oc},\ \varepsilon_{mc}$ and $\varepsilon_{cc}$ in Eq.(2) denotes limit values of those strains when the delamination of top coat happens at BC/TC interface. Two parameters $\varepsilon_{oc},\ \varepsilon_{mc}$ are determined by the heating test of small specimen. The test is very simple in comparison with thermal cycle test. it was found that parameter $\varepsilon_{oc}$ of first term in Eq.(2) was main parameter for the thermal fatigue damage of top coat from the results until now [1]-[3] and it may be set the value of a dotted line in Fig.5. On the other hand, parameter $\varepsilon_{cc}$ is determined from several mechanical tests of the specimen. However, it can solve the effect of the parameter $\varepsilon_{cc}$ in the design change of the shape of the parts.

## CONCLUSIONS

This paper examined the fracture mechanism of top coat in the experiments both thermal cycle test and heating test. Those specimens fracture in delamination near the BC/TC interface. The stress was calculated in both micro-model at the interface and specimen model in finite element analysis. As a results, it was found that the delamination is due to the growth of TGO layer, thermal stress, the shape of engine parts, the configuration of the interface and mismatch strain at the interface from the examinations of those fracture mechanisms. Thus, this paper investigated a simplified method to predict thermal fatigue life of top coat from those examinations. The method proposes an accumulated damage parameter D to predict the occurrence of delamination of top coat. The damage parameter D is composed three terms of the thickness of TGO layer, mismatch strain at BC/TC interface and the effect of the shape of part. Those parameters are determined both the calculations of finite element analysis and some specimens. A dotted line in Fig.5 shows a main parameter for the thickness of TGO layer. The proposed method in this paper is very simple, and also it is an effective method in the design of thermal fatigue life of top coat on engine part.

REFERENCES

[1]Y. Ohtake, N. Nakamura, N. Suzumura and T. Natsumura, "Evaluation for Thermal Cycle Damage of Thermal Barrier Coating," *Ceramic Engineering and Science Proceedings*, 24(3) 561-566 (2003).

[2]Y. Ohtake, T. Natsumura, "Investigation of Thermal Fatigue Life of Thermal Barrier Coating," *Ceramic Engineering and Science Proceedings*, 25(4) 357-362 (2004).

[3]Y. Ohtake, T. Natsumura, K.Miyazawa, "Investigation of Thermal Fatigue Life of Thermal Barrier Coating," *Ceramic Engineering and Science Proceedings*, 26(3) 89-93 (2005).

[4]R. A. Miller, "Oxidation-Based Model for Thermal Barrier Coating Life," Journal of the American Ceramic Society, 67 [8] 517-21 (1984).

[5]R. A. Miller, "Thermal Barrier Coatings for Aircraft Engines History and Directions," *Journal of Thermal Spray Technology*, 6 [1] 35-42 (1997).

[6]R. A. Miller, "Life Modeling of Thermal Barrier Coatings for Aircraft Gas Turbine Engines," *Journal of Engineering for Gas Turbines and Power*, 111 301-05 (1989).

[7]A. G. Evans, M. Y. He and J. W. Hutchinson, "Mechanics-based scaling laws for the durability of thermal barrier coatings," *Progress in Materials Science*, 46 249-271 (2001).

[8]A. G. Evans, D. R. Mumm, J. W. Hutchinson, G. H. Meier and F.S. Pettit, "Mechanisms controlling the durability of thermal barrier coating," *Progress in Materials Science*, 46 505-553 (2001).

[9]T. A. Cruse, S. E. Stewart and M. Ortiz, "Thermal Barrier Coating Life Prediction Model Development," Journal of Engineering for Gas Turbines and Power, 110 610-616 (1988).

[10]S. M. Meier, D. M. Nissley, K. D. Sheffler and T. A. Cruse, "Thermal Barrier Coating Life Prediction Model Development," *Journal of Engineering for Gas Turbines and Power*, 114 258-263 (1992).

[11]K. Nishimoto, K. Saida, D. Kim, Y. Nakao, "Transient Liquid Phase Bonding of Ni-base Singe Crystal Superalloy, CMSX-2," *The Iron and Steel Insitute of Japan*, **35[10]** 1298-1306 (1995).

[12]K. Sfar, J. Aktaa and M. F. Kanninen, "Analysing the Failure Behaviour of Thermal Barrier Coating Using the Finite Element Method," *Ceramic Engineering and Science Procedings*, 21(3) 203-211 (2000).

[13]R. G. Kerkhoff, R. Vaβen and D. Stover, "Numerlcally calculated thermal stresses in thermal barrier coatings on cylindrical substrates," *Proc. United Thermal Spray Conference*, 787-792 (1999).

[14]ABAQUS, "ABAQUS/Standard, Version5.8, " Hibbitt, Karlsson & Sorensen, Inc. (1999).

# Environmental Barrier Coatings for Si-Based Ceramics

# THE WATER-VAPOUR HOT GAS CORROSION BEHAVIOR OF $Al_2O_3$-$Y_2O_3$ MATERIALS, $Y_2SiO_5$ AND $Y_3Al_5O_{12}$-COATED ALUMINA IN A COMBUSTION ENVIRONMENT

Marco Fritsch*, Hagen Klemm

Fraunhofer-Institut for Ceramic Materials and Sintered Materials, IKTS Dresden, Winterbergstrasse 28, 01277 Dresden, Germany
*Corresponding author. Tel.:+49-(0)351-2553-869, Fax:+49-(0)351-2553-600,
E-mail address: marco.fritsch@ikts.fraunhofer.de

## ABSTRACT

The water vapour hot gas corrosion behavior of various oxide ceramics in the $Al_2O_3$-$Y_2O_3$ system ($\alpha$-$Al_2O_3$, $Y_3Al_5O_{12}$+$\alpha$-$Al_2O_3$, $Y_3Al_5O_{12}$ [yttrium aluminium garnet, or YAG], $YAlO_3$ [yttrium aluminium perovskite, or YAP], $Y_4Al_2O_9$ [yttrium aluminium monocline, or YAM], $Y_4Al_2O_9$+$Y_2O_3$) and of $Y_2SiO_5$ was determined. The tests were performed in a high temperature burner rig at temperatures between 1200 °C and 1500 °C, water vapour partial pressure of 0.24 atm, a total pressure of 1 atm, gas flow velocity of 100 m/s and test times about 100 .. 350 hours. YAG, YAP and YAM showed an order of magnitude lower corrosion rate in comparison to alumina. Nevertheless degradation and phase changes on surface were observed above 1350 °C. $Y_2SiO_5$ showed absolutely no degradation even after corrosion exposure at 1500 °C. The dependence of the temperature is discussed by compositional and microstructural investigations. Finally corrosion issues about environmental barrier coatings (EBC) on bulk are discussed. In case of EBC YAG coated $Al_2O_3$ improved hot gas stability was observed.

## INTRODUCTION

The demand of higher thermal efficiency and the decrease of emissions ($NO_x$, CO) of gas turbines lead to higher operating temperatures and to a reduction of cooling air for components in the hot gas path. That necessitates an increase of thermal loading of the components (combustor wall, gas turbine blade). The potential of metal based materials coated with thermal barrier coatings (TBC) seems to be limited, due to the low softening temperature of metals and a finite coating thickness. Ceramic based materials are promising candidates for structural components in a hot gas path of a gas turbine. Especially silicon based ceramics like silicon carbide (SiC) and silicon nitride ($Si_3N_4$) were subject numerous researches and led to a high level of mechanical properties of these ceramics at elevated temperatures [1]. A limitation for the long term application of these materials is the hot gas corrosion in combustion atmospheres at high temperatures and high gas flow velocity. The main structural ceramics like $Si_3N_4$, SiC [2-5] and also $Al_2O_3$ [6-8] show insufficient stability in high velocity combustion environments. Mainly the water vapour from the combustion gas attacks the ceramic surface by formation and evaporation of volatile hydroxides (e.g. $Si(OH)_4$, $Al(OH)_3$). The resulting degradation will be too high for long term applications (>10.000 hours). Therefore it is generally accepted to establish environmental barrier coatings (EBC) for these materials.

Materials in the $Al_2O_3$-$Y_2O_3$ (YAG, YAP and YAM) system are components of proposed oxide/oxide ceramic matrix composite (CMC), melt grown composite (MGC), environmental and/or thermal barrier coatings (EBC, TBC). Especially single crystal YAG shows a very low [9,10] and polycrystalline YAG [11] a superior creep resistance compared to other oxide materials and is therefore suitable as potential fibre or matrix constituent for ceramic composites. Single crystal and polycrystalline YAG shows an order of magnitude higher corrosion resistance compared to single crystal and polycrystalline alumina [8,12] respectively. With a similar thermal expansion coefficient to $Al_2O_3$, YAG is a potential environmental barrier coating material for alumina based fibre composites. Furthermore YAG shows a low thermal conductivity [13] and can host luminescence ions which make it suitable for thermal barrier sensor coatings [14]. Nevertheless limited information is known about the hot gas corrosion behavior of the materials in the $Al_2O_3$-$Y_2O_3$ phase system in high velocity combustion gases.

In this paper, we demonstrate the corrosion stability of polycrystalline $Al_2O_3$, YAG+$Al_2O_3$, YAG, YAP, YAM, YAM+$Y_2O_3$, $Y_2SiO_5$ and YAG coated alumina under high temperature and high gas flow conditions.

## EXPERIMENTAL

The corrosion behavior of oxide ceramics in the $Al_2O_3$-$Y_2O_3$ system and of $Y_2SiO_5$ (Table 1) was investigated. Dense alumina (AKP50 HC Stark – Goslar/Germany) was fabricated by sintering at 1300 °C in an air atmosphere. YAG+$Al_2O_3$, YAG, YAP, YAM, YAM+$Y_2O_3$ and $Y_2SiO_5$ were fabricated by stoichiometric mixture of the starting powders (AKP50 and $Y_2O_3$ HC Stark, $SiO_2$ Heraeus – Hanau/Germany) and hot pressing at 1700 °C for 2 hours at 30 MPa in a Ar atmosphere. Bending bars were cut and ground with a geometry of $3.8\times3\times36$ $mm^3$ as samples for the corrosion tests (polished surface quality).

Table 1: Summary of the materials investigated

| Material | Composition | Densification | Density g/cm$^3$ | $\alpha_{20..1450}$ / $K^{-1}$ $10^{-6}$ |
|---|---|---|---|---|
| $\alpha$-$Al_2O_3$ | $\alpha$-$Al_2O_3$ | Sintered | 3.98 | 9.06 |
| YAG+$\alpha$-$Al_2O_3$ | $\alpha$-$Al_2O_3$, 32.7 wt% $Y_2O_3$ | Hot pressed | 4.29 | 9.14 |
| YAG | $\alpha$-$Al_2O_3$, 57.1 wt% $Y_2O_3$ | Hot pressed | 4.55 | 9.23 |
| YAP | $\alpha$-$Al_2O_3$, 68.9 wt% $Y_2O_3$ | Hot pressed | 5.30 | 7.53 |
| YAM | $\alpha$-$Al_2O_3$, 81.6 wt% $Y_2O_3$ | Hot pressed | 4.55 | 6.13 (7.48 at 1400 °C) |
| YAM+$Y_2O_3$ | $\alpha$-$Al_2O_3$, 88,6 wt% $Y_2O_3$ | Hot pressed | 4.70 | 8.30 |
| $Y_2SiO_5$ | $SiO_2$, 79.0 wt% $Y_2O_3$ | Hot pressed | 4.45 | 7.00 |

YAG coatings on dense alumina were fabricated by dip coating. A slurry consisting on 42.9 wt% $Al_2O_3$ (AKP 50 HC Stark) and 57.1 wt% $Y_2O_3$ (HC Stark), in the composition of stoichiometric YAG, was milled together with high purity water in a ball mill for 4 hours. The coated alumina samples were heated with 1K/min to 1600 °C for 2 hours.

To consider the phase stability and to separate phase changes of YAG, YAP and YAM in comparison to the corrosion exposure in a combustion atmosphere, the oxide materials were exposed for 100 hours in a furnace at 1500 °C and stagnant air.

The corrosion tests were performed in a high temperature burner rig. The burner rig consisted of a combustor and a following test section with an inner diameter of 30 mm. The inner walls of the combustor consisted of SiC tubes. Natural gas (mainly $CH_4$) was used as fuel for the combustor. To obtain a higher water vapour partial pressure in the combustion gas, water vapour was introduced separately to the air/fuel gas stream using an evaporator. The temperature was controlled by a Pt-Rh thermocouple near to the sample holder. The test conditions established in the test section were:

Temperature: 1200 °C .. 1500 °C
Gas flow velocity: 100 m/s
Total Pressure: 1 atm
Test time: 100 hours .. 350 hours

The calculated composition of the combustion gas was 0.64 $N_2$, 0.24 $H_2O$, 0.08 $O_2$ and 0.04 $CO_2$ (partial pressure for each gas species) at $\Phi$=0.5 .. 0.6 fuel lean conditions. The calculation was obtained from the theoretical stoichiometric combustion equations and found to be in agreement to calculation performed with the computer program FactSage® (version 5.3.1, GTT-Technologies - Herzogenrath/Germany). This computer program is based on the principle of minimizing the Gibbs free energy of the system.

The recession of the specimens was obtained by measuring the weight difference before and after the corrosion test. It is expressed by weight change $\Delta W$ ($mg/cm^2$) relating to the area of the test specimen exposed to the gas flow. From the weight change $\Delta W$ dependence versus time the weight loss rate $K_w$ ($mg/cm^2h$) was determined. For linear corrosion kinetics the weight loss rate $K_w$ can be calculated with a linear fit. The tests were conducted 100..350 hours at temperatures between 1200 °C and 1500 °C. The weight loss rate was calculated after reaching equilibrium conditions (30..50 h).

In order to specify the corrosion attack, the samples surface were analysed before and after the furnace and the corrosion test by XRD. Based on these measurements the quantitative phase compositions were calculated with the computer program AutoQuan® (version 2.6.2.0, Rich. Seifert & Co. – Ahrensburg/Germany). Information about microstructure alterations were obtained through observations of the samples surface and polished cross sections with distributions of the elements in the SEM.

The linear thermal expansion of the oxide materials was determined by thermodilatometry (Netzsch – (Selb/Germany) Model 402 E) in an argon atmosphere up to 1450 °C with a heating rate of 2 K/min. The linear thermal expansion coefficient was calculated between 20 °C to 1450 °C (Table 1).

## RESULTS AND DISCUSSION

### Furnace Exposure

After furnace exposure for 100 hours at 1500 °C in a stagnant air atmosphere YAG and YAM showed no phase change in comparison to the manufactured state. Only YAP showed a small amount of decomposition into YAG and YAM phases. The results for YAG and YAM are in agreement with Mah et al. [17], who found no degradation for YAG, YAP and YAM even at 1650 °C for 100 hours in a stagnant air atmosphere.

### Burner Rig Test

The hot gas corrosion tests of the various oxide materials investigated, showed that significant corrosion attack below 1350 °C occurred only for $\alpha$-$Al_2O_3$ and YAG+$\alpha$-$Al_2O_3$. Measurable weight change or surface alteration were observed above 1350 °C for all oxide compounds, beside $Y_2SiO_5$ which showed absolutely no corrosion attack even at 1500 °C.

Figure 1 shows the corrosion kinetic at 1450 °C. The specimens showed a linear corrosion kinetic, indicating a surface controlled corrosion attack. The measured weight change and the calculated weight loss rate are summarized in Table 2.

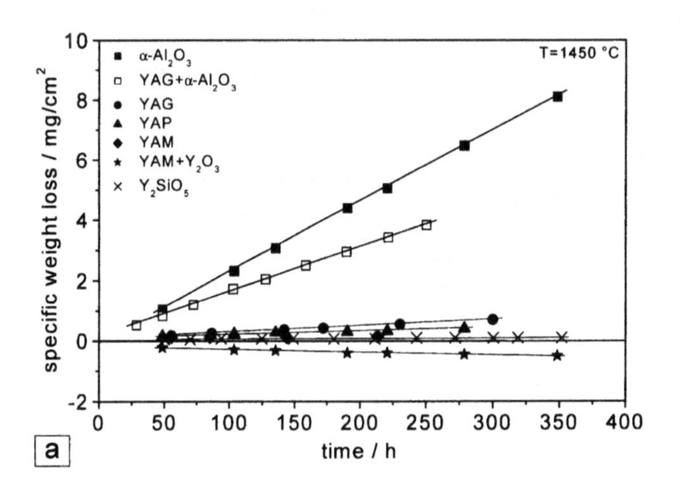


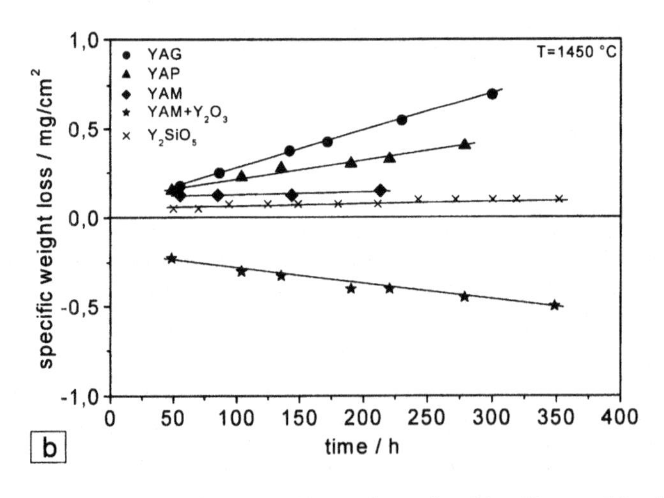


Fig.1 : Corrosion kinetic at 1450 °C: (a) all samples, (b) oxide materials with minor weight loss.

Table 2: Summary of corrosion test results

| Material | T=1450 °C | | | T=1500 °C | | |
|---|---|---|---|---|---|---|
| | Time / h | $\Delta m_t$ / mg | $K_w$ / $mg/cm^2 \cdot h$ | Time / h | $\Delta m_t$ / mg | $K_w$ / $mg/cm^2 \cdot h$ |
| $\alpha$-$Al_2O_3$ | 348.5 | -31.9 | $2.4 \cdot 10^{-2}$ | 161 | -26.9 | $4.7 \cdot 10^{-2}$ |
| YAG+$\alpha$-$Al_2O_3$ | 250.4 | -13.2 | $1.5 \cdot 10^{-2}$ | 76 | -8.8 | $3.4 \cdot 10^{-2}$ |
| YAG | 300 | -2.8 | $2.1 \cdot 10^{-3}$ | 109 | -5.4 | $5.4 \cdot 10^{-3}$ |
| YAP | 348.5 | -1.0 | $8.0 \cdot 10^{-4}$ | 33.5 | 0.0 | - |
| YAM | 213 | 0.0 | - *1 | 135.2 | +0,3 | $-7.1 \cdot 10^{-4}$ |
| YAM+$Y_2O_3$ | 348.5 | +1.1 | $-8.9 \cdot 10^{-4}$ *2 | 47 | +1.0 | $-4.3 \cdot 10^{-3}$ |
| $Y_2SiO_5$ | 351.8 | -0.4 | - | 96 | 0.0 | - |
| YAG EBC on $Al_2O_3$ | 360 | -12.4 | $9.4 \cdot 10^{-3}$ | 144.7 | -13.7 | $2.4 \cdot 10^{-2}$ |

*1 Results are within the uncertainty of the measurement
*2 negative weight loss rate according to weight gain

Figure 2 shows the corroded cross sections of the bulk oxide materials after corrosion at 1450 °C.

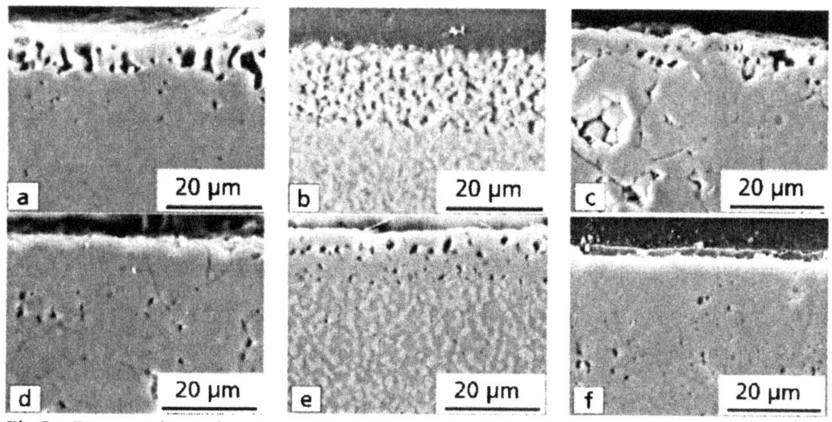


Fig.2 : Cross sections of corroded specimens at T=1450 °C : (a) YAG, (b) YAG+$\alpha$-$Al_2O_3$, (c) YAP, (d) YAM, (e) YAM+$Y_2O_3$, (f) $Y_2SiO_5$.

Table 3 summarizes the phase composition of the specimen surface before and after the hot gas corrosion test at 1450 °C. In general YAG, YAP and YAM decomposed in reaction with the hot gas atmosphere into lower stoichiometric $Al_2O_3$-$Y_2O_3$ phases (alumina volatility due to reaction with water vapour and building of volatile $Al(OH)_3$) and $Y_2O_3$. The $SiO_xH_y$ present in the combustion atmosphere (from the corrosion of the SiC combustor walls) reacted partially with the $Y_2O_3$ to form $Y_2SiO_5$. Hence the measured weight change of the specimens is a consequence of weight loss (alumina volatility) and weight gain (absorption of Si to build $Y_2SiO_5$ on surface).

Table 3: Phase composition of the surface before and after hot gas exposure at 1450 °C

| Material | Phase composition after manufacturing / wt % | Phase composition of surface after hot gas corrosion / wt % |
|---|---|---|
| $\alpha$-$Al_2O_3$ | 100 $\alpha$-$Al_2O_3$ | 100 $\alpha$-$Al_2O_3$ |
| YAG+$\alpha$-$Al_2O_3$ | 56 $Y_3Al_5O_{12}$+44 $\alpha$-$Al_2O_3$ | 91 $Y_3Al_5O_{12}$+9 $\alpha$-$Al_2O_3$ |
| YAG | 100 $Y_3Al_5O_{12}$ | 49$Y_3Al_5O_{12}$ + 33$Y_2SiO_5$ + 18$Y_4Al_2O_9$ |
| YAP | 100 $YAlO_3$ | 62 $Y_4Al_2O_9$+21 $Y_2SiO_5$ + 17 $YAlO_3$ |
| YAM | 100 $Y_4Al_2O_9$ | 59 $Y_4Al_2O_9$ + 41 $Y_2SiO_5$ |
| YAM+$Y_2O_3$ | 61 $Y_4Al_2O_9$+39 $Y_2O_3$ | 61 $Y_2SiO_5$ + 34 $Y_4Al_2O_9$ + 5 $Y_2O_3$ |
| $Y_2SiO_5$ | 100 $Y_2SiO_5$ | 100 $Y_2SiO_5$ |
| YAG EBC on $Al_2O_3$ | 85 $Y_3Al_5O_{12}$+15 $\alpha$-$Al_2O_3$ | 64 $Y_3Al_5O_{12}$ + 19 $\alpha$-$Al_2O_3$ + 17 $Y_4Al_2O_9$ |

**Corrosion of $Al_2O_3$ and YAG+$\alpha$-$Al_2O_3$**

Alumina and YAG+$\alpha$-$Al_2O_3$ composite showed above 1350 °C the highest corrosion rates in comparison to the other materials investigated. Bulk alumina corroded due to formation of volatile Al-hydroxides. Besides the alumina grains, the grain boundaries were attacked by corrosion, and the corroded surface appears thermally etched. The observed corrosion rates for 1500 °C are comparable to Yuri [6] (T=1500 °C, v=100 m/s, P=4 atm, $p_{H2O}$=0.45 atm, t=20 hours, $K_w=7\cdot10^{-2}$ mg/cm$^2$h) and Nakagawa [15] (T=1500 °C, v=250 m/s, P=1 atm, $p_{H2O}$=0.15 atm, t=10 hours, $K_w=2\cdot10^{-2}$ mg/cm$^2$h).

The YAG+$\alpha$-$Al_2O_3$ sintered composite showed lower corrosion rates in comparison to dense alumina. The microstructure consisted of polycrystalline YAG and alumina grains (in Figure 2b YAG is the bright phase and $Al_2O_3$ is the dark phase). After hot gas corrosion a corroded layer consisting mainly of porous YAG was observed. The alumina phase is chemically more unstable in comparison to YAG in water vapour and corroded due to build Al-hydroxides. The corrosion kinetic revealed that only a slight parabolic corrosion kinetic could be found (see Figure 2a). It is assumed, that water vapour can easily diffuse through the porous YAG layer and that this layer partially spall of.

**Corrosion of YAG, YAP and YAM**

The oxide materials in the $Al_2O_3$-$Y_2O_3$ system showed a magnitude lower corrosion rate in comparison to bulk alumina. Nevertheless a corrosion attack by the hot combustion gas was observed. The oxide materials reacted with the water vapour to lower stoiciometric $Al_2O_3$-$Y_2O_3$ phases (see Table 3: YAG, YAP→YAM→$Y_2O_3$) and Al-hydroxides. Opila showed that at elevated temperatures between 1300 °C and 1500 °C the $Al(OH)_3$ should be the dominant alumina volatile species [7, 16]. For example the following corrosion reaction for YAG can be deduced:

$$4Y_3Al_5O_{12}[YAG](s)+21H_2O(g)\leftrightarrow 3Y_4Al_2O_9[YAM](s)+14Al(OH)_3(g) \quad \text{(Eq.1)}$$

The $Y_2O_3$ finally formed, was found much more stable in the combustion environment [16]. Furthermore it reacted with the Si-hydroxides in the combustion environment to $Y_2SiO_5$. The corroded surfaces (Figure 2a, 2c and 2d) showed only a small or a negligible porous monosilicate

rich surface layer. A distinct influence on the corrosion kinetic to parabolic behavior was not found. It could be assumed that the mismatch in thermal expansion behavior between $Al_2O_3$-$Y_2O_3$ materials and $Y_2SiO_5$ (see also Table 1) led to cracking and to spallation of the surface layer.
For the YAP material a mixed corrosion degradation due to thermal induced phase transformation and to corrosion attack in combustion environment, could be assumed.

Exceptional corrosion kinetic was found for the YAM material. It showed no weight change at 1450 °C but a weight gain after corrosion at 1500 °C. A distinct phase change on surface could be observed with XRD. Comparable to YAG and YAP the YAM phase built out an $Y_2SiO_5$ layer. The weight gain due to absorbing silica from the atmosphere balanced the weight loss from alumina volatilization or exceeded it.

Corrosion researches of $Al_2O_3$/YAG eutectic composites (MGC) confirm the corrosion stability of these oxide materials [15, 18]. Regardless of the different microstructure (MGC consist of single crystal alumina and YAG without grain boundaries) the reported corrosion rate from Nakagawa [15] of $K_w=6\cdot10^{-3}$ mg/cm$^2$h (T=1500 °C, v=250 m/s, P=1 atm, $p_{H2O}=0.15$ atm, t=30 hours) is in the order of the measured corrosion rate of the polycrystalline YAG in Table 2. Further researches have to be done to explore the influence of grain boundaries of the amount of the corrosion degradation.

**Corrosion of YAM+$Y_2O_3$**

The YAM+$Y_2O_3$ composite approved that $Y_2O_3$ tends to react with silica from the atmosphere to form $Y_2SiO_5$. The composite showed a distinct weight gain (see Figure 1b). The corroded cross section in Figure 2e revealed that a porous double corrosion layer built out consisting of an outer $Y_2SiO_5$ and an inner YAM layer. It can be concluded that the $Y_2O_3$ diffuse to the surface and reacted to $Y_2SiO_5$ leaving behind a YAM rich phase.

**Corrosion of $Y_2SiO_5$**

The bulk yttrium monosilicate showed absolutely no corrosion attack even at 1500 °C. The outbuilding of the $Y_2SiO_5$ layer on surface of $Al_2O_3$-$Y_2O_3$ composites confirmed that this phase is highly stable in hot gas combustion atmospheres.

**YAG Environmental barrier coatings**

Figure 3 shows that the YAG coated dense alumina specimen showed a significant lower corrosion rate in comparison to the bulk alumina at 1450 °C. Nevertheless the corrosion rate of the coated sample is higher in comparison to the bulk YAG. The corrosion rate for the coated specimen at 1500 °C is comparable to bulk alumina (Table 2).

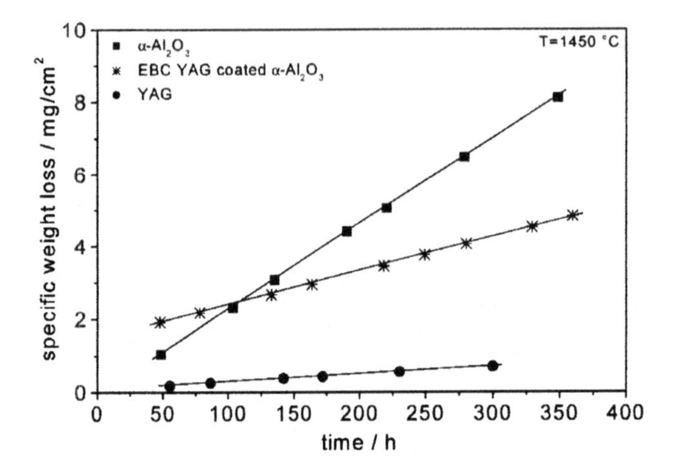


Fig.3 : Corrosion kinetic of YAG coated $\alpha$-$Al_2O_3$ compared to bulk $\alpha$-$Al_2O_3$ and YAG.

Figure 4a shows the cross section of the coating after manufacturing. It can be seen that after sintering the coating showed rest porosity. During hot gas exposure at 1450 °C and 1500 °C the coating tend to sinter forward, building a highly dense coating after corrosion exposure at 1500 °C (Fig.4e).

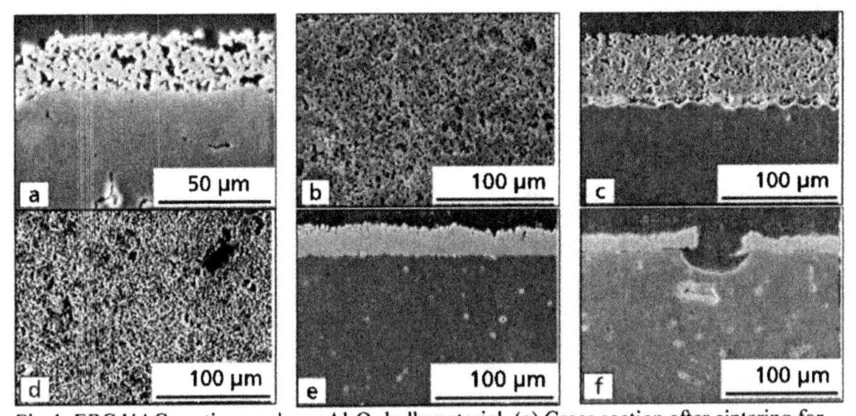


Fig.4: EBC YAG coating on dense $Al_2O_3$ bulk material. (a) Cross section after sintering for 2 hours at 1600 °C; (b) surface and (c) cross section after hot gas corrosion at 1450 °C for 360 hours; (d) surface and (e, f) cross sections after hot gas corrosion at 1500 °C for 145 hours.

In Figure 4c the surface of the alumina bulk seems to be rough, due to corrosion reaction with water vapour, which easily can diffuse through the porous YAG coating to the interface of YAG/$Al_2O_3$-bulk. In Figure 4e a surface corrosion attack of the YAG coating, comparable to the dense YAG specimen in Figure 2e, can be seen. Decomposition products like YAM can be found on surface after corrosion exposure (Table 2). Figure 4e shows the pitting of the bulk alumina at a location where the coating was interrupted. It demonstrates the noticeable corrosion protection of YAG coating in comparison to the uncoated alumina.

Nevertheless the high corrosion rates of YAG coated alumina at 1500 °C and the higher corrosion rates in comparison to the bulk YAG at 1450 °C could be attributed to the diffusion of alumina ions at high temperature. Due to the concentration gradient between the top of the coating (alumina volatilisation) and the interface (YAG/ $Al_2O_3$-bulk) it can be assumed that $Al^{3+}$ ions diffuse through the YAG coating to the surface and volatilizes, leaving a gap at the interface, which can result in spallation of the coating. A further decomposition of the YAG EBC to $Y_2O_3$ was not found (no $Y_2O_3$ or monosilicate in comparison to the bulk YAG were detected after corrosion exposure, Table 3). Probably the $Al^{3+}$ ion diffusion to the surface of the EBC YAG coating stabilizes the YAM phase.

Regarding the thermal expansion behavior of the oxide materials investigated (Table 1), the YAG and alumina showed similar thermal expansion behavior and is therefore a promising candidate for coatings on alumina based materials. YAP, YAM and $Y_2SiO_5$ showed significant lower thermal expansion. YAM and $Y_2SiO_5$ showed phase transitions and anisotropic thermal expansion comparable to literature values [19-21]. The phase change of YAM and with that the volume change (0.6% at ~1370°C) [19] can be avoided in a composite of YAM+$Y_2O_3$ with additions of $Y_2O_3$. The additions of $Y_2O_3$ led further to an increase of the thermal expansion coefficient.

The application of YAP, YAM and $Y_2SiO_5$ coatings on alumina based materials was not possible due to reaction of the bulk alumina into the alumina rich YAG phase starting at the interphase of the coating. This is consistent with the diffusion controlled phase development in $Al_2O_3$-$Y_2O_3$ powder mixtures starting from $Al_2O_3$ to build YAM→YAP→YAG with time and temperature [22].

## CONCLUSION

The hot gas corrosion behavior of oxide materials in the $Al_2O_3$-$Y_2O_3$ system, of $Y_2SiO_5$ and of YAG coated alumina was shown. YAG, YAP, YAM showed a higher corrosion stability in comparison to bulk alumina. With a similar thermal expansion of YAG and alumina, YAG is a promising EBC candidate for alumina based oxides. Nevertheless a certain corrosion attack, due to reaction with water vapour from the combustion atmosphere was observed, leading to a degradation and phase change on surface.

$Y_2SiO_5$ showed an outstanding corrosion stability, but will be difficult to use as an coating material, due to the different thermal expansion behavior and to phase reactions in alumina rich material systems.

## REFERENCES


1. D. W. Richerson, "Ceramic Components in Gas Turbine Engines: Why has it taken so long?", In Proceedings of the 28th International Conference on Advanced Ceramics and Composites held at Cocoa Beach, Florida, on January 25-30, 2004. Ed. Edgar Lara-Curzio and Michael J. Readey, issued as part of the Ceramic Engineering and Science Proceedings, Volume 25, Issues 3 and 4.
2. N. S. Jacobson, "Corrosion of Silicon-Based Ceramics in Combustion Environments", J.Am.Ceram.Soc., 76(1), 3-28, 1993.
3. E. J. Opila, "Oxidation and Volatilization of Silica formers in Water Vapor", J.Am.Ceram.Soc., 86(8), 1238-1248, 2003.
4. L. Yuri, T. Hisamatsu, Y. Etori, T. Yamamoto, "Degradation of Silicon Carbide in Combustion Gas Flow at High Temperature and Speed", ASME Turbo Expo 2000, Munich 2000, ASME Paper 2000-GT-664.
5. H. Klemm, "Corrosion of silicon nitride materials in gas turbine environment", J.Europ.Ceram.Soc., 22, 2735-2740, 2002.
6. I. Yuri, T. Hisamatsu, "Recession rate prediction for ceramic materials in combustion gas flow", In Proceedings of ASME Turbo Expo, June 16-19, Atlanta, Georgia, 2003, ASME Paper GT2003-38886.
7. E. Opila, D. L. Myers, "Alumina Volatility in Water Vapor at Elevated Temperatures", J.Am.Ceram.Soc., 87(9), 1701-1705, 2004.
8. H. Klemm, M. Fritsch, B. Schenk, "Corrosion of ceramic materials in hot gas environment", In Proceedings of the 28th International Conference on Advanced Ceramics and Composites held at Cocoa Beach, Florida, on January 25-30, 2004. Ed. Edgar Lara-Curzio and Michael J. Readey, issued as part of the Ceramic Engineering and Science Proceedings, Volume 25, Issues 3 and 4.
9. G. S. Corman, "High-Temperature Creep of Some Single Crystal Oxides", Ceram.Eng.Sci.Proc., 12(9-10), 1745-1766, 1991.
10. G. S. Corman, "Creep of yttrium aluminium garnet single crystals", J.Mater.Sci.Lett., 12, 379-382, 1993.
11. T. A. Parathasarathy, T. I. Mah, K. Keller, "Creep Mechanism of Polycrystalline Yttrium Aluminium Garnet", J.Am.Ceram.Soc., 75(7), 1756-1759, 1992.
12. Y. Harada, T. Suzuki, K. Hirano, N. Nakagawa, Y. Waku, "Effect of water vapour in degradation of in-situ single-crystal oxide eutectic composites", Proceedings of the 5th International Conference of High Temperature Ceramic Matrix Composites, Singh, M.(Ed.): High temperature ceramic matrix composites 5, Westerville, American Ceramic Society, 2005.
13. N. P. Padture, P. G. Klemens, "Low Thermal Conductivity in Garnets", J.Am.Ceram.Soc., 80(4), 1018-1020, 1997.
14. M. M. Gentleman, D. R. Clarke, "Concepts for luminescence sensing of thermal barrier coatings", Surface and Coatings Technology, Volumes 188-189, Pages 93-100, November-December 2004.
15. N. Nakagawa, H. Ohtsubo, K. Shibata, A. Mitani, K. Shimizu, Y. Waku, "High temperature stability of MGC's gas turbine components in combustion gas flow environments", Proceedings of GT2005, ASME Turbo Expo 2005, Power of Land, Sea and Air, June 6-9, 2005, Reno-Trahoe, Nevada, USA, GT2005-68658.

16. E. J. Opila, N. S. Jacobson, "Volatile hydroxide species of common protective oxides and their role in high temperature corrosion", Electrochemical Society Proceedings, Volume 96-26, 269-280, 1996.
17. T. I. Mah, K. A. Keller, S. Sambasivan, R. J. Kerans, "High-Temperature Environmental Stability of the Compounds in the $Al_2O_3$-$Y_2O_3$ System", J.Am.Ceram.Soc., 8[4], 874-878, 1997.
18. A. Otsuka, Y. Waku, K. Kitagawa, N. Arai, S. Hirano, "Hot corrosion resistance of the $Al_2O_3$/YAG eutectic composite", Ceram.Trans.Ceram.Proc.Sci., IV, Vol.112, 803-808, 2001.
19. H. Yamane, M. Omori, A. Okubo and T. Hirai, "High-Temperature phase transition of $Y_4Al_2O_9$", J.Am.Ceram.Soc., 76[9], 2382-84, 1993.
20. H. Yamane, K. Ogawara, M. Omori and T. Hirai, "Phase transition of rare-earth aluminates ($Re_4Al_2O_9$) and Rare-earth gallates ($Re_4Ga_2O_9$)", J.Am.Ceram.Soc., 78[9], 2385-90, 1995.
21. H.J. Seifert, S. Wagner, O. Fabrichnaya, H.L. Lukas, F. Aldinger, T. Ullmann, M. Schmücker, H. Schneider, „Yttrium Silicate Coatings on Chemical Vapor Deposition-SiC-Precoated C/C-SiC: Thermodynamic Assessment and High-Temperature Investigation", J.Am.Ceram.Soc., 88[2], 424-430, 2005.
22. K. M. Kinsman, J. McKittrick, E. Sluzky, K. Hesse, "Phase development and luminescence in chromium-doped yttrium aluminum garnet (YAG:Cr) phosphors", J.Am.Ceram.Soc., 77[11], 2866-72, 1994.

# EVALUATION OF ENVIRONMENTAL BARRIER COATINGS FOR SiC/SiC COMPOSITES

H. Nakayama
Japan Ultra-high Temperature Materials Research Institute, Ltd.
3-1-8 Higashi-machi, Tajimi-City 507-0801, Japan

K. Morishita
Graduate School of Engineering, Kyoto University, Sakyo-ku, Kyoto-City 606-8501, Japan

S. Ochiai
International Innovation Center, Kyoto University, Sakyo-ku, Kyoto-City 606-8501, Japan

T. Sekigawa and K. Aoyama
Mitsubishi Heavy Industries, Ltd. Nagoya Aerospace Systems,
10 Oye-cho, Minato-ku, Nagoya-City 455-8515, Japan

A. Ikawa
Japan Ultra-high Temperature Materials Research Institute, Ltd.
3-1-8 Higashi-machi, Tajimi-City 507-0801, Japan

## ABSTRACT

SiC fiber reinforced SiC matrix (SiC/SiC) composites are one of the most promising materials for high temperature structural applications such as power generation and propulsion systems. SiC/SiC composites are, however, susceptible to accelerated attacks in water vapor environments through oxidation and volatilization reaction, so Environmental Barrier Coatings (EBCs) are indispensable.

We have investigated some oxides and rare-earth silicates as topcoat candidates for EBCs. Topcoat materials must be stable in the high-water-vapor pressurized environments at high temperatures. Also, it is important that the thermal expansion coefficient of topcoat materials is similar to that of SiC/SiC composites.

In this study, rare-earth silicates such as lutetium silicates, 8YSZ and $ZrO_2$-$SiO_2$ were selected as topcoat candidates. They were exposed in the water containing atmosphere at a temperature 1673K for 100 hr under a total pressure 0.96 MPa and under an atmospheric pressure. Mass changes, structure of crystals and microstructures were investigated after the exposure experiments in order to evaluate the thermal stability of those materials. After this evaluation, lutetium silicates were considered to be promising for topcoat materials.

Lutetium silicates were coated on SiC/SiC composites, and their fracture toughness and microstructures were investigated and reported in this paper.

## INTRODUCTION

Many researches and developments are now conducted to improve durability of silicon-based ceramics and composites[1,2,3,4,5]. Development to improve oxidation resistance of SiC/SiC composites for the applications to power generation fields is being conducted in NEDO (New Energy and Industrial Technology Development Organization, Japan) project. In order to promote durability and oxidation resistance, we are now researching EBC systems on SiC/SiC composites[6,7].

This project started in the year 2002 and continues for 5 years. The goal of this project is to develop SiC/SiC composites with EBCs, which can be used in the high-water-vapor pressurized environments at high temperatures. Developed SiC/SiC composites with EBCs aim to retain residual strength of 80%, compared with the virgin SiC/SiC composites, after exposure in the water containing atmosphere at the temperature from 1573K to 1673K for 4000h under a total pressure 0.96MPa and water-vapor partial pressure 0.15MPa.

These development roles were divided among 5 companies. Development concerning topcoat and undercoat was assigned to MHI(Mitsubishi Heavy Industries,Ltd.) and KHI(Kawasaki Heavy Industries, Ltd.), respectively. Development for improving oxidization resistance of SiC/SiC matrix was assigned to IHI(Ishikawajima-Harima Heavy Industries co., Ltd.) and UBE(Ube Industries, Ltd.). JUTEMI(Japan Ultra-high Temperature Materials Research Institute) evaluated durability and oxidation resistance of SiC/SiC composites with EBCs.

One model of our concepts for developing EBC systems is shown in Fig.1. There are topcoat materials and undercoat materials in this system. Topcoat consists of rare-earth silicate and undercoat consists of mixed alumino-silicate and rare-earth silicate. Rare-earth silicate materials are mainly used in our concept of EBC system, because coefficient of thermal expansion of rare-earth silicate is similar to that of SiC/SiC composites and it seems to have good oxidation resistance in the high-water-vapor-pressurized environments at high temperatures.

The purpose of this study is to investigate the topcoat candidates and evaluate the possibility of selected materials as topcoat in order to establish EBC systems.

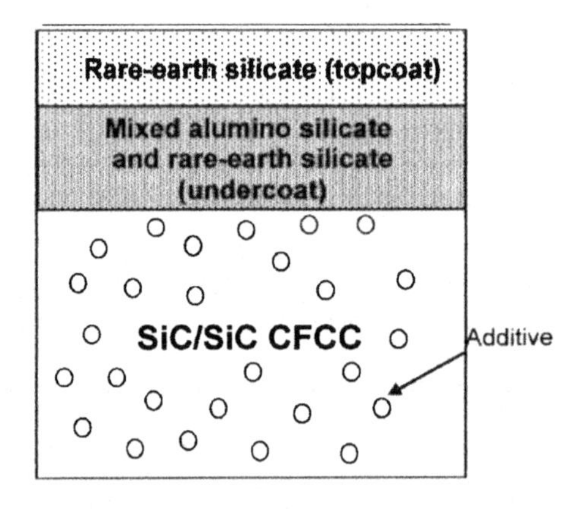


Fig.1 Concept of developing EBC system

## EXPERIMENTAL PROCEDURE

### Test Apparatus

The general outline of high-pressure water-vapor oxidizing test apparatus is shown in Fig.2. Environmental furnace consists of 2 main components of external pressure vessel and test chamber. External pressure vessel is made of stainless steel and test chamber is made of high-purity alumina. Pressure balance controller regulates the pressure difference between external pressure vessel and test chamber in order to prevent test chamber from being broken by tension stress.

Mixed gases of air, $N_2$, $O_2$, $CO_2$ and water-vapor can be fed into test chamber. But only air can be fed into external pressure vessel. The test zone size is 80mm diameter and 200mm length.

Maximum test temperature 1773K, maximum total pressure 0.96MPa and maximum gas-flow velocity $5x10^{-3}$ m/s can be obtained. This apparatus was fabricated by Toshin Kogyo co., Ltd. in Japan.

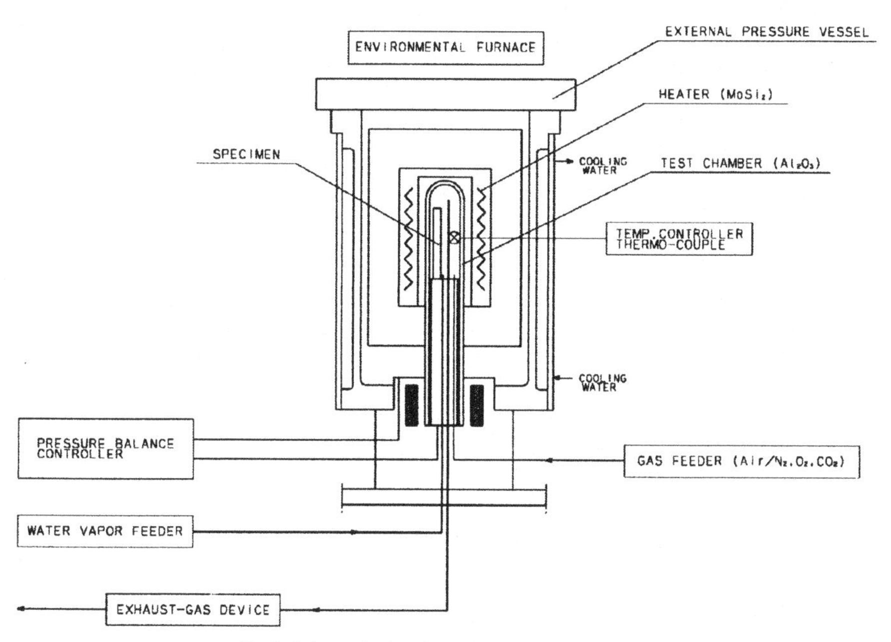


Fig.2 Schematic drawing of high-pressure water-vapor oxidizing test apparatus

We have another test apparatus, which can expose test specimens in the mixed gases of N2, O2 and water-vapor at the maximum temperature 2000K in an atmospheric pressure.

Investigation of Topcoat Candidate Materials

Some atmospheric plasma sprayed (APS) films of rare-earth silicate, such as (i)3 different composition powders of $Lu_2O_3$-$xSiO_2$, (ii) 33mol% $Y_2O_3$-67mol% $SiO_2$ and (iii)

oxide ceramics such as $ZrO_2$-8%$Y_2O_3$ and $ZrSiO_4$, were selected to evaluate the adaptability as topcoat materials. These APS films are stand alone films. Some films were heat-treated at 1373K in inert gas atmosphere for 10 hr after APS and others were as-sprayed.

These candidates were exposed in the high pressure steam environment by using high-pressure water-vapor oxidizing test apparatus and dry air atmosphere by using atmospheric pressure test apparatus, and their oxidation resistance was evaluated.

In case of high pressure steam environment, exposure condition of a temperature 1673K for 100hr under a total pressure 0.96MPa, water-vapor partial pressure 0.15MPa and gas-flow velocity $0.5\times10^{-3}$ m/s was adopted. On the other hand, exposure condition of a temperature 1673K for 100h under atmospheric pressure and gas-flow velocity $1\times10^{-3}$ m/s was adopted as dry air atmosphere oxidizing test.

After exposure of specimens under these oxidizing atmospheres, mass changes of these candidates were measured, and crystal structures were observed by XRD analysis.

In addition, thermal expansion of plasma sprayed films of 33mol% $Lu_2O_3$-67mol% $SiO_2$ were measured by using laser method in order to decide the heat treatment condition for stabilizing the structure of plasma sprayed films.

### Evaluation of Topcoat on SiC/SiC Composites

APS films of 33mol% $Lu_2O_3$-67mol% $SiO_2$ was applied as topcoat material of EBC system on SiC/SiC composites.

SiC/SiC used as substrate consists of 3D preform of Tyranno fiber SA grade and SiC matrix formed by PIP method. Rare-earth silicate was mixed in the SiC matrix as additive to improve oxidation resistance of matrix by IHI and Ube Industries, Ltd.

The coated specimens were heat-treated at 1373K in inert gas atmosphere for 10 h to stabilize the structure of topcoat layer and exposed at a temperature 1573K for 100h in the 3 kinds of oxidizing atmospheres ; (a)high pressure steam (total pressure 0.96MPa, water-vapor partial pressure 0.15MPa), (b) normal pressure steam (atmospheric pressure, 47% $H_2O$) and (c) dry air atmosphere.

After exposure of specimens under these oxidizing atmospheres, the fracture toughness of topcoat layers was measured by Indentation Fracture Method, and microstructures were also observed with SEM. The fracture toughness values were estimated by equation (1).

$$K_{Ic} = 0.036 E^{0.4} P^{0.6} a^{-0.7} \left( \frac{c}{a} \right)^{-1.5} . \qquad (1)$$

Where E, P, 2a and 2c refer to the Young's modulus, load(4.9N), diagonal length and crack length, respectively.

## RESULTS AND DISCUSSION

### Investigation of Topcoat Candidates

Fig.3 shows the mass changes of topcoat candidates after exposure under oxidizing atmosphere. In case of high pressure steam exposure, almost all plasma-sprayed films of topcoat candidates show mass gains, except that zircon composition film (F-ZS) shows mass losses in both as-sprayed and heat-treated conditions. It is considered that almost all plasma-sprayed films will form hydroxides after high pressure steam exposure. This may be the reason why the candidates except F-ZS shows the mass gain after exposure test. But the conclusive evidence of forming hydroxides is not obtained yet. These mass gain values are very small compared with sintered SiC. We conducted the same high pressure steam exposure test of sintered SiC formerly. And sintered SiC showed the large mass gain of about $10mg/cm^2$. In this case, sintered SiC formed $SiO_2$ after exposure test. From the XRD analysis, it seems that the structure of F-ZS consists of $ZrO_2$ crystals and small amount of amorphous silica even after heat treatment. Amorphous silica is susceptible to accelerated attacks under high pressure steam[8,9,10]. This may be the reason why F-ZS shows the mass loss.

On the other hand, all plasma sprayed films show mass losses after dry air atmosphere exposure. The mass loss was, however, very small.

Heat-treated plasma sprayed films of 33mol% $Lu_2O_3$-67mol% $SiO_2$ and $ZrO_2$-8wt%$Y_2O_3$ show little mass changes after both high pressure steam and dry air atmosphere exposures.

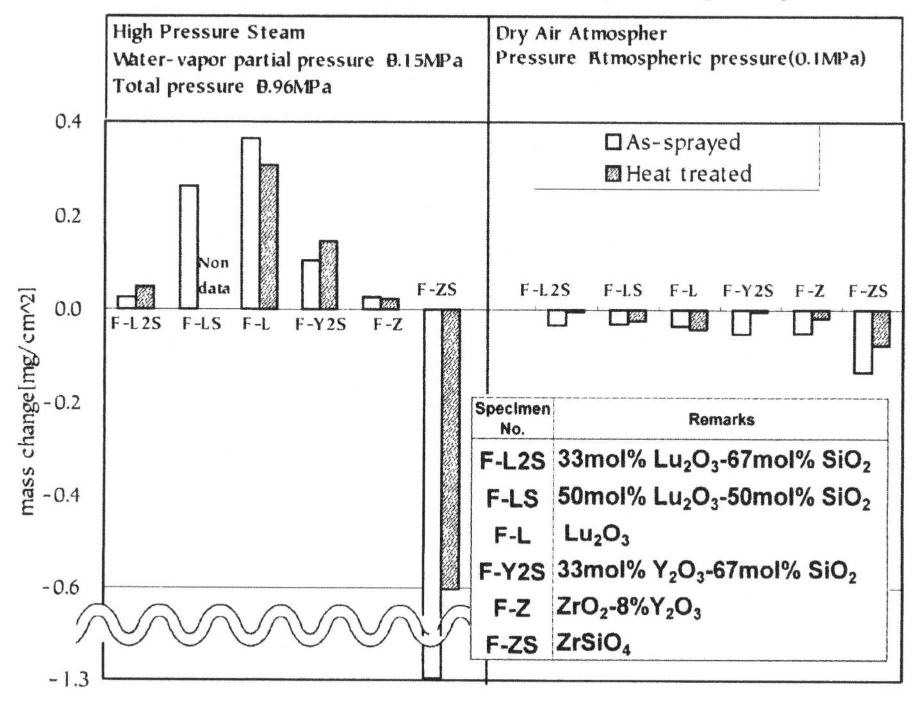


Fig.3 Mass changes of topcoat candidates (plasma sprayed stand alone films) after exposure in steam or dry air at 1673K for 100hr

In these experiments, gas-flow velocity is small. It is clarified that the influence of gas-flow velocity on the mass loss and recession rate of ceramics is large[11]. But it is reported that sintered lutetium silicates have good recession resistance in the high gas-flow velocity, compared with sintered alumina, $Si_3N_4$ and $SiO_2$[12].

Considering the thermal expansion coefficient, lutetium silicate is regarded as the most promising topcoat materials for SiC/SiC composites.

Fig.4 shows XRD analysis, which was conducted on the plasma sprayed films of 33mol% $Lu_2O_3$-67mol% $SiO_2$. From the results of XRD analysis, it appears that the structure of as-sprayed films changes from amorphous-like state to the mixed structures of $Lu_2SiO_5$, $Lu_2Si_2O_7$ and Lu2O3 after heat treatment at 1373K in inert gas atmosphere, and $Lu_2SiO_5$ structure is predominant. Those structures after heat treatment are very stable, and do not change after dry air atmosphere exposure and high pressure steam exposure.

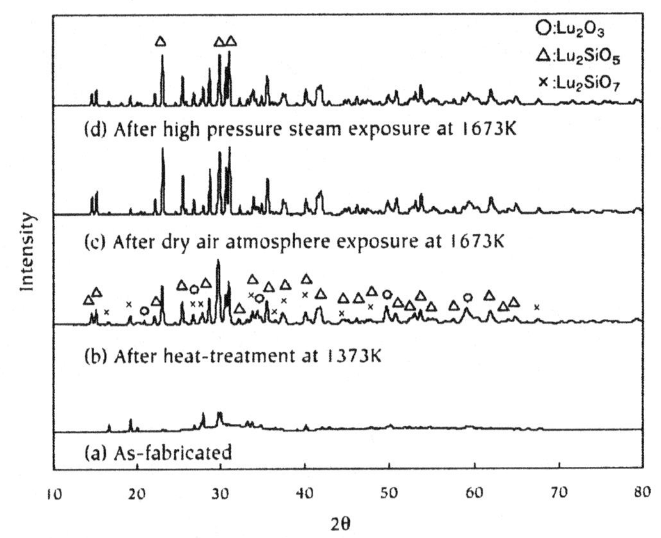


Fig.4 XRD (Cu K) patterns of APS film of 33mol% $Lu_2O_3$-67mol% $SiO_2$

Fig.5 shows the thermal expansion of plasma sprayed films of 33mol% $Lu_2O_3$-67mol% $SiO_2$ up to the temperature 1773K. As-sprayed film shows the abrupt expansion above the temperature 1300K. On the other hand, heat-treated film at 1373K for 10hr after APS shows the gradual expansion with the temperature. The results of this experiment confirm the XRD analysis that the structure of as-sprayed film crystallized and became stable after heat treatment at 1373K for 10hr. And mean thermal expansion coefficient value from ambient temperature to 1773K was $5.6 \times 10^{-6}$. Its value is small enough compared with 8YSZ, CTE value of which is about $10.0 \times 10^{-6}$.

Evaluation of Topcoat on SiC/SiC Composites

SiC/SiC composites with EBCs were exposed under high pressure steam oxidizing atmosphere. And it proved that residual flexural strength of SiC/SiC did not decrease, compared with that of as-fabricated (heat-treated) SiC/SiC composites.

Fig.6 shows the fracture toughness of lutetium silicate topcoat of EBC system on SiC/SiC composites after various exposure tests. Fracture toughness values show the large scatter, but it seems that the mean values show the ranking among specimens after various exposure tests. Fracture toughness of as-fabricated (heat-treated) condition shows the lowest value, and that of the specimen after high pressure steam exposure shows the highest value.

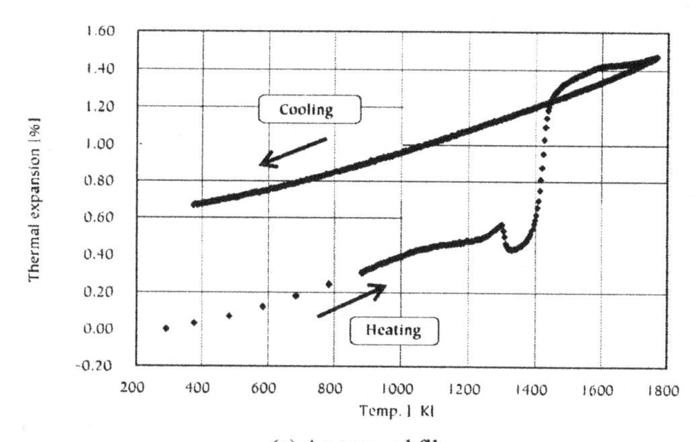


(a) As-sprayed film

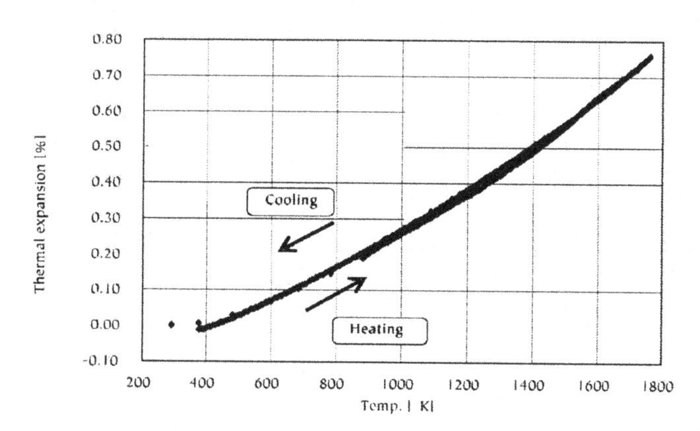


(b) Heat-treated film at 1373K for 10hr after APS

Fig.5 Thermal expansion of plasma sprayed films of 33mol% $Lu_2O_3$-67mol% $SiO_2$

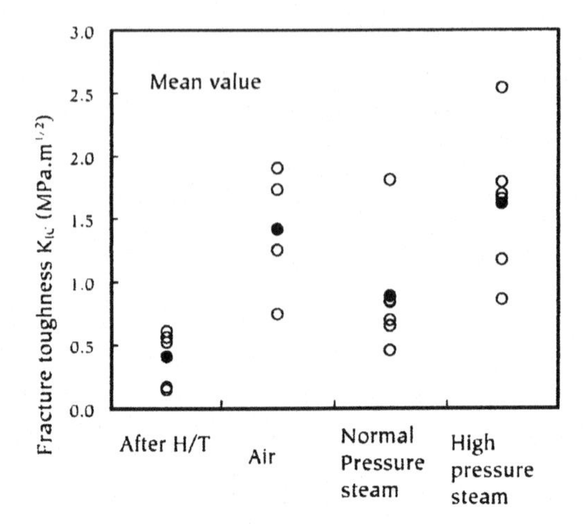


Fig.6 Fracture toughness of lutetium silicate topcoat on SiC/SiC composite after various xposure tests

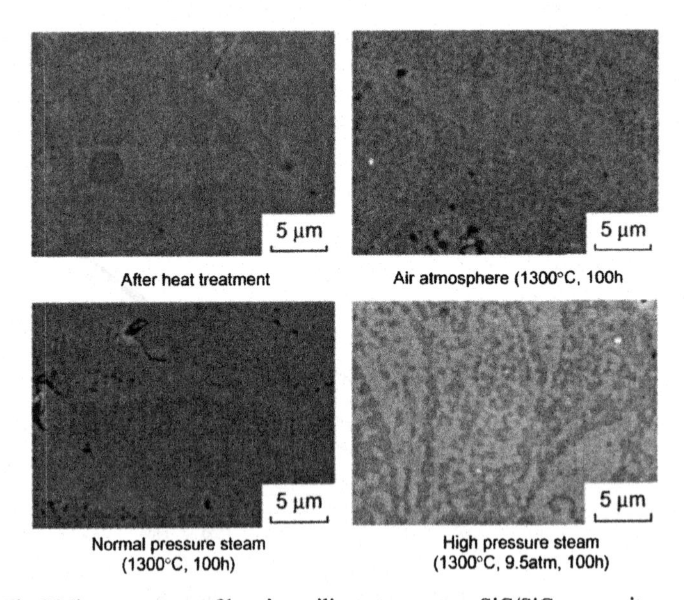


Fig.7 Microstructures of lutetium silicate topcoat on SiC/SiC composite after various exposure tests (SEM)

Fig.7 shows the microstructures of lutetium silicate topcoat on SiC/SiC composite after various exposure tests. Small amount fine particulate dispersed structures are observed in the topcoat layer before exposure and those structures increase and grow after exposure. The microstructures after high pressure steam exposure show the most growth of the dispersed particles structures. It seems that the change of fracture toughness is related to the change of microstructures.

## CONCLUSION

In order to improve durability and oxidation resistance of SiC/SiC in the applications of power generation fields, investigation of topcoat candidates and evaluation of topcoat on SiC/SiC composites were conducted and following results were obtained.

1) Oxidation resistance of stand alone plasma sprayed films of rare-earth silicate was evaluated for topcoat candidates and lutetium silicate of composition ratio of 33mol% $Lu_2O_3$-67mol% $SiO_2$ has good oxidation resistance compared with other rare-earth silicate.

2) The structure of plasma sprayed films of 33mol% $Lu_2O_3$-67mol% $SiO_2$ crystallized and became stable by heat treatment at 1373K for 10 hr in inert gas atmosphere.

3) EBC systems including lutetium silicate topcoat layer on SiC/SiC composites were exposed under various oxidizing atmospheres, and fracture toughness of topcoat layer was measured. And the fracture toughness of topcoat layer increases after exposure under oxidizing atmospheres.

4) As the results of these evaluation tests, lutetium silicates are considered to be promising for topcoat materials. And it is the next subject to evaluate lutetium silicates under high pressure steam environment with more high gas-flow velocity.

## ACKNOWLEDGMENTS

These researches and developments are supported by NEDO.

The authors want to thank all persons of each company, such as UBE, IHI and KHI, who are engaged in this project.

## REFERENCES

[1]J. A. Haynes, M. J. Lance, K. M. Cooley, M. K. Ferber, R. A. Lowden and D. P. Stinton, "CVD Mullite Coatings in High-Temperature, High-Pressure Air-$H_2O$," *J. Am. Ceram. Soc.*, **83**(3)657-59(2000).

[2]K. N. Lee, D. S. Fox, R. C. Robinson and N. P. Bansal, "Environmental Barrier Coatings for Silicon-Based Ceramics," *High Temp. Ceram. Mat. Comp.* 224-29(2001)

[3]A. K. Misra, "Development of Advanced Engine Materials in NASA's Ultra Efficient Engine Technology Program," *ISABE-2001-1106*(2001)

[4]H. Klemm, C. Taut and G. Wotting, "Long-term stability of nonoxide ceramics in an oxidative environment at 1500C," *J. Euro. Ceram. Soc.*, **23,** 619-27 (2003)

[5]K. N. Lee, D. S. Fox, J. I. Eldridge, D. Zhu, R. C. Robinson, N. P. Bansal and R. A. Miller, "Upper Temperature Limit of Environmental Barrier Coatings Based on Mullite and BSAS," *J. Am. Ceram. Soc.*, **86**(8)1299-306(2003)

[6]H. Nakayama, K. Aoyama, M. Yamamoto, H. Sumitomo, K. Okamura, T. Yamamura and M.Sato, "Evaluation of Environmental Barrier Coatings for CFCC in the High-Water-Vapor-

Pressurized Environments at High Temperatures.“ *High Temp. Ceram. Mat. Comp.* **5**, 613-618 (2004)

[7]H. Nakayama, K. Morishita, S. Ochiai, T. Sekigawa, K. Aoyama, T. Oi, M. Yamamoto, K. Okamura, and M.Sato, “Quest and Evaluation of Topcoat Materials for Environmental Barrier Coatings of SiC/SiC Composites“ *Proc.Encera 04/ The 3rd Int. Sympo. Sci. of Eng. Ceram.* (2004)

[8]E. J. Opila and R. E. Hann Jr., “Paralinear Oxidation of CVD SiC in water Vapor,” *J. Am. Ceram. Soc.*, **80**(1)197-205(1997)

[9]E. J. Opila, J. L. Smialek, R. C. Robinson, D. S. Fox and N. S. Jacobson, “SiC Recession Caused by $SiO_2$ Scale Volatility under Combustion Conditions: 2, Thermodynamics and Gaseous-Diffusion Model,” *J. Am. Ceram. Soc.*, **82**,1826-34(1999)

[10]E. J. Opila, “Oxidation and Volatilization of Silica Formers in Water Vapor,” *J. Am. Ceram. Soc.*, **86**(8)1238-48(2003)

[11]I. Yuri and T. Hisamatsu, “Recession Rate Prediction for Ceramic Materials in Combustion Gas Flow,” *ASME TURBO EXPO* (2003)

[12]I. Yuri, T. Hisamatsu, S. Ueno and T. Ohji, “Recession Characteristics of $Lu_2Si_2O_7$ in Combustion Gas Flow at High Temperature and High Speed,” *J. Soc. Mat. Sci., Japan,* **54**(10), 1075-79(2005)

# LIFE LIMITING PROPERTIES OF UNCOATED AND ENVIRONMETAL-BARRIER COATED SILICON NITRIDE AT HIGHER TEMPERATURE

Sung R. Choi,* Dongming Zhu, and Ramakrishna.T. Bhatt
NASA Glenn Research Center, Cleveland, Ohio 44135

ABSTRACT

Life limiting properties of a commercial gas-turbine grade silicon nitride were determined with and without environmental barrier coatings (EBCs) at 1371°C in air. Uncoated silicon nitride, AS800, exhibited a significant life limiting phenomenon with a relatively low life prediction parameter $n \approx 10$ either in dynamic fatigue or in stress rupture under flexure. This enhanced susceptibility to life limiting was also consistent in tensile stress rupture loading. By contrast, despite some limited data, AS800 with plasma-sprayed rare-earth multi-component hafnia or yittria-hafnia EBCs showed much improved resistance to life limiting over the uncoated AS800, resulting in four to seven-fold life extension.

## INTRODUCTION

Advanced ceramics are candidate structural materials for high temperature aeroengine applications. One of the major limitations involved with structural ceramics in such applications is a life limiting phenomenon associated with mechanisms such as slow crack growth, creep, material degradation due to combustion environments, all accelerating with increased operating temperatures. In general, combination of development of better base material, design methodology, and use of appropriate environmental barrier coatings (EBCs) seems to be a common approach to ensure structural integrity/life of engine components. Despite its importance, little studies have been done on the subject on life limiting behavior of advanced gas turbine grade silicon nitrides with and without EBCs particularly at higher temperatures >1350 °C. A previous study showed that a silicon nitride (SN88) with CVD-mullite coatings exhibited little improvement in long-term performance over the uncoated counterpart [1]. The testing, however, was conducted at a relatively low temperature of 850 °C with a main focus on the intermediate-temperature stability of the material.

This paper reports life limiting behavior of a gas-turbine grade silicon nitride (AS800) with and without EBCs at a very high temperature of 1371 °C in air. Both dynamic fatigue and stress rupture testing were performed with uncoated AS800 test specimens in flexure. Also, additional stress rupture testing was conducted at 1371 °C in pure tension using some limited number of uncoated tensile dog-boned specimens. Two different EBC systems were obtained through air plasma-spraying EBCs onto AS800 flexure specimens, and their lifetimes were determined in stress rupture at a specified level of applied stress of 250 MPa.

## EXPERIMENTAL PROCEDURS

* Now with Naval Air Systems Command, Patuxent River, MD 20670; sung.choi1@navy.mil

Table 1. Basic mechanical and physical properties of AS800 and SN282 silicon nitrides at ambient temperature [7,8]

| Material | Elastic modulus E (GPa) | Poisson's ratio ν | Density ρ (g/cm³) | Hardness H (GPa) | Flexure Strength | | | Fracture toughness $K_{Ic}$ (MPa√m) |
|---|---|---|---|---|---|---|---|---|
| | | | | | Mean strength (MPa) | Weibull modulus | Characteristic strength (MPa) | |
| AS800 $Si_3N_4$ | 309 | 0.27 | 3.27 | 13.6(1.4) | 775(45) | 21 | 795 | 8.1(0.3) |

Notes:

1. For detailed test methods, refer to references [12,13]
2. The numbers in the parentheses indicate ±1.0 standard deviations.

Material

The material used in this work was a commercially available gas-pressure sintered silicon nitride, AS800 (fabricated by Honeywell Ceramic Components, Torrance, CA, '99 vintage, gel-cast). This silicon nitride has been considered one of the strong candidate ceramics for gas-turbine applications in view of its substantially improved elevated-temperature properties [2-4]. AS800 is an in-situ toughened silicon nitride, with microstructures tailored to achieve elongated grain structures (see Figure 1). AS800 silicon nitride has been used at the NASA Glenn Research Center in lifing and FOD (foreign object damage) programs [5-8]. The billets were machined into flexure target specimens measuring 3 mm by 4 mm by 25 or 50 mm, respectively, in depth, width and length in accordance with a machining procedure detailed in a test standard ASTM C 1211 (size "B") [9]. The final finish of target specimens was completed with a 600-grit diamond wheel. The basic mechanical and physical properties of AS800 are shown in Table 1 [7,8].

The EBCs investigated in this study were multi-component, $HfO_2$ and rare earths (yttria, gadolinia and ytterbia) doped alumino-silicate (designated here "EBCs-1"), and the yttria-stabilized $HfO_2$ (designated "EBCs-2") which were further doped with silicon for improved bonding performance. Depending on the composition optimizations, the both coatings can also be designed and used as ceramic-based EBC bond coats and/or interlayers for Si-based ceramics, without the traditional Si bond coat or the EBC systems for better high temperature stability. The plasma-spray technique was used for processing the two EBC systems. The pre-mix composition oxide powders were first spray dried, and then plasma-reacted and spheroidized. The coatings were then sprayed using the advanced doped alumino-silicate and $HfO_2$ powders in a furnace at elevated temperature, directly onto AS800 flexure bar substrates. The total EBC thicknesses were approximately 100 μm. Figure 2 shows the high-temperature plasma-spray facility at the NASA Glenn Research Center for the environmental barrier coating processing.

Life Prediction Testing: Dynamic Fatigue and Stress Rupture

Both dynamic fatigue and stress rupture testing for uncoated AS800 test specimens were carried out in flexure with 20-mm inner and 40-mm outer spans at 1371 °C in air. The dynamic fatigue test method, as specified in ASTM C 1465 [10], determines flexure strengths as a function of applied test rate. From the determined strength as a function of applied test rate, slow crack growth or life prediction parameters of a material can be evaluated using a relevant relation. Four

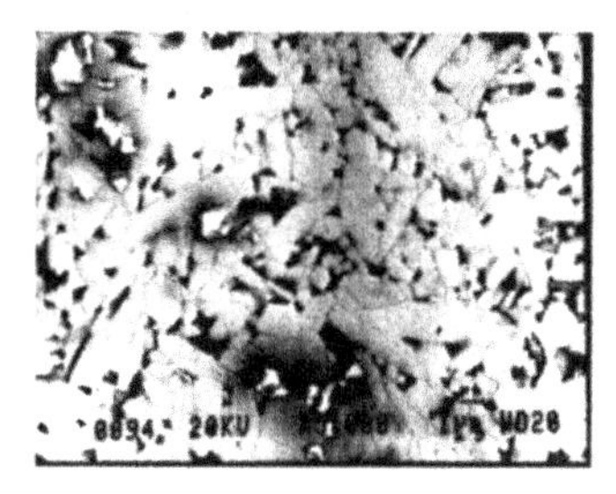

Figure 1. Microstructure of AS800 silicon nitride

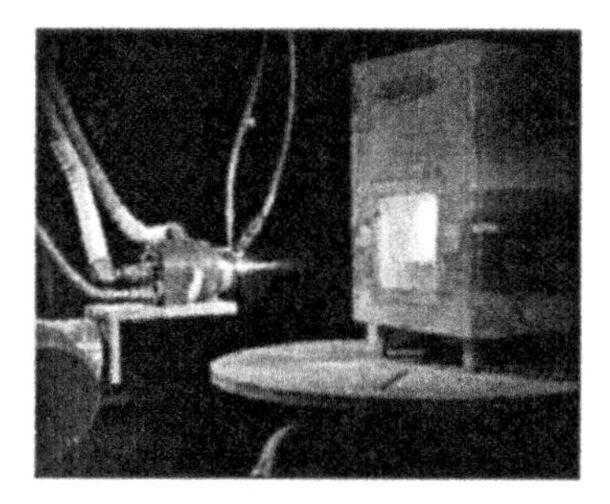

Figure 2. Plasma-spray processing of environmental barrier coatings using a high temperature plasma-spray system.

different test rates of 50, 0.5, 0.05, and 0.005 MPa/s were used in load control using an electromechanical test frame (Model 8562, Instron, Canton, MA) or a servohydraulic test frame (Model 8501, Instron). A total of 10 to 15 test specimens, depending on test rate, were used at each test rate. Stress rupture testing for uncoated AS800 was also conducted using the same test fixture, specimen configuration, and test temperature that were used in dynamic fatigue testing. Stress rupture testing determines life as a function of applied stress, with which life prediction parameters of a material can be estimated. Eight different levels of applied stresses, ranging from 200 to 500 MPa, were used. A total of 21 test specimens were used in stress rupture testing.

Stress rupture testing for AS800 flexure test specimens plasma-sprayed with two different EBC systems, EBCs-1 and EBCs-2, was carried out at a particular level of applied stress of 250 MPa at 1371 $^{o}$C in air. Since it has been observed that some amount of strength degradation of test specimens was inevitable upon plasma coating, a relatively low applied stress of 250 MPa was chosen in this study to eliminate any premature failure in stress rupture testing. Due to limited availability of EBCs, five coated specimens were used for each EBC system.

## RESULTS AND DISCUSSION

### Life Limiting Behavior of Uncoated AS800

Results of dynamic fatigue testing for AS800 silicon nitrides at 1371 °C in air are shown in Figure 3. Between the applied stresses of 50 MPa/s and 0.5 MPa/s, no strength degradation occurred. However, at the applied stress rates lower than 0.5 MPa/s, strength degradation took place with further decreasing applied stress rate, showing a susceptibility to slow crack growth. Weibull strength distributions, determined at different stress rates, are shown in Figure 4. Weibull modulus ($m$), despite some limited number of test specimens used due to *higher-temperature* life testing, ranged from $m$ = 10 to 40 showing a decrease in $m$ with decreasing test rates. Figure 5 presents typical fracture surfaces of specimens tested at the fast test rate of 50 MPa/s and the slowest test rate of 0.005 MPa/s, showing a difference in degree of slow crack growth. Failure was in many cases associated with volume flaws (large grains and/or pores) particularly at lower test rates. It was also observed that more enhanced oxidation occurred in the surfaces of test specimens subjected to lower test rates, because of increased test time therein. A typical example showing the extension of oxidation is shown in Figure 6, where bubbling was manifest from the specimen surface tested at 0.005 MPa/s in which an average test time was about 20 h.

Figure 7 shows the results of stress rupture testing at 1371 °C for uncoated AS800 flexure test specimens. Life limiting was apparent, showing a significant decrease in life with increasing applied stress. A typical fracture surface subjected to an applied stress of 250 MPa with a life of 80 h is shown in Figure 8, where a region of enhanced slow crack growth is evident, associated with a volume flaw. Severe oxidation took place for the specimen tested at 200 MPa with a lifetime of 550 h, as also shown in Figure 8. As a consequence, the limit of use of this uncoated AS800 material at 1371 °C in air for a purpose of long-term life/stability is self-evident. Also note from Figure 8 that the specimen was surrounded by the *dark band*, which is indicative of the enhanced oxidation probably due to the material's sintering additive systems [1,11].

The basic formulation of slow crack growth (SCG) for advanced monolithic and composite (reinforced with particulates, platelets or whiskers) ceramics at elevated temperatures follows an empirical power-law form

$$v = A[K_I / K_{Ic}]^n \qquad (1)$$

where $v$, $K_I$ and $K_{Ic}$ are crack velocity, mode I stress intensity factor, and mode I fracture toughness, respectively. $A$ and $n$ are material/environment dependent SCG parameters. The important life prediction (or SCG) parameter is $n$, as reflected from the functional form of Eq. (1), and this parameter can be evaluated from an appropriate relation in dynamic fatigue between fracture strength ($\sigma_f$) and applied stress rate ($\dot{\sigma}$) [10]. Using this relation, the parameter was estimated to be $n$ = 11, when the range of applied stress rates showing strength degradation was considered. For brittle materials, it is generally categorized that significant life limiting occurs for $n < 30$, intermediate for $30 < n < 70$, and insignificant life limiting for $n > 70$. Hence, it can be stated that

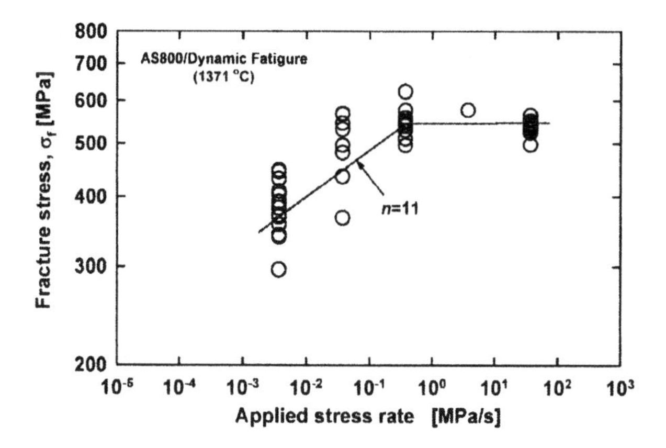


Figure 3. Results of dynamic fatigue testing for uncoated AS800 flexure test specimens at 1371 °C in air. Life prediction parameter $n$ is included with a solid line.

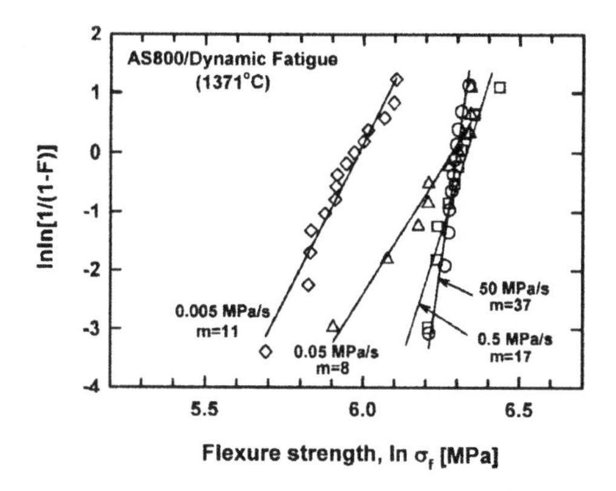


Figure 4. Weibull strength distributions of uncoated AS800 flexure test specimens in dynamic fatigue at 1371 °C in air for different applied stress rates. $m$: Weibull modulus.

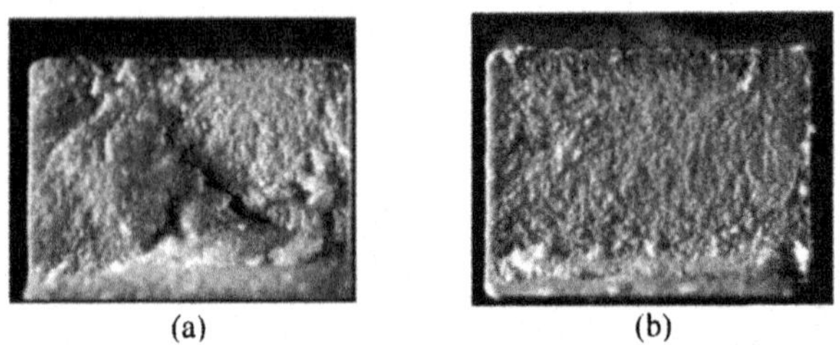


Figure 5. Typical appearances of fracture surfaces of uncoated AS800 flexure specimens in dynamic fatigue at 1371 °C in air, tested at (a) 50 MPa/s and (b) 0.005 MPa/s.

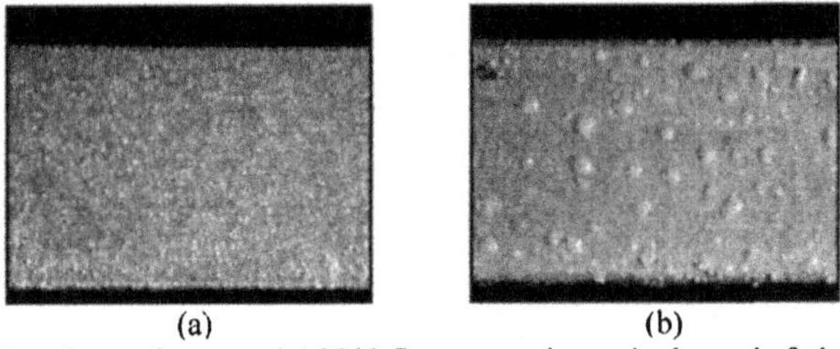


Figure 6. Typical surfaces of uncoated AS800 flexure specimens in dynamic fatigue at 1371 °C in air, tested at (a) 50 MPa/s and (b) 0.005 MPa/s.

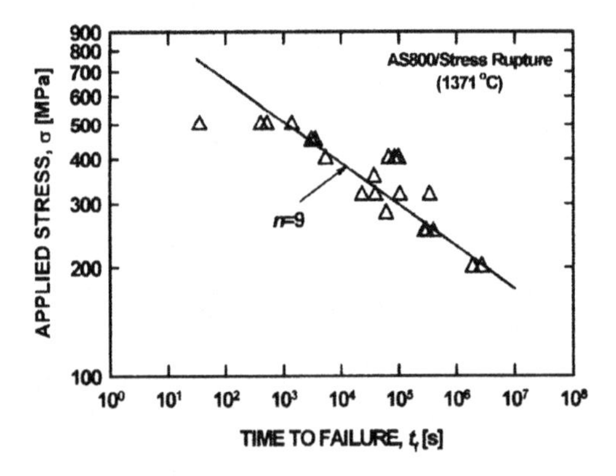


Figure 7. Results of stress rupture testing for uncoated AS800 flexure test specimens at 1371 °C in air. Prediction of life from the dynamic fatigue data of Figure 3 is included as a solid line.

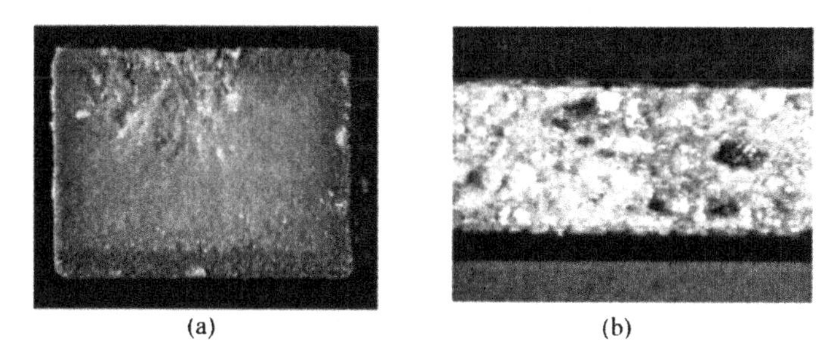


Figure 8. (a) Typical fracture surface of uncoated AS800 flexure test specimen in stress rupture at 250 MPa at 1371 °C in air; (b) typical surface of specimen at 200 MPa in stress rupture with a test duration of 530 h.

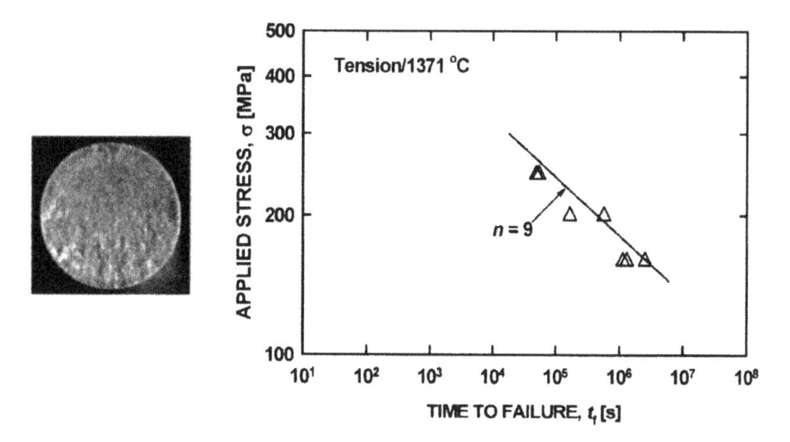


Figure 9. Results of stress rupture testing for uncoated AS800 tensile dog-boned test specimens at 1371 °C in air. A typical fracture surface of a specimen tested at 250 MPa is also shown.

the susceptibility to life limiting was significant to uncoated AS800 at 1371 °C. The results of dynamic fatigue testing (Figure 3) can be converted to stress rupture using the following relation

$$t_f = f(m, \sigma_i, n, D, F, \sigma) \tag{2}$$

where $t_f$ is life, $F$ is the failure probability, $D$ is the intercept in dynamic fatigue curve, $\sigma_i$ is the inert strength, and $\sigma$ is the static applied stress. For simplicity, a failure probability of approximately $F$ = 50 % was used in prediction. The prediction, as can be seen from a solid line in Figure 7, was in good agreement with the stress rupture data. The results of stress rupture [illegible] pure tension also showed a similar parameter $n$ = 9, as presented in Figure 9 (However, the use of limited number of tensile dog-boned test specimens should be noted.). All of these results imply that similar failure mechanism(s) might have been operative, irrespective of loading configurations, either dynamic fatigue or stress rupture in flexure or pure tension. Of course, more detailed microstructural examinations are needed to better assess the governing mechanism(s) associated with failure, although the governing mechanism is thought to be related to the damage zone developed due to oxidation.

Life Limiting Behavior of Environmental Barrier Coated AS800

The results of stress rupture testing for AS800 coated with two different EBC systems, subjected to 250 MPa at 1371 °C in air, are shown in Figure 10. This figure compares life between the uncoated AS800 and the coated AS800 specimens. Some of the EBCs-1 specimens exhibited life equivalent to or shorter than that (≈ 100 h) of the uncoated. This could be due to some open channels/crackings within the coatings, possibly attributed to inhomogeneity involved in plasma-spray process. Theses open channels or crackings would have degraded the effectiveness of coatings as an environmental barrier. By contrast, other specimens with both EBC systems showed significantly increased lives greater than 400 h. The runout time for EBCs-1 and EBCs-2, in which test was interrupted if the test time was over the prescribed time, could amount to 700 h and 500 h, respectively. This indicates that the use of the two EBC systems could achieve up to five to seven -fold life extension, as compared to that of the uncoated AS800. Also the test specimens' surfaces appeared to be very smooth (see Figure 11), unlike the one shown in Figure 8(b) due to severe oxidation. The specimens tested in this long duration exhibited significant creep deformation, estimated from the magnitude of deformed curvature of crept specimens. Therefore, creep was not an immediate cause of material failure for the uncoated specimens. Rather, slow crack growth and/or new flaw populations generated due to loading and oxidation might have been a predominant source of premature failure of the uncoated specimens. Since the scatter in stress rupture data was appreciable for both EBC systems, the use of more test specimens are required to ensure statistical reproducibility, commonly practiced in any strength and/or life prediction testing.

CONCLUSIONS

1. Life limiting phenomenon of uncoated AS800 at 1371 °C in air was significant with a low life prediction parameter of $n \approx 10$, irrespective of dynamic fatigue or stress rupture loading in flexure or pure tension.

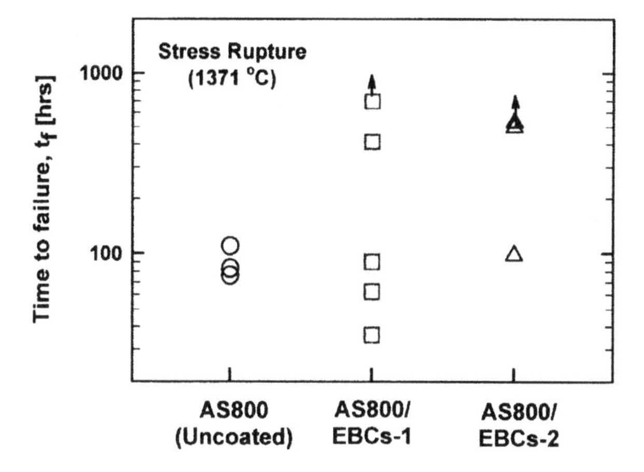


Figure 10. Results of stress rupture testing for AS800 flexure specimens with EBCs-1 and EBCs-2 systems at 1371 °C in air. Arrows indicate the specimens not failed (the runout). The data of uncoated AS800 flexure test specimens are included for comparison.

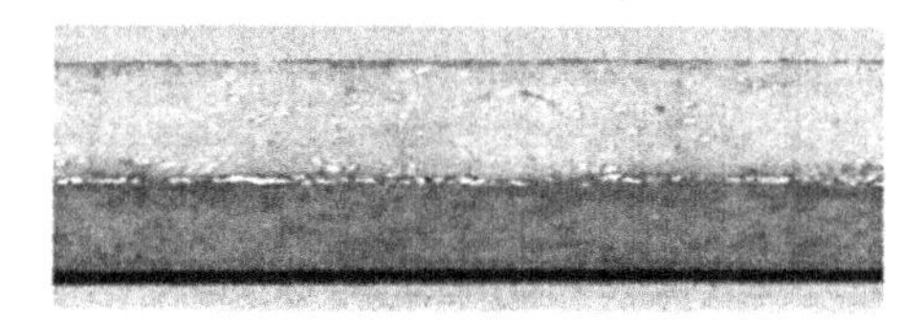

Figure 11. Typical appearance of an AS800 specimen with EBCs-1 system, subjected to stress rupture at 1371 °C in air at 250 MPa, showing a smooth surface as compared with the uncoated specimen in Figure 8(b). Test duration ≈ 700h.

2. Despite their premature failure in some cases, AS800 plasma-sprayed with two different EBC systems showed a significant life extension when tested at 250 MPa at 1371 °C in air. The life extension amounted to four to seven-fold, as compared to the life (≈ 100 h) of the uncoated AS800 specimens.
3. Statistical reproducibility of stress rupture data needs to be evaluated using more coated test specimens, considering the scatter of the current stress rupture data. This would allow one to better assess material/processing parameters in response to long-term life/stability.

ACKNOWLEDGEMENTS

The authors are grateful to Ralph Pawlik for mechanical testing and George Leissler for processing EBCs. This work was supported by Higher Operating Temperature Propulsion Components (HOTPC) and Ultra-Efficient Engine Technology (UEET) programs, NASA GRC, and the Space Act Agreement with Honeywell Engine Systems. Helpful discussion of Laura Lindberg of Honeywell is appreciated.

REFERENCES

1. H. T. Lin, M. K. Ferber, T. P. Kirkland, and S. M. Zemskova, "Dynamic Fatigue of CVD-Mullite Coated SN88 Silicon Nitride," Proceedings of ASME Turbo Expo, Power for Land, Sea, and Air, June 16-19, 2003, Atlanta, Georgia; ASME paper No. GT2003-38919.
2. T. Ohji, "Long Term Tensile Creep Behavior of Highly Heat-Resistant Silicon Nitride for Ceramic Gas Turbines," *Ceram. Eng. Sci. Proc.*, **22**[3] 159-166 (2001).
3. F. Lofaj, S. M. Wiederhorn, and P. R. Jemian, "Tensile Creep in the Next Generation Silicon Nitride," *Ceram. Eng. Sci. Proc.*, **22**[3] 167-174 (2001).
4. H. T. Lin, S. B. Waters, K. L. More, J. Wimmer, and C. W. Li, "Evaluation of Creep Property of AS800 Silicon Nitride from As-Processed Surface Regions," *Ceram. Eng. Sci. Proc.*, **22**[3] 175-182 (2001).
5. S. R. Choi and J. P. Gyekenyesi, "Elevated-Temperature "Ultra" Fast Fracture Strength of Advanced Ceramics: An Approach to Elevated-Temperature "Inert" Strength", Trans. of ASME, *J. Eng. Gas Turbines & Power*, **121** 18-24 (1999).
6. S. R. Choi, J. M. Pereira, L. A. Janosik, and R. T. Bhatt, "Foreign Object Damage Behavior of Two Gas-Turbine Grade Silicon Nitrides by Steel Ball Projectiles at Ambient Temperature," NASA/TM-2002-211821, National Aeronautics and Space Administration, Glenn Research Center, Cleveland, Ohio (2002).
7. S. R. Choi, J. M. Pereira, L. A. Janosik, and R. T. Bhatt, "Foreign Object Damage in Flexure Bars of Two Gas-Turbine Grade Silicon Nitrides," *Mat. Sci. Eng.*, **A379** 411-419 (2004).
8. S. R. Choi, J. M. Pereira, L. A. Janosik, and R. T. Bhatt, "Foreign Object Damage in Disks of Gas-Turbine Grade Silicon Nitrides by Steel Ball Projectiles at Ambient Temperature," *J. Mater. Sci.*, **39** 6173-6182 (2004).
9. ASTM C 1211, "Test Method for Flexural Strength of Advanced Ceramics at Elevated Temperatures," *Annual Book of ASTM Standards*, Vol. 15.01, American Society for Testing & Materials, West Conshohocken, PA (2005).
10. ASTM C 1465, "Standard Test Method for Determination of Slow Crack Growth Parameters of Advanced Ceramics by Constant Stress-Rate Flexural Testing at Elevated Temperatures," *Annual Book of ASTM Standards*, Vol. 15.01, American Society for Testing and Materials, West Conshohocken, PA (2005).
11. S. R. Choi and R. T. Bhatt, "Delayed Failure of Two Gas-Turbine Grade Silicon Nitrides at Elevated Temperatures," presented at the 28$^{th}$ Cocoa Beach Conference on Advanced Ceramics and Composites, January 25-30, 2004, Cocoa Beach, FL; Paper No. CB-S4-28-2004.

MULTILAYER EBC FOR SILICON NITRIDE

C.A. Lewinsohn, Q. Zhao, B. Nair
Ceramatec Inc.
Salt Lake City, UT, 84119

The lifetime of silicon-based ceramic components in turbine engine applications are limited by hydrothermal corrosion.[1,2] Materials with adequate resistance to hydrothermal corrosion do not possess the required mechanical properties to be used as engine components, but may perform satisfactorily as coatings.[3] Reliable coatings of many candidate materials are not possible due to significant property mismatches with potential substrate materials. Likewise, interlayers, and methods of applying them, to accommodate property mismatches, must meet stringent requirements. This paper describes multilayer, environmental barrier coatings (EBC) with sufficient adhesion, environmental and thermal stability, and processing robustness to provide hydrothermal corrosion resistance to silicon nitride components .

## INTRODUCTION

Silicon based ceramics and composites have the high temperature thermomechanical properties required for use in gas turbine engine hot-section components and sensors. Although silicon carbide (SiC) has high strength and good thermal conductivity, it suffers from low fracture toughness and, hence, reliability. Therefore, components consisting of silicon nitride ($Si_3N_4$), which can be manufactured with higher values of fracture toughness than silicon carbide, and silicon carbide- or silicon nitride–matrix composites are currently under development for components that will be subject to appreciable stresses in operation. These materials are stable under purely oxidizing conditions, due to the formation of passivating oxide layers. They can be significantly corroded, however, by $H_2O$ and CO, which are common components in gas turbine systems.[4,5,6]

Materials that exhibit good hydrothermal corrosion resistance typically have much higher coefficients of thermal expansion than silicon nitride[7], or mismatched elastic properties, such that unacceptably high residual stresses develop in the substrate or coating that subsequently lead to failure after processing or during operation. One approach to mitigate these residual stresses has been to insert materials with intermediate properties between the coating and the substrate.[8] The choice of interlayer material, however, has been limited by the requirements that it adheres to both top coat and substrate materials, have good high-temperature stability, does not exhibit any deleterious reactions with either the top coat or substrate, and has acceptable thermoelastic properties. Recently, amorphous, non-oxide ceramics derived from preceramic polymers (polymer-derived ceramics, PDC) have demonstrated remarkable oxidation stability and mechanical properties at elevated temperatures.[9] Furthermore, these materials show excellent adherence to a wide range of materials, including non-oxide ceramics, oxide ceramics, and metals. For example, in an earlier project at Ceramatec, a method of forming high-temperature, gas tight seals between solid oxide fuel cell electrolytes (zirconia) and metal interconnects (steel) was demonstrated.[10,11] The coatings described in this paper also could be applied to lightweight oxide materials that are susceptible to hydrothermal corrosion. In this work, interlayers comprised of amorphous, non-oxide PDC materials for duplex, environmental barrier coatings were developed for resistance to hydrothermal corrosion.

Extensive research on EBC coating materials has been conducted and several oxide materials with low silica activity are being investigated for their hydrothermal corrosion resistance. These oxides include ytterbium silicate ($Yb_2Si_2O_7$)[7,12,13], lutetium silicate ($Lu_2Si_2O_7$)[13], yttria-stabilised zirconia (8mol% yttria + 92mol% $ZrO_2$, 8YSZ)[12], strontium-stabilised celsian ((1-x)BaO-xSrO-AlO2-SiO2, 0<x<1), BSAS)[8], and mullite ($3Al_2O_3$-$2SiO_2$).[14] Ceramatec is also investigating other oxides with low silica activity using a "geomimetic" approach, since these materials may have excellent resistance to hydrothermal corrosion.

Based on promising results from mullite and calcium aluminosilicate coatings, NASA scientists developed a state-of-the-art EBC for silicon carbide, which consists of three layers: a silicon bond coat, a mullite or a mullite+BSAS intermediate coat, and a BSAS top coat by modified plasma spray process[8]. This EBC had been scaled up and applied on SiC/SiC CMC combustor liners used at ~1250°C for over 24 000 hours. The upper temperature limit for this EBC system is around 1300°C (EBC surface temperature) because of significant recession of BSAS via volatilization in water-vapor-containing atmospheres[8]. Systems with similar properties are required for silicon nitride ceramics to allow their use in turbine engine components such as blades, disks and rotors.

## METHODS

A flow chart illustrating the overall process to coat silicon nitride substrates, using two bond coat layers, is shown in Figure 1. The bond coat can be applied as several layers, with the same or different composition, or as a single layer. A flow chart illustrating the preparation of the bond coat layer is shown in Figure 2. One of the unique features of the coatings described in this work is the combination of partially pyrolysed preceramic precursor (aHPCS, Starfire Systems, Inc) with unpyrolysed polymer. The partially pyrolysed material reduces the shrinkage during processing of the coatings.

A second unique feature of the multilayer coatings described in this paper is the use of geomimetic top coat materials for hydrothermal corrosion resistance. The top coat materials were selected from a database generated by geochemists, hence the name geomimetic, for the stability of materials at high temperature and pressure. The geomimetic materials used in this work are multi-component oxides, of proprietary composition, that are extremely stable at high temperature environments containing moisture.

Coatings were applied to bend bars made from various commercial types of silicon nitride. In some cases the bend bars had as-processed surfaces, but in most cases the surfaces were machined and ground. Cross-hatch, adhesive tape peel tests, similar to ASTM D3359 Method A, were used to evaluate adhesion of the bond coat to the substrate, The flexural strength of the bars was tested according to ASTM C-1160. Hydrothermal corrosion of the top coat material was tested in a furnace containing high-pressure (18 bar) water vapor at The Oak Ridge National Laboratory. Coated specimens were also subjected to thermal cycle testing at 1300°C in an environment of 90% $H_2O$, 10% $O_2$ flowing at 2.2 cm/sec. The thermal cycles were performed by shuttling the specimens in and out of the hot zone of a furnace held at 1300°C. The specimens were cycled between room temperature and 1300°C with a heating time of 20 seconds to temperature, a 1 h hold at 1300°C, a cooling time of several minutes, and a 20 minute hold at room temperature. The specimens were contained in platinum crucibles and the were contained within an alumina furnace tube. The testing was performed at the NASA Glenn Research Center.

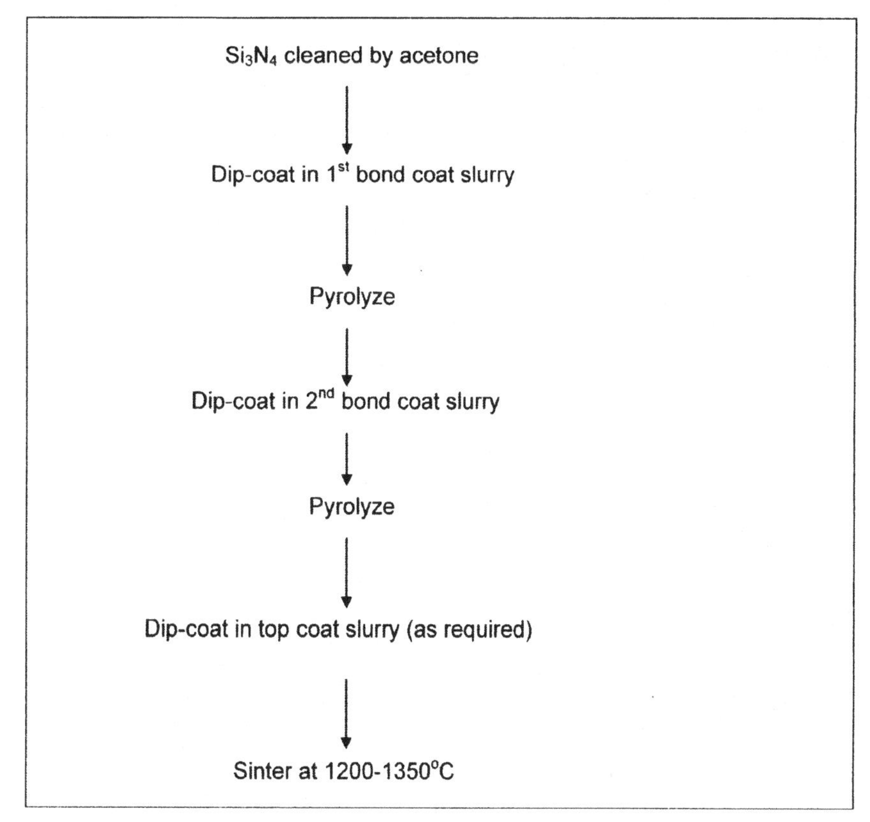


Figure 1 Overall process flow chart for applying bond coat(s) and outer coatings, as required.

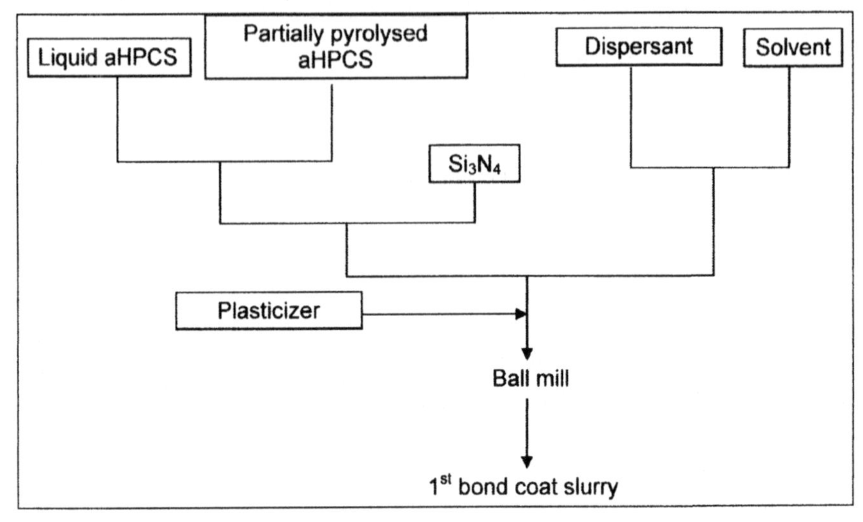


Figure 2 Preparation of a bond coat layer.

## RESULTS AND DISCUSSION

Cross-hatch, adhesive tape peel tests, similar to ASTM D3359 Method A, were used to demonstrate good adhesion of the bond coat to silicon nitride substrate (Figure 3). In addition, to good adhesion to substrates, as demonstrated by the cross-hatch results, the bond coats also demonstrated good adhesion to other, relevant, outer coating materials such as mullite and ytterbium silicate. Furthermore, the bond coats also exhibited good adhesion to both the substrates and the outer coatings after thermal cycle testing.

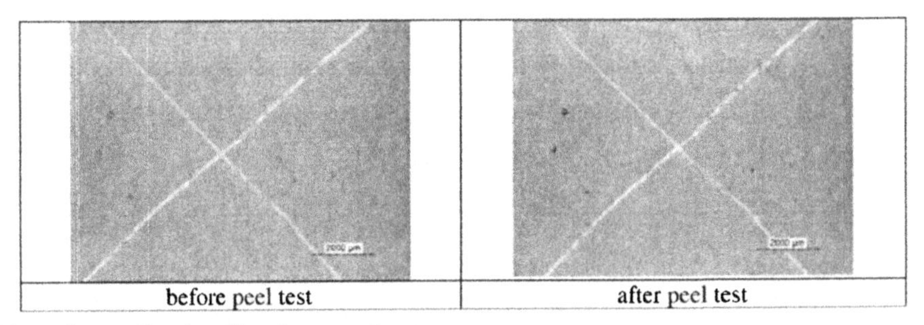


Figure 3 Results of bond coat peel test

The feasibility of a multilayer EBC was demonstrated by applying a bond coat of PDC and silicon nitride, a second bond coat of PDC and 3mol% yttria-stabilized zirconia (3YSZ), and an outer coat of 3mol% zirconia (these materials were readily available at Ceramatec), shown in Figure 3. Although each layer was quite thin, and 3mol% zirconia is not known to have good hydrothermal corrosion resistance, the results demonstrate that these types of coatings can be

Figure 4 Micrographs of a representative coating system: 3YSZ (top layer) on NT-154 $Si_3N_4$ (bottom layer).

applied to silicon nitride. Furthermore, the fact that thin layers could be deposited may be beneficial to the achievement of graded coatings with small property gradients.

The results in Table I summarise flexural testing of untreated bars of silicon nitride compared with bars coated with bond coat layers demonstrate that the coating process does not degrade the strength of the substrate. It was demonstrated that both as-processed and machined NT154 $Si_3N_4$ bars retained nearly 100% of their strength after bond coat application. The results in Table II, show that bend bars coated with a multilayer EBC, with a geomimetic top coat composition, retain >75% of their flexural strength after 70 h of exposure to flowing water vapor in air at 1100°C.

The weight change, measured after exposure for 2 000 h at 1250°C in 10% steam/90% $H_2O$ at a total approximately 18 bar, of dense top coat material coupons was slightly negative for the most promising geomimetic composition (Figure 5). The weight change of the best composition was between that of strontium aluminosilicate materials. Additional exposure testing and microscopy of these compositions is underway.

Table I
Effect of bond coat on strength of unmachined silicon nitride substrates

| Type | Ave. strength (MPa) | Std. dev. (MPa) |
|---|---|---|
| Un-coated (baseline) | 596 | 153 |
| 1 bond coat | 654 | - |

Table II
Effect of geomimetic EBC system on strength of silicon nitride substrates

| Type | # tested | Average Strength (MPa) | 95% CI (MPa) |
|---|---|---|---|
| as-processed | 5 | 596 | 134 |
| machined | | 935* | |
| as-processed, 70 h exposure | 6 | 688 | 40.5 |
| machined, 70 h exposure | 3 | 626 | 144 |

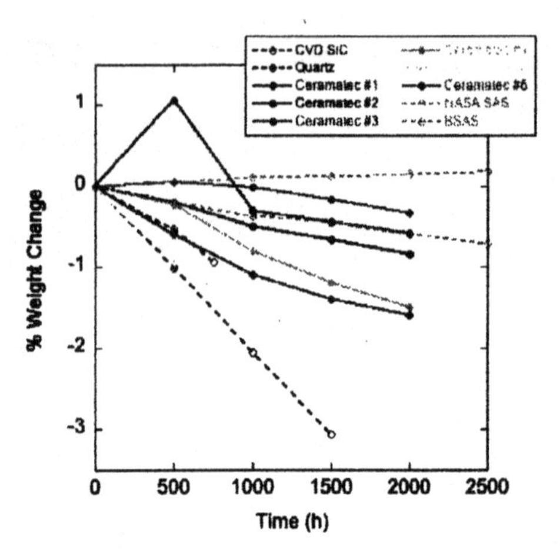


Figure 5 Weight change of top coat materials after exposure in 10% steam/90% air environment at 1250°C.

## SUMMARY

In summary, the multilayer EBCs described in this paper appear promising for the protection of silicon-based ceramics from hydrothermal corrosion. Bond coats, derived from preceramic polymers, provide good adhesion to silicon nitride substrates and relevant top coat materials. Geomimetic compositions provide hydrothermal corrosion protection. The coating system does not degrade the strength of desired substrates excessively and survives numerous thermal cycles.

## ACKNOWLEDGEMENTS

The authors are grateful to a subcontract from the University of Tennessee – Battelle for support of this work and to Dr. D. Stinton and Dr. T. Tiegs, at The Oak Ridge National Laboratory, for administering this support. The authors are also grateful to Dr. V. Pujari and Mr. A. Varabedian for supplying samples of NT-154 and for performing hydrothermal screening tests. The authors are grateful to Dr. K. More and Dr. P. Tortorelli, at the Oak Ridge National Laboratory, for performing high temperature hydrothermal corrosion testing on top coat specimens. Finally, the authors are also grateful to Mr. H. Anderson, at Ceramatec, Inc., for preparing samples.

## REFERENCES

[1] Hagen Klemm, "Corrosion of Silicon Nitride Materials in Gas Turbine environment", Journal of the European Ceramic Society, 22, pp. 2735–2740 (2002).

[2] R.C. Robinson and J.L. Smialek, "SiC Recession Caused by $SiO_2$ Scale Volatility under Combustion Conditions: I, Experimental Results and Empirical Model," J. Am. Ceram. Soc., 82 [7], 1817-1825 (1999).

[3] H. E. Eaton and G. D. Linsey, "Accelerated oxidation of SiC CMC's by water vapor and protection via environmental barrier coating approach", Journal of the European Ceramic Society, 22, pp. 2741–2747 (2002).

[4] E.J. Opila and N. S. Jacobson, "SiO(g) formation from SiC in Mixed Oxidizing/reducing Gases," *Oxidation of Metals* **44** [5/6] 527-544 (1995).

[5] E.J. Opila, J.L. Smialek, R.C. Robinson, D.S. Fox, N.S. Jacobson, "SiC Recession Caused by SiO2 Scale Volatility under Combustion Conditions: II, Thermodynamics and Gaseous-Diffusion Model," *J. Am. Ceram. Soc..*, **82** [7] 1826-1834 (1999).

[6] E. J. Opila, "Oxidation and Volatilization of Silica Formers in Water Vapor", *J. Am. Ceram. Soc.*, 86 [8] 1238–48 (2003).

[7] S. Ueno, D. Jayaseelan, N. Kondo, T. Ohji, and H.-T. Lin, "Development of EBC for Silicon Nitride," presented at ISASC-2004/International Symposium of New Frontier of Advanced Si-based Ceramics and Composites, Gyeongju, Kore, June 2004.

[8] K. N. Lee, "Current Status of Environmental Barrier Coatings for Si-Based Ceramics," *Surf. Coatings Tech.* **133-134** 1-7 (2000).

[9] R.Riedel, A. Kienzle, W. Dressler, L. Ruwisch, J. Bill, F. Aldinger, "Silicoboron carbonitride ceramic stable to 2,000 °C," *Nature*, **382**, 796 (1996).

[10] C.A. Lewinsohn, S. Elangovan, "Development of amorphous, non-oxide seals for solid oxide fuel cells," Ceram. Engng. & Sci. Proc., **24** [3], 317-322, 2003.

[11] C. Lewinsohn, S. Elangovan, M. Quist, "Durable Seal Materials for Planar Solid Oxide Fuel Cells," Cer. Engng. & Sci. Proc., vol. 25, *in press.*

[12] T. Fukudome, S. Tsuruzono, T. Tatsumi, Y. Ichikawa, T. Hisamatsu, I. Yuri, "Developments of Silicon Nitride Components for Gas Turbine," presented at ISASC-2004/International Symposium of New Frontier of Advanced Si-based Ceramics and Composites, Gyeongju, Kore, June 2004.

[13] S. Ueno, N. Kondo, T. Ohji, S. Kanzaki, J. Doni, "High Temperature Hydro Corrosion Resistance of Silica Based Oxide Ceramics", American Society of Mechanical Engineers, International Gas Turbine Institute, Turbo Expo (Publication) IGTI, v 1, 2003, p 625-632.

[14] K. N. Lee, "Key Durability Issues With Mullite-Based Environmental Barrier Coatings for Si-Based Ceramics", Journal of Engineering for Gas Turbines and Power, Vol. 122, 632-636, October (2000).

# Non-Destructive Evaluation of Thermal and Environmental Barrier Coatings

# CHARACTERISATION OF CRACKS IN THERMAL BARRIER COATINGS USING IMPEDANCE SPECTROSCOPY

Lifen Deng, Xiaofeng Zhao, Ping Xiao*
Materials Science Centre, School of Materials, Grosvenor Street, University of Manchester, Manchester, M1 7HS, UK

## ABSTRACT

Finite element method has been developed to calculate impedance spectra of a model thermal barrier coating (TBC) with presence of cracks. Impedance spectra have been calculated to examine the effect of the crack size, total length of cracks, crack thickness, and location of the cracks on impedance spectra of TBCs. The calculated results indicate that both crack size and total length of cracks have significant effect on the impedance spectra, whereas the crack thickness and crack location have little influence on the impedance spectra. The initial impedance measurement results have confirmed some of the calculated results.

## INTRODUCTION

Crack formation and propagation in thermal barrier coatings (TBCs) due to thermal treatment leads to spallation of TBCs[1, 2]. However, failure mechanisms of TBCs are not clear yet although there have been extensive studies on this important topic[3-6]. Therefore, monitoring crack propagation in TBCs is important in predicting lifetime of TBCs in service and will become critical in further development of TBCs for wide applications. Experimental studies of TBC degradation suggested that cracks often occur at either TGO/BC[7] or YSZ/TGO[6] interfaces, and the accumulation of cracked regions led to a final spallation of the TBCs. Several techniques such as ultrasonic imaging coupled with acoustic emission[8, 9] and thermal wave[10] have been used to directly monitor the crack propagation in TBCs. Fluorescence spectroscopy has been used to examine the stress distribution in TGO, and used the non-uniformity of stresses in cracked region to detect crack formation[11, 12]. However, other factors than cracks could affect the stress distribution, so that uncertainty in measurements would be introduced. Previous works tried to link impedance measurements of TBCs to degradation and crack propagation of TBCs[13, 14]. However, it is difficult to quantify the crack size and location based on measured impedance spectra. We have done extensive studies on impedance spectra of TBCs[14, 15], and studied the TGO growth[16], TBC sintering and phase transformation[17, 18] using impedance spectroscopy.

The present study is to simulate impedance spectra of TBCs based on the finite element method and to examine the effect of the cracks on the impedance spectra. A two dimensional model of TBC structure has been established to calculate impedance spectra of the TBCs. The

* To Whom correspondence should be addressed (ping.xiao@manchester.ac.uk)

calculated Bode plots indicate that the presence of cracks contributes to the increase and widening of the peak at frequency of ~$10^2$Hz, which is also attributed to the presence of the TGO layer. Impedance measurements of TBCs with and without cracks further confirm the overlapping effect of the cracks and TGO. Further modelling results suggest that impedance spectra of TBCs change significantly with an increase of total crack length, ie. the sum of crack sizes of all cracks, while the impedance spectra varies moderately with the variation of crack size when the total crack length is constant. However, little change appears in impedance spectra with the change of crack thickness and location of cracks. This study provides a further understanding of impedance spectra of TBCs and indicates that cracks in TBCs can be characterised using impedance measurements.

## IMPEDANCE MODELLING

Figure 1(a) shows that the EB-PVD TBC system consists of three layers, YSZ top coat with a columnar microstructure, a thermally grown oxide (TGO) layer and a metallic bond coat (BC) layer. Fig. 1(b) shows a two dimensional schematic structure of the TBC prepared by electron beam physical vapor deposition (EB-PVD). The gb_1 represents vertical gap between columns, and the gb_2 represents grain boundaries inside the columns. The sample size is 5 mm width while the YSZ column width was set as 250 μm to save computer memory and computing time. Although the approximation of the column size would lead to some error in calculated impedance spectra at a frequency higher than $10^3$ Hz, the effect of cracks on impedance spectra at a frequency lower than $10^3$ Hz would be very similar. The YSZ layer thickness is 150 μm and the TGO layer thickness is 5 μm. The width of grain boundaries is set as 2 μm. The electrode size is 1mm. Fig. 2 shows the presence of cracks at either the YSZ/TGO or TGO/BC interface. The parameters of crack numbers, size and thickness are given in Table 1. Seven series of parameters have been used for calculation. At first, a total length of cracks is constant as 4000μm, and individual crack size decreased and the crack number increased in order to investigate the effect of crack size distribution on impedance spectra (No.1-3). Secondly, the total crack length was reduced by decreasing the number of uniform cracks with a fixed crack size of 200 μm to study the influence of crack total length on impedance spectra (No. 4-6). With the same crack size and number of cracks, crack thickness decreased from 5μm to 2μm (No.3 and No.7 series of parameters) or change crack location from TGO/BC to YSZ/TGO interface, to identify the effect of crack thickness and location on impedance spectra.

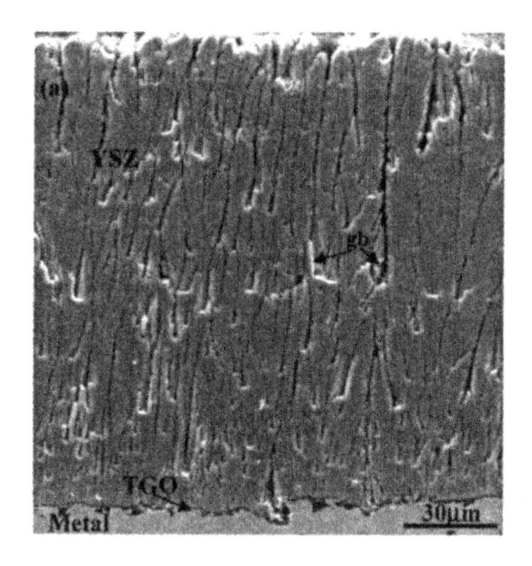


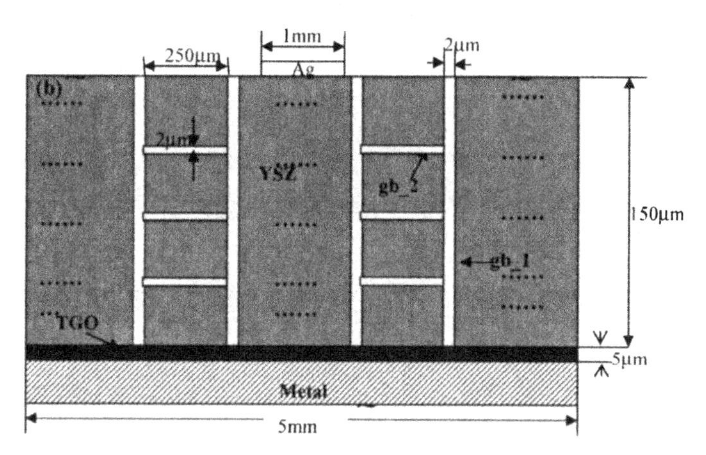


Figure 1. (a) SEM image of a EB-PVD TBC cross section; (b) A schematic figure of cross section of EB-PVD TBCs.

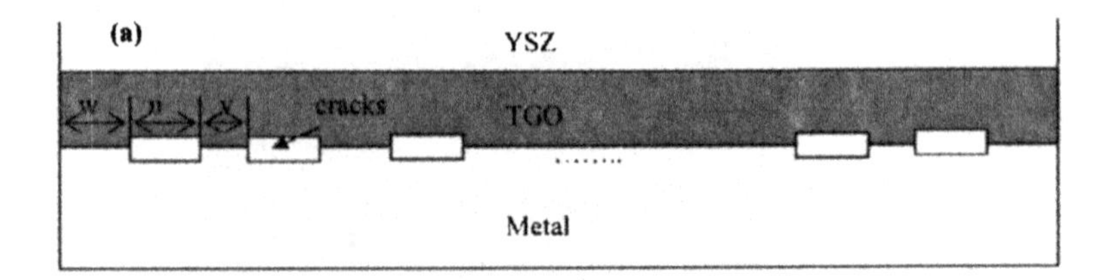


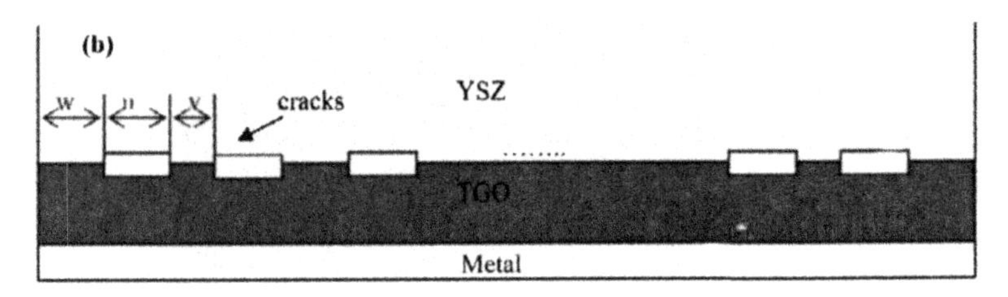


Figure 2. Schematic figures of TBC showing cracks at (a) YSZ/TGO interface and (b) TGO/BC interface

Table 1. The geometric parameters of cracks at TGO/BC interface in models.

| No. | Thickness /μm | Size u/μm | Number of cracks | Separation distance μm | Total length of cracks μm |
|---|---|---|---|---|---|
| 1. | 5 | 500 | 8 | 200 | 8X500=4000 |
| 2. | 5 | 400 | 10 | 100 | 10X400=4000 |
| 3. | 5 | 200 | 20 | 40 | 20X200=4000 |
| 4. | 5 | 200 | 15 | 100 | 15X200=3000 |
| 5. | 5 | 200 | 10 | 300 | 10X200=2000 |
| 6. | 5 | 200 | 5 | 800 | 5X200=1000 |
| 7. | 2 | 200 | 20 | 40 | 20X200=4000 |

The electric potential in TBC satisfied the Poisson's equation, as stated by Fleig [19],

$$\text{div}\cdot\text{grad}\,\Phi(r,t)=-\frac{1}{\varepsilon}\rho(r,t) \quad (1)$$

where $\Phi$ is AC electric potential, $\rho$ is charge density, ε is the absolute dielectric constant and ρ is the charge density, r and t are space and time respectively. There is no applied magnetic field during impedance measurements and the diffuse charge on the surface and interface can be ignored, therefore, the charge density in a uniform domain of TBC samples is equal to zero,

$\rho(r,t)=0$, thus Eq(1) can be reduced to Laplace's equation

$$\frac{\partial^2\hat{\Phi}}{\partial x^2}+\frac{\partial^2\hat{\Phi}}{\partial y^2}=0 \qquad (2)$$

Three boundary conditions were used to solve Laplace's equation (2): i) There is no potential drop occurred at the electrode/electrolyte interface; ii) No current goes out of the sample, therefore, the normal current density at free edges of the model is equal to zero: grad $\hat{\varphi}\bullet n=0$, where n represents a normal vector of relevant side; iii) At interfaces such as the Ag/YSZ interface, YSZ/gb1 interface, YSZ/gb2 interface, YSZ/TGO interface, TGO/gb1 interface and TGO/BC interface, the normal component of complex current density is continuous. For example, at the interface between the 1st sub-domain and the 2nd sub-domain: $\hat{\kappa}_1 grad\hat{\varphi}\bullet n=\hat{\kappa}_2 grad\hat{\varphi}\bullet n$. Here, complex conductivity $\hat{\kappa}=\sigma+i\omega\varepsilon$, $\sigma$ is the electric conductivity, $\varepsilon$ is the absolute dielectric constant. With the three boundary conditions, the equation (2) has been solved, and then the complex current density was calculated according to

$$\hat{j}=-\hat{\kappa}grad\hat{\varphi} \qquad (3)$$

Further, the current at the electrode/electrolyte interface was obtained by integrating the current density along the electrode boundary:

$$I=\int_s j\bullet ds=\int_s \hat{\kappa}grad\hat{\varphi}\bullet ds \qquad (4)$$

With applied voltage $U_0$ and the gained current I, impedance can be gained by:

$$\hat{Z}=\frac{U_0}{I}=\frac{U_0}{\int_s \hat{\kappa}grad\hat{\varphi}\bullet dA}. \qquad (5)$$

Table 2 shows material properties used for calculations. The electrical properties of YSZ were calculated from the experimental data by measuring YSZ bulk samples with standard shape and size at 400 °C. The conductivity of grain boundary is much less than that of bulk and assumed as $\sigma_{gb}\approx10^{-3}\sigma_{YSZ}$[20], while the permittivity of YSZ grain boundary can be regarded as the same with that of YSZ bulk, $\varepsilon_{gb}=\varepsilon_{bulk}$[21]. TGO layer is mainly composed of $Al_2O_3$, so that set its electric conductivity $\sigma_{TGO}=1.0\times10^{-8}$ /Ω and permittivity $\varepsilon_{r,TGO}=10$ [22]. Cracks are filled with air, electric conductivity of air $\sigma_{air}=0$, and relative permittivity of air $\varepsilon_{r\,air}=1$. The electric properties of Ag and substrate metal electrodes are set as $\sigma = 1.0\times10^7$ /Ω, $\varepsilon_r$=1. For the two dimensional model, the thickness of the 3rd dimension is assumed as infinite. The unit of electric conductivity is $\Omega^{-1}$. The calculation was carried out using FEMLAB with MATLAB software.

Table 2. The electric properties of materials used in models.

| | $\sigma/\Omega^{-1}$ | $\varepsilon_r$ |
|---|---|---|
| YSZ grain | 1e-3 | 28 |
| YSZ Grain boundary | 1e-6 | 28 |
| TGO | 1e-8 | 10 |
| air | 0 | 1 |
| Ag / Metal | 1e+7 | 1 |

## EXPERIMENTAL

EB-PVD TBCs, provided by Rolls-Royces, Derby, UK, consisted of Ni-superalloy, Pt diffusion bond coat and 8 wt.% YSZ topcoats. 5×5 $mm^2$ plate samples were cut from engine components and isothermally heated at 1200°C for 60 hours and 90 hours.

Impedance measurements were carried out using a Solartron SI 1255 HF frequency response analyser (FRA) coupled with a computer controlled Solartron 1296 dielectric interface, with a horizontal tube furnace at 400 °C. During measurements, an a.c. amplitude of 100 mV was used, over a frequency range of $10^6$–$10^{-2}$ Hz. 1 $mm^2$ silver (Ag) was painted on the YSZ surface and then fired at 690 °C for 30 min. The fired Ag acted as an electrode. For TBCs served at temperature higher than 1000°C, microstructure change of TBCs caused by impedance spectroscopy measurement 400 °C and 690 °C for 30 min Ag annealing can be ignored. The polished base of the superalloy substrate acted as the other electrode. Samples were placed in a jig between $Al_2O_3$ plates; Pt wires was attached the electrodes for impedance measurements.

SEM/EDX (JEOL 6300 and Philips 525M) was used to examine the cross-section of the TBC samples. The samples were prepared using a precision diamond wheel saw (Struers Accutom-50) at low speed to minimise mechanical damage to the coating. The samples were placed in epoxy resin, before grinding and polishing using successively finer SiC paper. Polished samples were carbon coated (Edwards E306A coater) for SEM examination.

## RESULTS

### Modelling results

Figure 3 shows a Bode plot (minus phase angle (θ) vs. frequency (Log (f)) and a Nyquist plot (imaginary impedance (Zi) vs. real impedance (Zr)) of TBCs with cracks (using No. 2 parameters in table 1) and without crack. The Bode plot shows three peaks, corresponding to three relaxation processes in TBCs, YSZ grain (g, approx. $10^{6\sim7}$Hz), YSZ grain boundary (gb, approx. $10^4$Hz), and TGO (approx. $10^{2\sim3}$Hz) which were confirmed experimentally in our previous study without consideration of cracks[17]. The Nyquist plots show large semicircles of the TGO layer, whereas the semicircles for g and gb at high frequencies are too small to be seen in figure 3b. In comparison with the TBC without cracks, the cracks in the TBC induced a significant change in impedance spectra, i.e. the presence of cracks at the TGO/BC interfaces led to higher and broader "TGO" peak in the Bode plot. The semicircle corresponding to the "TGO" in the Nyquist plot became much larger with the presence of cracks.

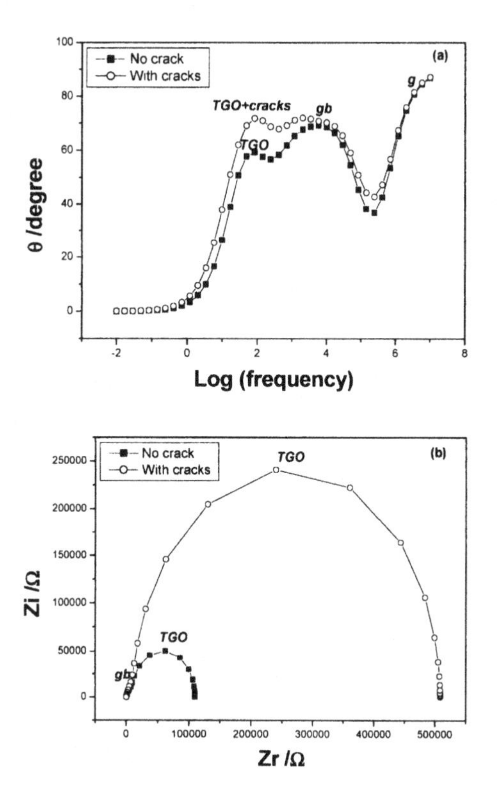


Figure 3. (a) Bode plots (minus θ vs. Log(f)) and (b) Nyquist plots (Zi vs. Zr) of models with cracks (No 2 in table 1) and without crack.

## Experimental Results

Fig. 4 shows the cross section of EB-PVD TBC samples after isothermal treatments at 1200 °C for 60 hours and 90 hours. With the treatment time increasing from 60 hr to 90 hr, there is little change in the TGO thickness, but a large crack appeared at the TGO/BC interface in the 90 hr treated sample whereas no apparent crack appears in the 60 hr treated sample. Although the TGO is separated from BC metal in Fig. 4b, the TGO layer is still attached to the BC in other region. The large crack in Fig. 4b was partially induced by polishing during sample preparation for SEM, but difference in thermal treatment certainly lead to difference in crack size in the two samples.

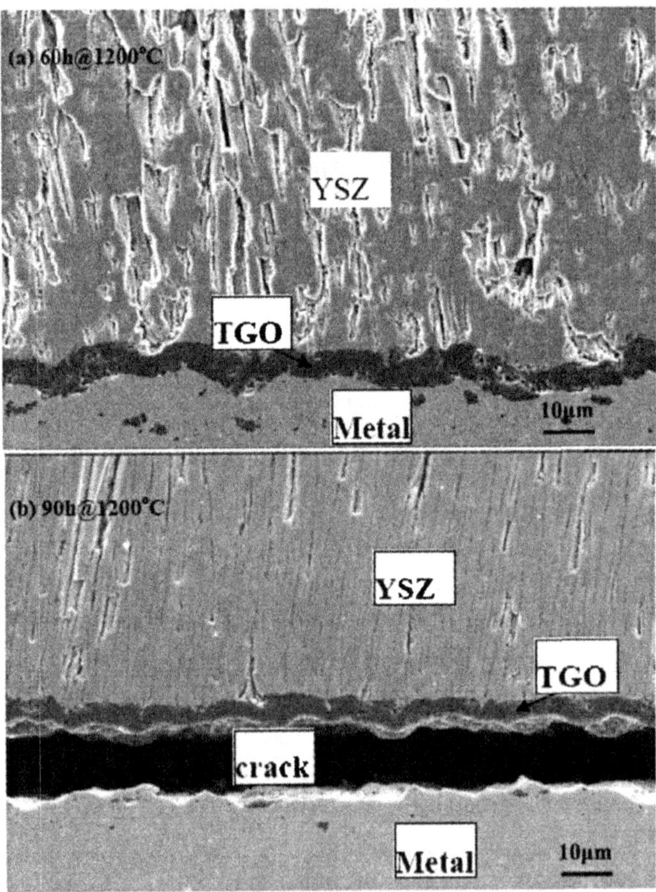


Figure 4. SEM images of cross section of EB-PVD TBC samples isothermally heated at 1200 °C for (a) 60 hours and (b) 90 hours.

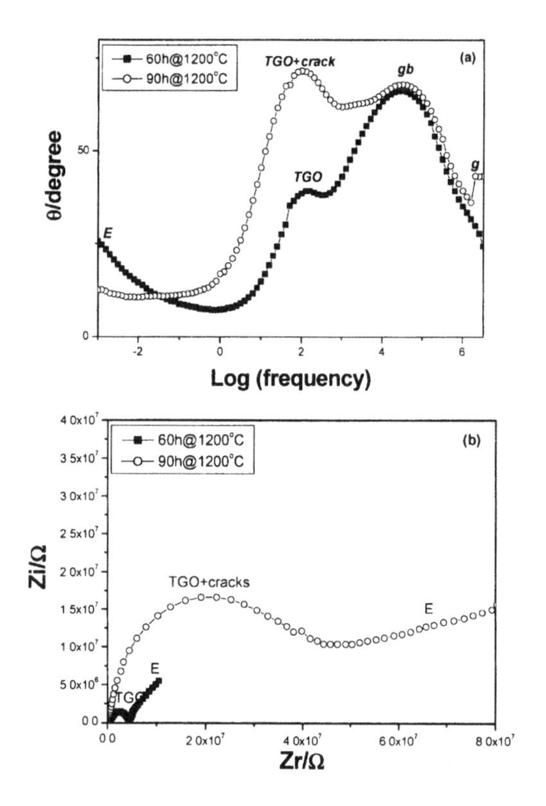


Figure 5. The Bode (a) and Nyquist (b) plots measured from EB-PVD TBCs after isothermal treatments at 1200 °C for 60 hours and 90 hours.

Fig. 5 shows impedance spectra of EB-PVD TBC samples with microstructure shown in figure 4. The Bode plots indicate grain effect (approx. $10^{6\sim7}$ Hz), grain boundary effect (approx. $10^4$ Hz), TGO effect (approx. $10^{2\sim3}$ Hz) and electrode effect (E, approx. $10^{-2\sim-3}$ Hz). Because the spectrum measured at frequencies higher than $10^6$ Hz is not reliable[23], the grain effect, ie. g peak in the Bode plot is not clear in figure 5. The Nyquist plots show a semicircle contributed by TGO+cracks, and a tail corresponding to electrode effect (E). However, the significant increase in the "TGO" effect in both the Bode plot and the Nyquist plot confirmed the effect of the cracks. In addition, the crack effect on the Bode plot and the Nyquist plot is the same as in the calculated spectra (Fig. 3). It should be noted that there is no consideration of electrode effect in calculation of the impedance spectra.

Further Modelling

Experimental measurements (Fig. 5) are in good agreement with modelling results (Fig. 3) about the effect of cracks at the TGO/BC interfaces on the impedance spectra of TBCs. Then, the effects of crack size, the total length of cracks, crack location and crack thickness have been examined and discussed as follows:

Total Crack Length Effect

The size of each crack was fixed as 200μm, the number of cracks changed from 20 to 15, 10, and 5 (No.4-6 in table 1). Fig. 6 shows calculated impedance spectra where the Bode plots show that the TGO+cracks peak near $10^2$Hz increased with increasing total crack length. In addition, semicircles in Nyquist plots become larger with an increase in the total crack length. When the total crack length equal to or less than 5×200 μm=1000 μm (No.6 parameters in table 1), it is difficult to differentiate impedance spectra with cracks from those without crack. This suggests that for the 1000μm Ag electrode, the minimum total length of cracks which can be monitored is about 1000μm. Measurements with smaller Ag electrode size might increase sensitivity of impedance spectra to the total crack length. Further investigation need to be carried out to examine this issue.

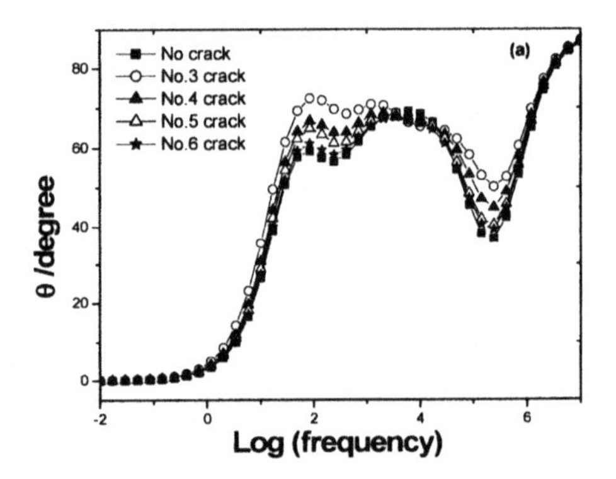

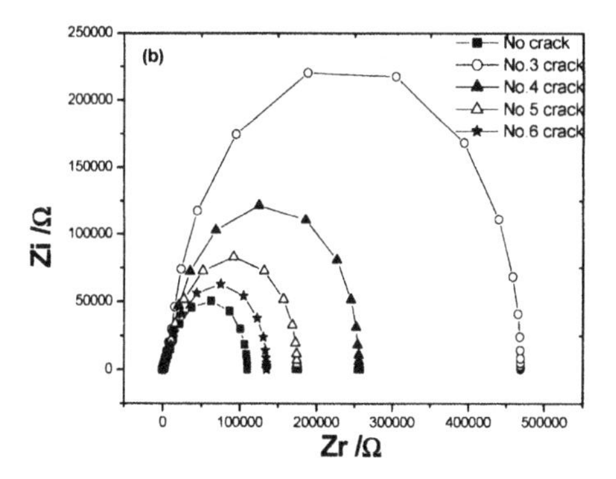


Figure 6. Calculated Bode and Nyquist plots of model with No.3, No.4, No.5 No.6 cracks in table 1 (the crack size is constant while the number of cracks changes) and without crack .

Crack Size

With the total length of cracks fixed, e.g. 4000 μm, the size of individual crack changing from 1000 μm, to 500 μm, to 200 μm, (No.1 to No.3 parameters in table 1), the calculated impedance spectra are shown in Fig. 7. The Bode plots indicate that the "TGO+cracks" peak near $10^2$ Hz shows little change with the change of crack size. However, the semicircles on Nyquist plots increased slightly with the increase of crack size.

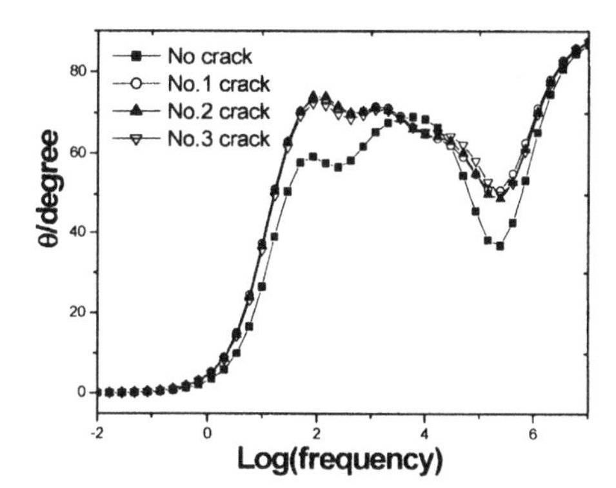

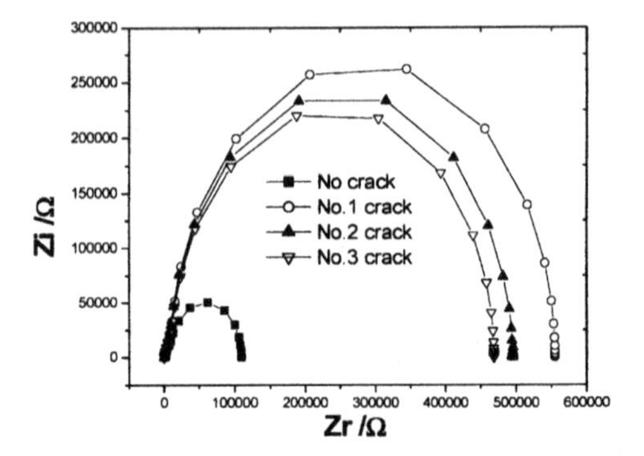


Figure 7. Calculated Bode and Nyquist plots of model with No.1, No.2 and No.3 cracks in table 1 (the total length of cracks is same as 4000um, the crack size is different) and without crack.

Crack thickness and location

For TBC degradation, cracks appeared sometimes at the TGO/BC interfaces[7] and sometimes occurred at the YSZ/TGO interfaces[24]. Fig. 2 shows the cracks at different locations in TBCs. With No. 2 cracks in Table 1, Fig. 8 shows impedance spectra of TBCs with different crack locations. No visible difference appears on impedance spectra due to difference in the crack location.

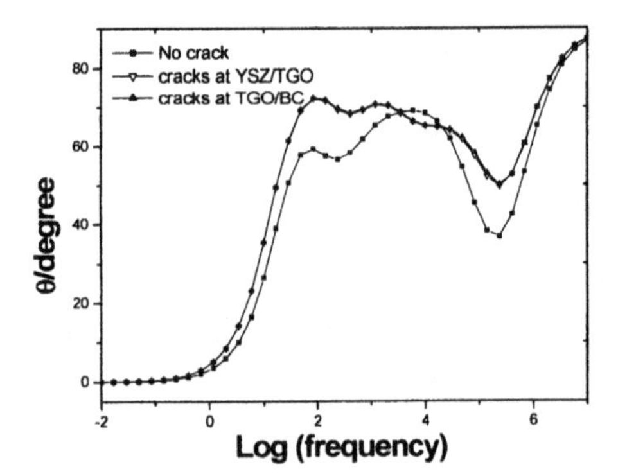

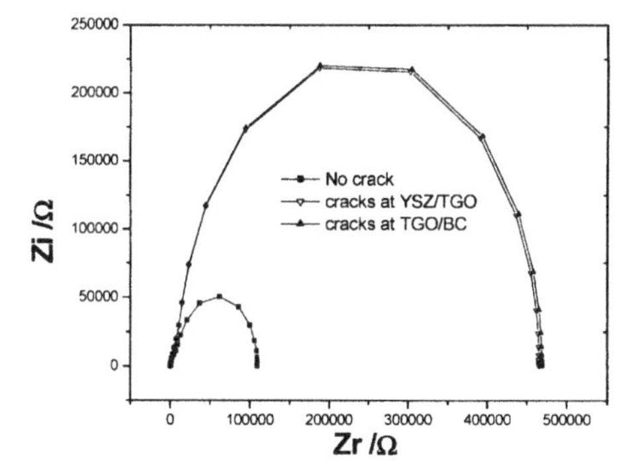


Figure 8. Impedance spectra calculated from models with cracks (No.2 crack parameter in the table 1) at the YSZ/TGO or TGO/BC interface and without cracks.

With the thickness of cracks changing from 5 μm to 2 μm, calculated impedance spectra in Fig. 9 show no apparent difference. It should be noted that the calculation was made with arrangement of 1mm Ag electrode at surface of TBC and the metallic substrate as the other electrode. The crack thickness effect and crack location effect on impedance spectra may be different with different arrangements of the electrodes. Further research is required to study the effects of crack thickness and location on impedance spectra of TBCs.

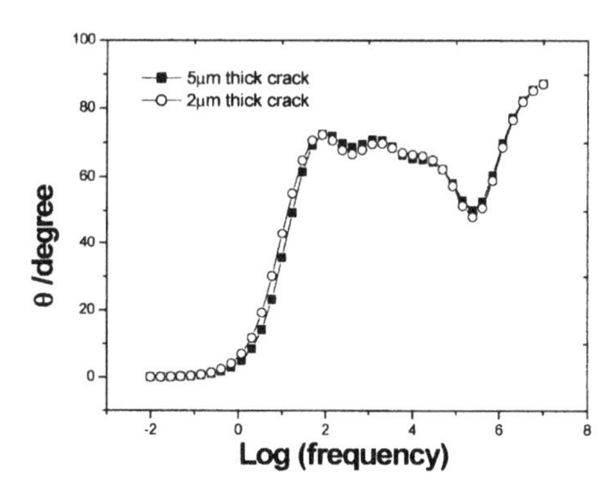

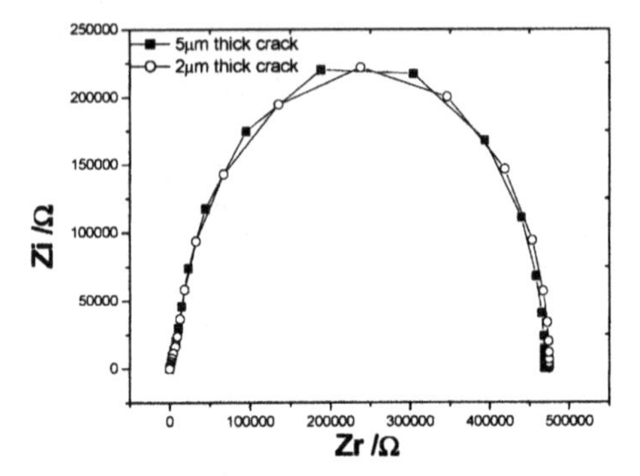


Figure 9 Bode and Nyquist plots calculated from models when the thickness of cracks was changed from 5μm to 2μm (No.3 and No.7 cracks in table 1).

## CONCLUSION

A 2-dimensional model of TBC has been built to calculate impedance spectra of the TBCs using the finite element method. The effect of cracks in TBCs has been investigated by modelling and experimental results. Both results show that the higher TGO peak near $10^2$ Hz in the Bode plot and larger semicircles on Nyquist plots were induced with crack formation. An increase in the total length of cracks led to significant change in both the Nyquit plot and the Bode plot. With the total length of cracks being constant, an increase in crack size induced change in the Nyquist plot but little change in the Bode plot. The crack thickness change and change in crack location has negligible effect on impedance spectra with arrangement of 1mm Ag electrode at surface of TBC and the metallic substrate as the other electrode.

## REFERENCES

[1] S. Nusier,G. Newaz, "*Analysis of interfacial cracks in a TBC/superalloy system under thermal loading,*" Eng. Fract. Mech. **60**, 577-81 (1998).

[2] D. R. Clarke, R. J. Christensen,V. Tolpygo, "*The evolution of oxidation stresses in zirconia thermal barrier coated superalloy leading to spalling failure,*" Surf. Coat Tech. **94-95**, 89-93 (1997).

[3] A. G. Evans, D. R. Mumm, J. W. Hutchinson, G. H. Meier,F. S. Pettit, "*Mechanisms controlling the durability of thermal barrier coatings,*" Prog. Mater. Sci. **46**, 505-53 (2001).

[4] D. R. Mumm,A. G. Evans, "*On the role of imperfections in the failure of a thermal barrier coating made by electron beam deposition,*" Acta Mater. **48**, 1815-27 (2000).

[5] G. Qian, T. Nakamura,C. C. Berndt, "*Effects of thermal gradient and residual stresses on thermal barrier coating fracture,*" Mech. of Mater. **27**, 91-110 (1998).

[6] A. Rabiei,A. G. Evans, "*Failure mechanisms associated with the thermally grown oxide in plasma-sprayed thermal barrier coatings,*" Acta Mater. **48**, 3963-76 (2000).

[7] D. R. Clarke,S. R. Phillpot, "*Thermal barrier coating materials,*" Mater. Today. **8**, 22-29 (2005).

[8] Y. C. Zhou,T. Hashida, "*Thermal fatigue failure induced by delamination in thermal barrier coating,*" Int. J. Fatigue. **24**, 407-17 (2002).

[9] N. Mesrati, Q. Saif, D. Treheux, A. Moughil, G. Fantozzi,A. Vincent, "*Characterization of thermal fatigue damage of thermal barrier produced by atmospheric plasma spraying,*" Surf.& Coat. Tech. **187**, 185-93 (2004).

[10] G. Newaz,X. Chen, "*Progressive damage assessment in thermal barrier coatings using thermal wave imaging technique,*" Surf Coat Tech. **190**, 7-14 (2005).

[11] M. M. Gentleman,D. R. Clarke, "*Concepts for luminescence sensing of thermal barrier coatings,*" Surf Coat Tech. **188-189**, 93-100 (2004).

[12] V. K. Tolpygo, D. R. Clarke,K. S. Murphy, "*Evaluation of interface degradation during cyclic oxidation of EB-PVD thermal barrier coatings and correlation with TGO luminescence,*" Surf. Coat. Tech. **188-189**, 62-70 (2004).

[13] B. Jayaraj, S. Vishweswaraiah, V. H. Desai,Y. H. Sohn, "*Electrochemical impedance spectroscopy of thermal barrier coatings as a function of isothermal and cyclic thermal exposure,*" Surf Coat Tech. **177-178**, 140-51 (2004).

[14] M. S. Ali, *A novel technique for evaluating the degradation of engine components non-destructivtely*, in *Department of Mechanical Engineering*. 2002, Brunel University: London.

[15] M. S. Ali, S. Song, P. Xiao, "*Evaluation of degradation of thermal barrier coatings using impedance spectroscopy,*" J. Eur. Ceram. Soc. **22**, 101-07 (2002).

[16] X. Wang, J. Mei, P. Xiao, "*Determining oxide growth in thermal barrier coatings (TBCs) non-destructively using impedance spectroscopy,*" J. Mater. Sci. Lett. **20**, 47-49 (2001).

[17] P. S. Anderson, X. Wang,P. Xiao, "*Impedance spectroscopy study of plasma sprayed and EB-PVD thermal barrier coatings,*" Surf. Coat. Tech. **185**, 106-19 (2004).

[18] P. S. Anderson, X. Wang, P. Xiao, "*Effect of isothermal heat treatment on plasma-sprayed yttria-stabilized Zirconia studied by impedance spectroscopy,*" J. Am.Ceram.Soc. **88**, 324-30 (2005).

[19] J. Fleig,J. Maier, " *Finite element calculations of impedance effects at point contacts,*" Electrochemica Acta. **41**, 1003-09 (1995).

[20] J. Fleig,J. Maier, "*The impedance of ceramics with highly resistive grain boundaries: validity and limits of the brick layer model,*" J. Eur. Ceram. Soc. **19**, 693-96 (1999).
[21] J.-S. Lee, J. R. Fleig,J. Maier, "*Conventional and microcontact impedance studies of Mn–Zn ferrite ceramics,*" J. Mater. Res. **19**, 864-71 (2004).
[22] D. R. Lide, *CRC Handbook of Chemistry and Physics*. 85th ed. 2004: CRC press LLC. 12-52.
[23] X. Wang, *Non-Destructive Characterisation of Structural Ceramics Using Impedance Spectroscopy*, in *Department of Mechanical Engineering*. 2001, Brunel University: London. p. 8.
[24] M. Okazaki,H. Yamano, "*Mechanisms and mechanics of early growth of debonding crack in an APSed Ni-base superalloy TBCs under cyclic load,*" Int. J. Fatigue. **27**, 1613-22 (2005).

# NONDESTRUCTIVE EVALUATION METHODS FOR HIGH TEMPERATURE CERAMIC COATINGS

William A. Ellingson, Rachel Lipanovich, Stacie Hopson, Robert Visher
Argonne National Laboratory
9700 S. Cass Avenue
Argonne, IL 60439

ABSTRACT

Various high temperature ceramic coatings are under development for components for the hot gas path of gas turbine engines. These coatings are being developed for both metal and ceramic substrates. Regardless of the coating system, environmental barrier coatings for ceramics or thermal barrier coatings for metals, there is a need to determine the condition of the coating as well as determine physical parameters such as thickness. Two laser-based, non-contact, nondestructive methods are being developed that provide information regarding the condition or "health" of the coating, including thickness variations. The elastic optical scatter (EOS) method has been demonstrated to correlate to spall conditions for thermal barrier coatings however there has been a question about sensitivity to surface features. The laser-based optical coherence tomography (OCT) method has been demonstrated to provide thickness measurements for thermal and environmental barrier coatings. Several well-controlled sample sets have been examined with these two NDE methods and extensive data have been acquired. The test methods, test samples, and results obtained to date showing correlations with destructive measurements are presented.

## INTRODUCTION

Two very different high temperature coating systems are being addressed by NDE technology: 1)-Environmental barrier coatings mainly for ceramic composites being used in high temperature oxygen-rich environments and 2)-thermal barrier coatings mainly for metals used in the hot gas flow path of turbine engines. For gas-turbine combustor liners made of ceramic matrix composites-or any high temperature application where oxygen is present, there is a need to reduce oxidation-induced recession in silicon-based ceramic composites [1] and a need to reduce the operating temperature to reduce creep in oxide-based composites [2]. These two negative effects on composite performance can be mitigated by application of environmental barrier coatings (EBC) [2]. For SiC/SiC materials, present EBCs, typically 100-200 micron thick, are composed of some form of barium-strotium-alumino-silicate (BSAS). For oxide-based composites, a special oxide-based functionally graded insulator (FGI) has recently been developed and patented by Siemens-Westinghouse [1]. For either of these EBCs, what is desired to be measured (determined) by NDE technology are: uniformity of thickness, adherence quality and detection of any flaws in the CMC under the EBC. Several processing steps are necessary to fully fabricate a CMC liner with an EBC. It is possible that initial flaws within the CMC material could affect the EBC adhesion or provide an increased potential to spall. Thermal barrier coatings (TBCs), on the other hand, are necessary to reduce the operating temperature of metal components in the hot gas path of gas turbines. Components such as the first stage blades and vanes down stream of the combustor as well as the combustor itself are examples of hot gas path components. Failure of a TBC could lead to a costly unplanned outage, and could lead to catastrophic events. It is therefore necessary to monitor the condition of the TBC so as to avoid such failures - and, if possible, provide pre-cursor information that would suggest a spallation is immanent. Significant work has been done by many investigators ([3-6]) and this work has shown that failure of TBCs depend upon the coating type EB-PVD or APS. Use of NDE technology for detection of regions where spallation might occur in the future therefore depends upon the coating type because the failure mechanism is different. There is also a need to develop NDE technology that can determine changes in the thermal conductivity of these TBC materials. Changes in thermal conductivity can arise from a diffusion of combustion constituents into the coating, especially for turbines burning "dirty" fuels; the conductivity can also change from microstructural changes in the TBC such as in-situ sintering. From the NDE point of view, the microstructural differences between an EB-PVD and an APS TBC are significant. An EB-PVD TBC has a better defined columnar microstructure ([7]) and is less optically scattering than the "splat" microstructure of an APS. NDE technologies that can accommodate these microstructural differences and measure the desired TBC parameters are important for long-term reliable operation of advanced, high efficiency and low-emission gas turbines.

## TEST METHODS

Two optical NDE test methods have been under development for application to these coating systems. One is optical coherence tomography (OCT) ([8]) and the other is elastic optical back scatter (EOS)

([9-10]). OCT is a fairly new NDE technique that can be used to take cross-sectional images of various materials providing that the material is optically translucent. Our work on OCT has been to investigate this as a tool for evaluating the thickness of TBCs and EBCs as well as perhaps detecting any disbond. The OCT technique is described more fully in reference ([8]) and a block diagram of the Argonne OCT system is shown in Figure 1. In an OCT set up, light from an optical source is split into two paths, a sample path and a reference path. Light in the reference path is reflected from a fixed-plane mirror whereas light in the sample path is reflected from surface and subsurface features of the ceramic sample. The reflected light from the sample path will only be detected if it travels a distance that closely matches the distance traveled by the light in the reference path; this constraint incorporates depth resolution into the technique. Thus, data can be obtained from a cross-sectional plane perpendicular or parallel to the surface of the sample.

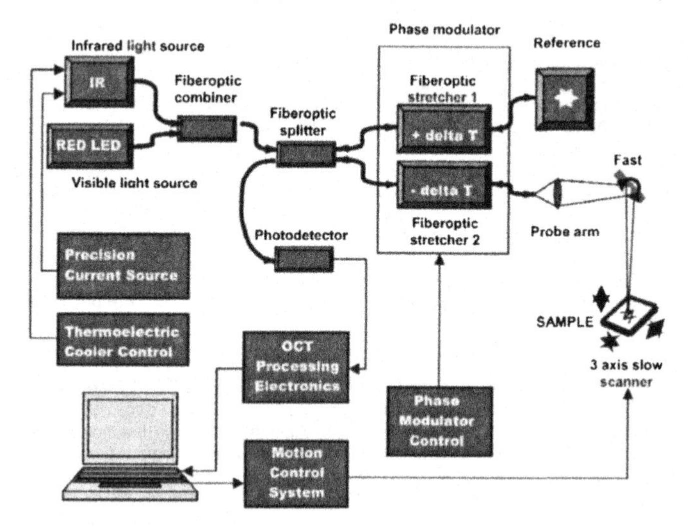


Figure 1: Block diagram of ANL OCT system.

The EOS or laser backscatter, technique ([9-10]), has been extensively discussed previously. In this method, polarized laser light is used to investigate the surface and subsurface characteristics ceramic coatings such as TBCs and EBCs. The underlying physical principle behind laser backscatter is that light incident on the surface of the coating will be partially reflected and partially transmitted. The polarization state of any surface- reflected light will not change, whereas the polarization state of a portion of the transmitted light that is subsequently reflected will change. The characteristics of the reflected light can be used to distinguish between light reflected from the surface and light reflected from a sub-surface feature. A two-detector EOS system is shown in Figure 2. In this system, the sample under investigation is mounted on the translational motion stages. The laser beam is incident on the sample at a given coordinate and a measurement is taken. The sample is moved so that the laser beam is now incident on a new location, adjacent to the previous location, and the measurement is repeated. This process continues until the area of the sample being studied has been covered. Typically, the locations are separated by a distance on the order of 5-10 microns. The two-dimensional array of collected measurements is then normalized with respect to the minimum and maximum measured values to create a gray-scale image.

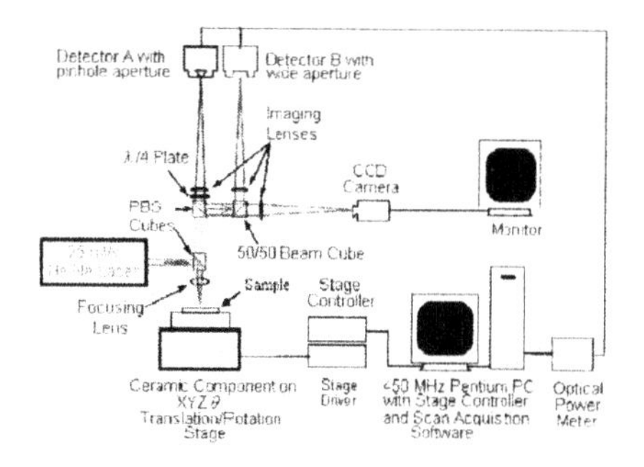


Figure 2: Schematic representation of ANL-developed EOS system.

**RESULTS**

Thermal Barrier Coatings

Previous data has resulted in empirically correlated results between the laser back scatter and spallation of both EB-PVD and APS TBC sample sets [9]. An example of what the EOS data look like for an EB-PVD coating and one example of the correlation with spallation is given in Figure 3.

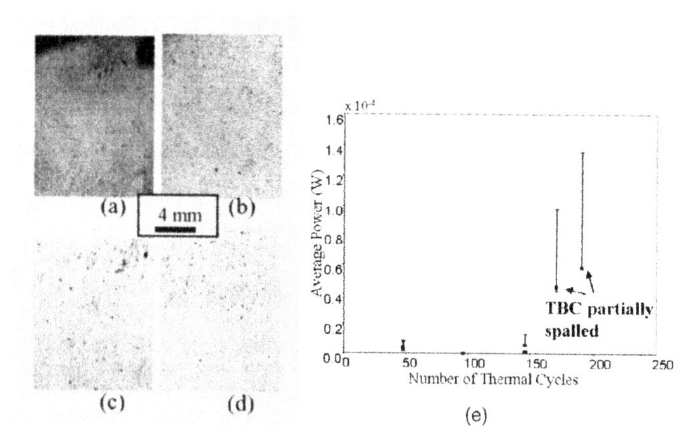


Figure 3: Laser backscatter sum images for: EB-PVD TBC sample after exposure to (a) 48, (b) 146, (c) 171, and (d) 191 thermal cycles and (e) average detected power of back scattered light as a function of the number of thermal cycles

However, there has been some concern over the sensitivity of this method to small features on the surface of the samples. To address this issue a special TBC sample was prepared and then polished. An

EB-PVD TBC sample was prepared at the German Aerospace Institute in Koln, Germany, DLR. The test sample, 10 mm by 28 mm, consisted of an EB-PVD TBC on a metal substrate. For reference purposes, three indents were placed on the sample. First, the sample was examined for surface features using a microscope and recording the images. Then data were acquired with the laser backscatter system. In these laser scatter tests the sample was raster scanned in 10 μm steps at a velocity of 20 mm/s. Using the two detectors in the set up, the first detector detected the light directly reflected by the sample, while the second detected the scattered light. One half of the sample was then polished to reduce surface effects and was scanned using the same data acquisition parameters. The total reflected light is the sum of detector 1 and detector 2. Figure 4 shows the resulting sum images obtained before and after polishing.

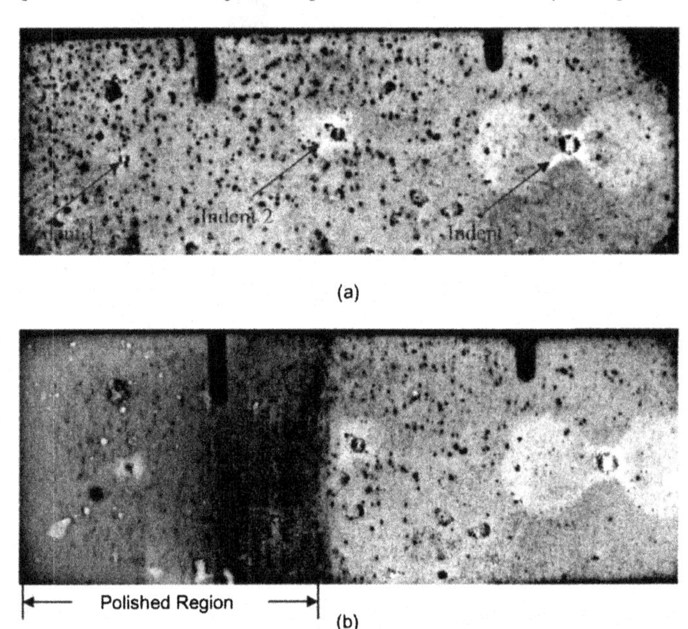


Figure 4: EOS sum image data

(a) before polishing and (b) after polishing.

Examination of Figure 4b reveals that there are fewer "dark spots" in the EOS data in the polished area after polishing than before polishing. This suggests that some "dark spots" that were present prior to polishing in fact are a result of surface features. The correlation between "dark spots" before and after polishing was confirmed by comparing microscope images and scan images. The digital microscope images were overlaid on the digital EOS images. The opacity of the microscope image was digitally adjusted from zero to 100 percent, where zero shows only the EOS data scan and 100 shows only the microscope image. Results of these overlays with opacities of 0%, 50%, and 100% are shown in Figure 5, with corresponding "dark spots" labeled A through G. The "'butterfly" region to the right side is a point where an indent was placed.

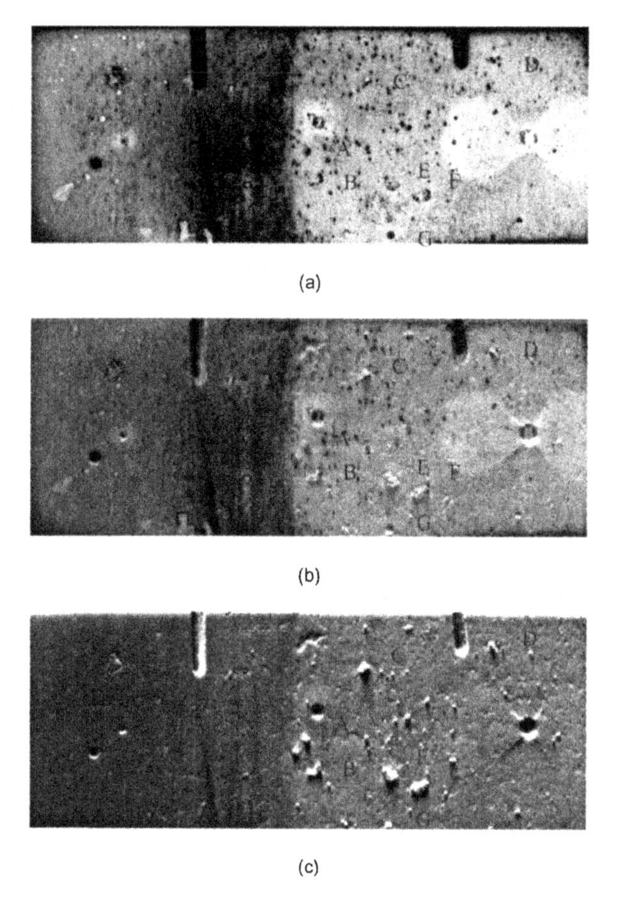

(a)

(b)

(c)

Figure 5: Results of overlaying digital microscope images over EOS images with opacities of (a) 0%, (b) 50%, and (c) 100%.

Figure 5 however shows that the lighter areas around the three indents are not surface features. Instead, these sections are areas of probable spallation as a result of the indentations.

Further analysis of TBCs has been done using OCT to obtain direct thickness measurements on an actual airfoil section. An example is shown in Figure 6.

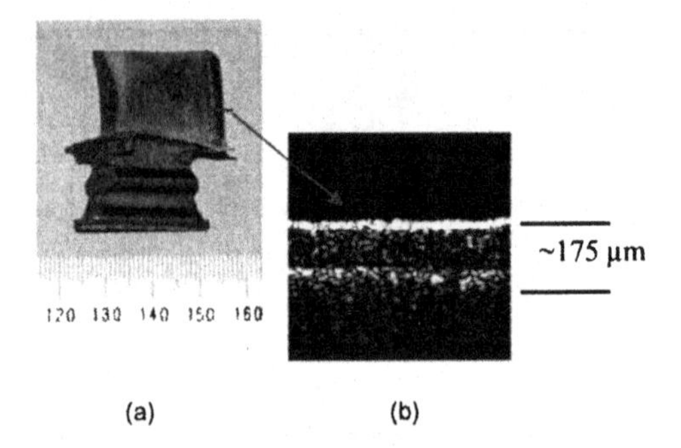


(a) (b)

Figure 6: (a) Photograph of turbine blade with TBC and (b) OCT cross sectional image showing TBC thickness of ~ 175 um

Environmental Barrier Coatings

Use of OCT to measure thickness of EBCs has recently been demonstrated on both plasma sprayed BSAS EBCs and slurry applied EBCs made of a proprietary oxide material. The first example to be shown is of a BSAS EBC applied on a melt infiltrated (MI) SiC/SiC. Figures 7(a) shows a diagram where several OCT cross sections were acquired from the sample shown in Figure 7(b). Figure 7(c) show resulting cross sectional images that correspond to the various locations noted in 7a. Clearly the OCT data provide reasonable thickness data within a few 10s of microns.

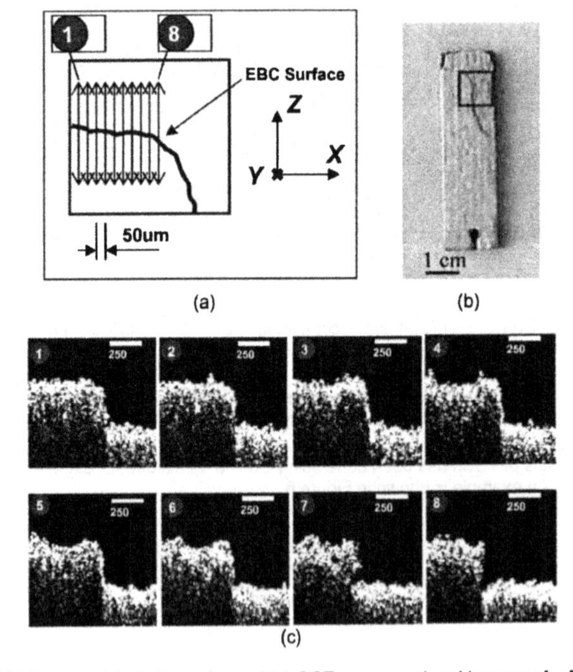


(a) (b)

(c)

Figure 7: (a) Diagram, (b) photograph , and (c) OCT cross-sectional images of a EBC sample

The second example of the use of OCT to measure EBC thickness of a new slurry dipped coating. Optical transmission characteristics for the slurry coating in the "green" state suggested that there were poor optical properties. However, at 1038 um of the OCT there was small transmission and therefore attempts were made to use the OCT. A special sample was prepared by using a standard quartz microscope slide so that the slurry coating could be readily observed. Figure 8(a) shows a photograph of the slide with the slurry coating applied only on one end. Figure 8(b) shows a resulting OCT cross sectional image that clearly shows the thickness of the slurry coated EBC on the top as well as the bottom. By focusing in on the top section, see Figure 8(c), one can estimate the thickness of the coating very well. Data from the OCT using a calibrated setting, suggests that the coating is 42um um thick. By placing the sample under the microscope the actual thickness was estimated to be 40 um . Thus, the comparison is very good.

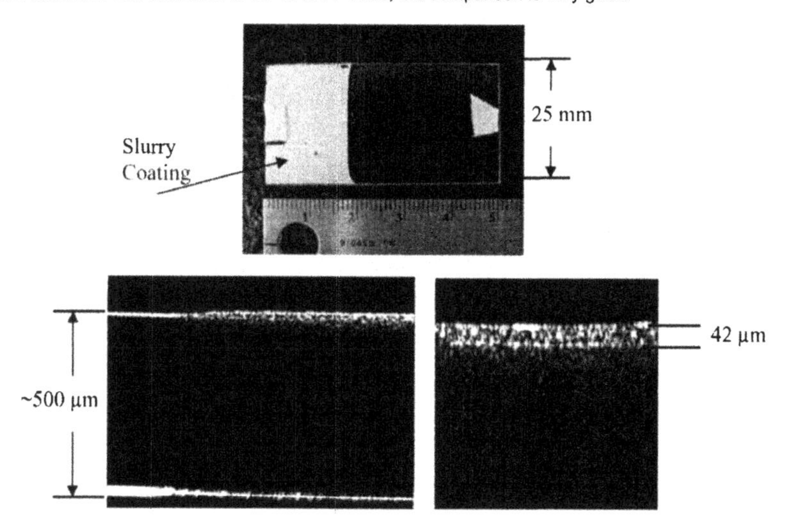


Figure 8: (a) Photograph of a slurry coating on a quartz slide and (b, c) the resulting cross-sectional images

## CONCLUSIONS

Two optical NDE methods have been studied for characterizing high temperature ceramic coatings: environmental barrier coatings and thermal barrier coatings. Optical coherence tomography (OCT) has been shown to be able to determine thickness of both TBCs and EBCs including green state slurry coated EBCs. Elastic Optical Scatter (EOS) has been shown to provide one possible way to estimate spallation of TBCs and that, indeed, while there is more sensitivity to surface features than initially considered, there is a dominant effect of other factors that out weigh the effects of surface features.

## REFERENCES

1. A. Szweda, T. E. Easler, R. A. Jurf and S. C. Butner, "Ceramic Matrix Composites for Gas Turbine Engines", in Ceramic Gas Turbine Component Development and Characterization, eds., M. van Roode, M. K. Ferber and D. W. Richerson, ASME press, 2003, pgs 277-289.

2. K. N. Lee, H. Fritze and Y. Ogura, "Coatings for Engineering Ceramics", in Ceramic Gas Turbine Component Development and Characterization, eds., M. van Roode, M. K. Ferber and D. W. Richerson, ASME press, 2003, pgs 641-664.

3. Karlsson. A. M., Hutchinson, J. W., and Evans, A. G. "The Displacement of the Thermally Grown Oxide in Thermal Barrier Coating Systems upon Temperature Cycling", in Mat. Sci. and Eng. A, 351 1-2, pgs 244-257, 2003

4. Karlsson. A. M., Hutchinson, J. W., and Evans, A. G. "A Fundamental Model of Cyclic Instabilities in Thermal Barrier Systems", Journal of the Mechanics and Physics of Solids, 50, 1565-89.

5. Mumm, D. R., and Evans, A. G., 2000, "On the Roles of Imperfections in the Failure of a Thermal Barrier Coating Made by Electron-Beam Physical Vapor Deposition," Acta Materiala, 48, 1815-27.

6. D. R. Clarke, J. R. Christensen, V. Tolpygo, "The Evolution of Oxidation Stresses in Zirconia Thermal Barrier Coated Superalloy Leading to Spalling Failure," *Surf. Coat. TechnoL*, **94/95**, 89-93 (1997).

7. Advisory Group for Aerospace Research and Development (AGARD)/NATO, "Thermal Barrier Coatings", 1997, AGARD report 823, 1997, North Atlantic Treaty Organization

8. B. E. Bouma and G. J. Tearney, Handbook of Optical Coherence Tomography, Marcel Dekker, New York (2002).

9. W. A. Ellingson, R. J. Visher, R. S. Lipanovich and C. M. Deemer, "Optical NDT Techniques for Ceramic Thermal Barrier Coatings", in Mat. Eval. Vol. 64, No. 1, pgs 45-51, 2006

10. Visher, R. J., Gast, L., Ellingson, W. A. and Feuerstein, A. "Health Monitoring and Life Prediction of Thermal Barrier Coatings using a Laser-Based Method", ASME paper GT2005-68252, 2005

# NONDESTRUCTIVE EVALUATION OF ENVIRONMENTAL BARRIER COATINGS IN CFCC COMBUSTOR LINERS

J. G. Sun, J. Benz, W. A. Ellingson
Argonne National Laboratory
Argonne, IL 60439

J. B. Kimmel, and J. R. Price
Solar Turbines Inc.
San Diego, CA 92186

## ABSTRACT

Advanced combustor liners fabricated of SiC/SiC continuous fiber-reinforced ceramic composite (CFCC) and covered with environmental barrier coatings (EBCs) have been successfully tested in Solar Turbines Inc. field engines. The primary goal for the CFCC/EBC liners is to reach a 30,000-h lifetime. Because the EBCs, when applied on the hot surfaces of liners, protect the underlying CFCC from oxidation damage, their performance is critical in achieving the lifetime goal. To determine CFCC/EBC liner condition and assess operating damage, the liners were subjected to nondestructive evaluation (NDE) during various processing stages, as well as before and after the engine test. The NDE techniques included pulsed infrared thermal imaging, air-coupled ultrasonic scanning, and X-ray computerized tomography. It was found that EBC damage and spallation depend on the condition of the CFCC material. The NDE results and correlations with destructive examination are discussed.

## INTRODUCTION

The main objective of the Ceramic Stationary Gas Turbine and Advanced Material Program sponsored by the U.S. Department of Energy (DOE) is to utilize ceramic technology by selectively replacing cooled metallic hot-section components with ceramic parts (blades, nozzles, and combustor liners). In field tests, a team led by Solar Turbines Inc. has successfully demonstrated the use of SiC/SiC CFCC liners in turbines designed to reduce $NO_x$ and CO emissions. Because the recession rate of CFCC liners in the high steam environment of the gas turbine combustor is high, and because the goal for these liners is a 30,000-h lifetime, EBCs were applied to their hot surfaces (Miriyala et al., 2001).

Several sets of CFCC/EBC liners have been tested in Solar Turbines gas engines. One set of CFCC/EBC inner and outer liners was field tested at a Malden Mills facility (Lawrence, MA, USA). The test was terminated when, after 15,144 hours of engine operation that included 92 starts/stops, extensive cracking was observed in the inner liner during routine borescope inspection. Goodrich Corp. (formerly BF Goodrich Aerospace) fabricated the inner liner and GE Power Systems Composites (formerly Honeywell Advanced Composites Inc.) fabricated the outer liner used in this test. The inner liner was a Tyranno ZM/SiC-Si composite made by the melt-infiltration (MI) process. The outer liner was an enhanced Hi-Nicalon/E-SiC composite made by the chemical-vapor-infiltration (CVI) process. A SiC seal coat was applied on both liners by chemical vapor deposition for additional environmental protection. Both liners were 20-cm long and •3.5-mm thick. The diameter of the inner liner was 33 cm, and that of the outer liner, 76 cm. A 3-layer EBC coating was subsequently applied to the gas-path surfaces of the two liners by United Technology Research Center by a thermal spray process. The EBC coating consisted of a top layer of barium strontium aluminum silicate (BSAS), an intermediate layer of mullite and BSAS mixture, and a bottom layer of silicon. The total thickness of the coatings was •450 μm. Fabrication details and analysis results for the coatings can be found in Kimmel et al. (2003). Figure 1 shows a typical CFCC/EBC liner set after the EBC application.

Fig. 1. Photograph of CFCC liners after EBC application.

To determine the condition of and assess operating damage to the liners, they were subjected to nondestructive evaluation (NDE) at Argonne National Laboratory during various processing stages, as well as before and after the engine test. The NDE techniques included pulsed infrared thermal imaging, air-coupled ultrasonic scanning, and X-ray computerized tomography (CT) (Sun et al., 1999, 2002; Sun, 2005). This paper presents the NDE results for the CFCC liners after EBC application and after the engine test. NDE analyses were also performed to determine post-test EBC damage.

## RESULTS

### NDE Results from As-Processed CFCC/EBC Liners

Through-thickness infrared thermal imaging and through-transmission air-coupled ultrasonic scanning were conducted to assess the condition of the CFCC liners before and after EBC application. The NDE data were presented as thermal diffusivity and ultrasonic transmission (UT) images that mapped the entire liners (100% inspection). The thermal diffusivity was affected slightly by local thickness variations, because an average thickness was used for each liner to derive the local thermal diffusivity (Sun et al., 1999).

Figure 2, thermal diffusivity and air-coupled UT images of the inner liner after the EBC application, reveals several defect regions. When compared with the surrounding areas, thermal diffusivity of these defect regions is lower, and the UT intensity is abnormal. The defects were within the CFCC material because they were found within the CFCC inner liner before EBC application. To determine the defect type, one-sided thermal imaging and X-ray CT were performed for the two major defects indicated in Fig. 2.

Figure 3a shows the depth map predicted from one-sided thermal imaging for a portion of Defect 1 in Fig. 2. The map indicates that Defect 1 is a plane-type defect, with a constant depth of 1.1 mm from the inside surface of the liner. Defect 1 is, therefore, a delamination. Figure 3b shows a 0.25-mm-thick CT slice of the Defect 2 region at an average depth of 1.2 mm from the inside surface of the liner. The defect exhibits lower density and appears to be planar, so it is also a delamination. Therefore, both Defects 1 and 2 were delaminations at a depth •2.4 mm from the outside-diameter (OD) surface of the inner liner, where EBC was applied.

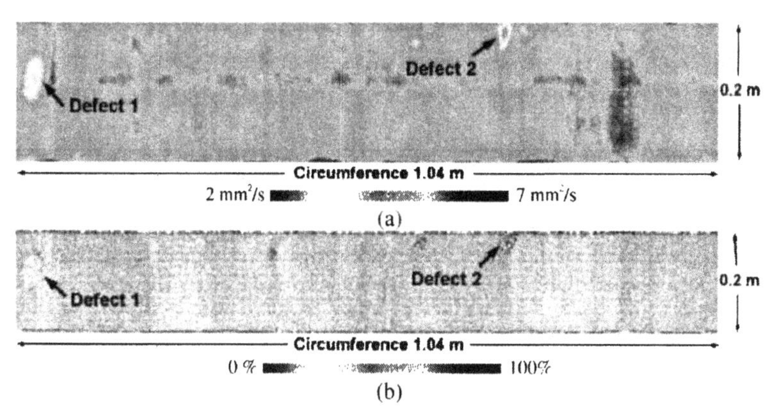


Fig. 2. (a) Thermal diffusivity image and (b) air-coupled UT image of Tyranno ZM/SiC-Si MI inner liner after EBC application.

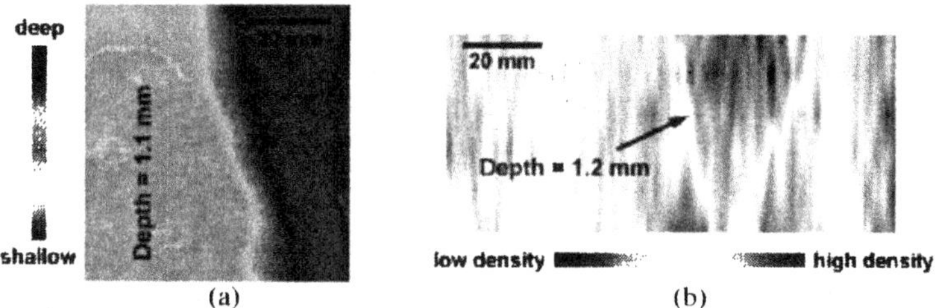


Fig. 3. (a) Depth map predicted by one-sided thermal imaging for Defect 1 in Fig. 2 and (b) X-ray CT image of Defect 2 in Fig. 2. Depths are from the inside surface of the liner.

Both thermal diffusivity and air UT tests showed that the outer liner was generally uniform and free of major defects. The thermal diffusivity image of the outer liner after EBC application is shown in Fig. 4. The image indicates some variation in thermal diffusivity. The regions with lower diffusivity are likely to have lower density. Several observed axial lines were seams that showed on liner surfaces. As indicated in Fig. 4, several thermocouple holes were machined into the liner. In addition, two SiC seal-coat chips were present on the top edge. Nevertheless, none of these variations were considered significant.

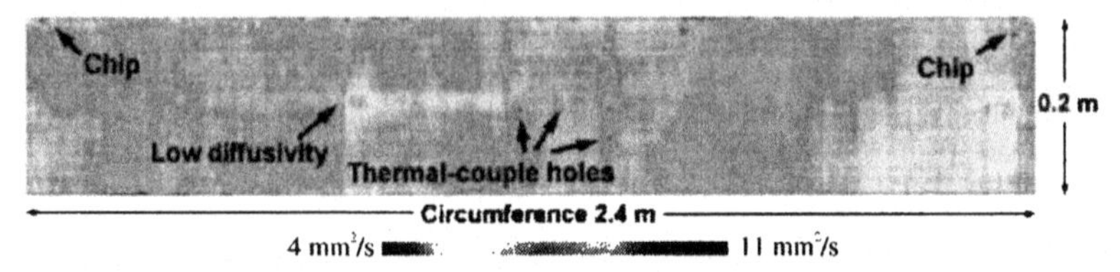


Fig. 4. Thermal diffusivity image of enhanced Hi-Nicalon/E-SiC CVI outer liner after EBC application.

NDE Results from Field-Engine-Tested Liners

Digital images of the EBC-coated (hot-side) surfaces of the inner and outer liners after field engine tests are shown in Fig. 5. The engine test was stopped after 15,144 hours because circumferential and axial cracks in the inner liner had progressed through the CFCC thickness. A thin crack along the circumference of the outer liner was also observed. It is evident that EBC was still present on large sections of the liners, and that EBC loss was concentrated at the aft-end edge (lower edge) and in the middle of both liners. The EBC loss in the middle of the liners occurred in areas where fuel injectors were located, i.e., the hottest areas.

In Fig. 6, the cracks in the field-tested inner liner are superimposed on the thermal diffusivity image of the liner before the engine test. The circumferential cracking of the inner liner matched well the areas of high thermal diffusivity. Higher diffusivity may be the result of higher density or higher silicon concentration as compared to surrounding areas, and was associated with CFCC liner processing. The axial cracking occurred at locations of seams that can be matched to the NDE data in Fig. 2.

Figure 7, the thermal diffusivity image of the outer liner after the engine test, showed that the diffusivity of regions around the EBC spalled areas was lower than it was in other areas. Close examination of these regions indicated that the EBC had debonded from the CFCC liner and could be easily scratched off. A comparison of the thermal diffusivity images shown in Figs. 4 and 7 revealed that the circumferential cracking occurred within a region of lower thermal diffusivity.

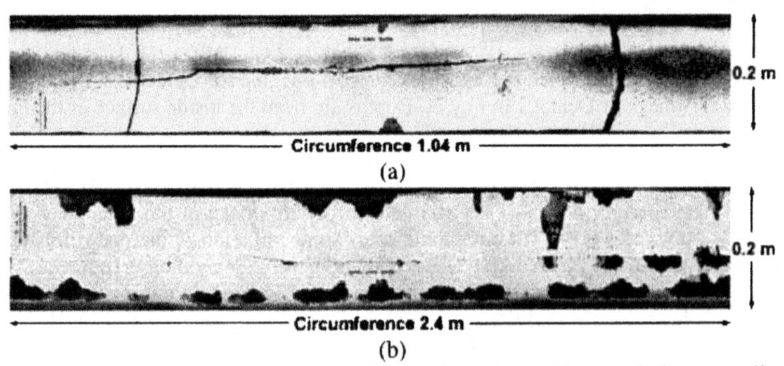


Fig. 5. Digital images of EBC-coated surfaces of (a) inner liner and (b) outer liner after 15,144-h field-engine test.

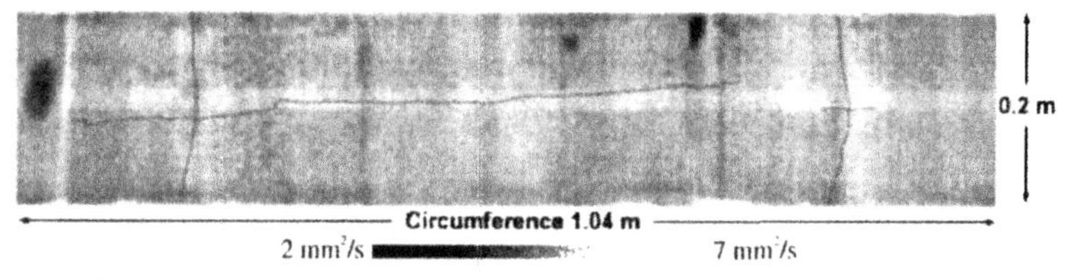


Fig. 6. CFCC cracking superimposed on thermal diffusivity image of Tyranno ZM/SiC-Si MI inner liner taken before field-engine test.

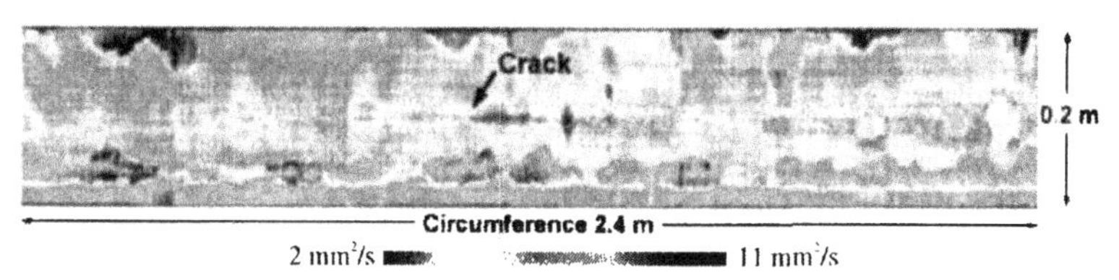


Fig. 7. Thermal diffusivity image of enhanced Hi-Nicalon/SiC CVI outer liner after 15,144-h field test.

NDE Evaluation of EBC

From the digital image of the EBC-coated surface of the inner liner in Fig. 5a, it is evident that the EBC on top of the CFCC delaminated regions had not spalled after the 15,144-h field-engine test. In comparison, EBC spallations over delaminated areas of an inner liner that was eventually field tested for 13,937 hours were observed within 900 hours of engine test (Sun et al., 2002). To understand the relationship between CFCC delamination and EBC spallation, NDE data for these two liners were closely examined. Figure 8 summarizes the characteristics of the two major delaminations in the 15,144-h field-tested liner. The thermal diffusivities of the two delaminations, from left to right, were 64 and 55% of the average diffusivity of the liner, and their depths were 2.4 and 2.3 mm from the OD or hot surface of the liner. In comparison, the thermal diffusivity of the delamination that caused EBC spallation in the 13,937-h field-tested liner was 26% of the average liner diffusivity and the depth was 0.9 mm from the OD surface. These results indicate that delaminations with minor separation (less diffusivity reduction) and near the cold side of the liner are not detrimental to the top EBC layer during engine testing.

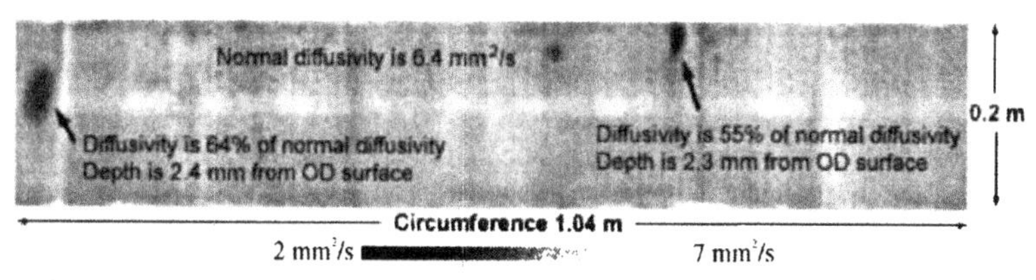


Fig. 8. Characteristics of delamination defects detected in thermal diffusivity image of Tyranno ZM/SiC-Si MI inner liner taken before field-engine testing.

Both inner and outer liners were sectioned for microstructural analyses after the 15,144-h engine test. The analyses showed that a layer of silica had formed at the silicon/mullite+BSAS interface within the EBC and that extensive cracking that extended into the mullite+BSAS layer had occurred at the silica peaks (Kimmel et al., 2003). The cracking was hypothesized to be the cause of the EBC spallation across most of the liner surfaces. Thermal imaging was carried out to detect EBC cracking and debonding in a cut section of the outer liner (Fig. 9a). Figures 9b and 9c show one-sided thermal imaging data, the logarithmic temperature-decay slope $d(\ln T)/d(\ln t)$ images, where $T$ is surface temperature and $t$ is time (Sun, 2005), at 0.05 and 0.4 s after the thermal flash. In the early-time image, Fig. 9b, all regions with EBC cracking and debonding display lower temperature-decay slopes (darker grayscale) whereas normal EBC regions have relatively higher slopes. In the later-time image, Fig. 9c, regions with EBC cracking now exhibit higher temperature-decay slopes (brighter grayscale), and regions with EBC debonding still display a lower slope than the normal regions. The change of temperature-decay slope in EBC cracking regions is due to the smaller lateral size and thinner separation gap of the cracking, while the slope in EBC debonded regions will change at a much later time because of larger lateral size and thicker separation gap of EBC debonds from substrate. Therefore, the temperature-decay slope can be used to monitor the EBC damage.

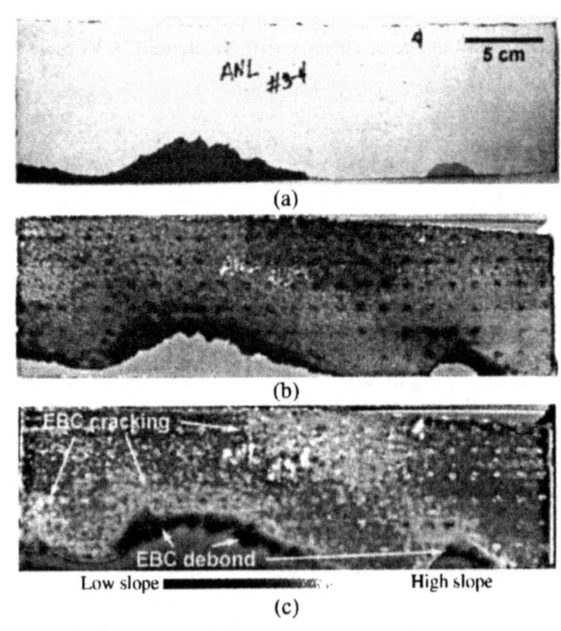


Fig. 9. (a) Photograph of cut section of enhanced Hi-Nicalon/SiC CVI outer liner after 15,144-h field test, and temperature-decay slope images at (b) 0.05 s and (c) 0.4 s after one-sided thermal flash.

## CONCLUSIONS

Infrared thermal imaging, air-coupled UT scanning, and X-ray CT NDE techniques have been successfully used to characterize full-scale EBC-coated CFCC combustor liners that were tested for 15,144 hours in a Solar Turbines engine. Extensive cracking occurred in the CFCC of the inner liner. The cracks were located at regions where the thermal diffusivities were higher than in surrounding areas. Two major delamination defects were found in the CFCC of the inner liner. However, these delaminations were not detrimental to the EBC top layer during engine testing because the separation was minor (less diffusivity reduction) and near the cold side of the liner. In addition, examination of the condition of the EBC on the field-tested liners by one-sided thermal imaging showed that the temperature-decay slope can be used to monitor EBC cracking and determine EBC debonding.

## ACKNOWLEDGMENTS

This work was partially funded by the U.S. Department of Energy, Energy Efficiency and Renewable Energy, Office of Industrial Technologies, under Contract W-31-109-ENG-38, and partially funded under DOE Contracts DE-AC02-92CE40960 and DE-FC02-00CH11049.

## REFERENCES

Kimmel, J., Price, J., More, K., Tortorelli, P., Sun, E., and Linsey, G., 2003, "The Evaluation of CFCC Liners after Field Testing in a Gas Turbine - IV," ASME Paper GT-2003-38920, presented at ASME Turbo Expo 2003, Power for Land, Sea, and Air, Atlanta, Georgia, June 11-19, 2003.

Miriyala, N., Fahme, A., and van Roode, M., 2001, "Ceramic Stationary Gas Turbine Program - Combustor Liner Development Summary," ASME Paper 2001-GT-512, presented at the International Gas Turbine and Aeroengine Congress and Exposition, New Orleans, Louisiana, USA, June 2001.

Sun, J. G., 2005, "Analysis of Pulsed Thermography Methods for Defect Depth Prediction," accepted for publication in *J. Heat Transfer.*

Sun, J. G., Deemer, C., Ellingson, W. A., Easler, T. E., Szweda, A., and Craig, P. A., 1999, "Thermal Imaging Measurement and Correlation of Thermal Diffusivity in Continuous Fiber Ceramic Composites," in *Thermal Conductivity 24*, Eds. P. S. Gaal and D. E. Apostolescu, pp. 616-622.

Sun, J. G., Erdman, S., Russel, R., Deemer, C., Ellingson, W. A., Miriyala, N., Kimmel, J. B., and Price, J. R., 2002, "Nondestructive Evaluation of Defects and Operating Damage in CFCC Combustor Liners," in *Ceramic Engineering and Science Proceedings*, Eds. H-T Lin and M. Singh, Vol. 23, Issue 3, pp. 563-569.

# Ceramic Coatings for Spacecraft Applications

# CHARGING OF CERAMIC MATERIALS DUE TO SPACE-BASED RADIATION ENVIRONMENT

Jennifer L. Sample, Ashish Nedungadi, Jordan Wilkerson, Don King, David Drewry, Ken Potocki, and Doug Eng

Johns Hopkins University Applied Physics Laboratory
11100 Johns Hopkins Road
Laurel, MD 20723

## ABSTRACT

Radiation-induced charging of spacecraft coatings depends on material properties such as resistivity, dielectric constant, and photoelectron and secondary electron emission. Surface charging of a spacecraft is a complex phenomenon that depends on these material properties, as well as spacecraft geometry, orientation, sunlight intensity and distribution, temperature, and radiation environment. Ceramics can be problematic as spacecraft coatings due to their potential to store charge and develop differential voltage potentials. This paper examines the relative charging between a bare carbon-carbon heat shield and one coated with $Al_2O_3$ in near solar (0.3 AU), Jovian, and deep space (2.5 AU, 5 AU) radiation environments, corresponding to points along a notional Solar Probe trajectory.

Findings indicate that when modeled as a simple conical heat shield, the carbon structure alone does not develop a differential potential, whereas the $Al_2O_3$ does, primarily in the cold Jovian environment. However, differential charging appears to be mitigated by photoelectron and secondary electron emission from the $Al_2O_3$. The electrical conductivity of the ceramic increases near the sun due to increasing temperature, thus enabling equilibration of charge. The results of this study reject the initial hypothesis that electrically insulating coatings will create insurmountable charging issues for solar missions. Trajectory-dependent spacecraft illumination and temperature can be used to exploit desirable characteristics of the ceramic such as emission and conductivity, thus mitigating the problem. Integrated spacecraft charging simulations are planned.

## INTRODUCTION

Spacecraft charging is a complex phenomenon derived from several factors, including radiation environment to which the spacecraft is exposed, material properties such as electrical resistivity, dielectric constant, photoemission constants, and secondary emission behavior. Mission-specific details such as attitude and rotation of the spacecraft can also affect charging. Thus, spacecraft charging simulations are typically performed in order to assess and quantify risk to the mission. Risk areas are twofold: buildup of absolute potential over the entire spacecraft, possibly distorting the integrity of science experiments, and buildup of differential potentials between different parts of the spacecraft, possibly leading to arcing and resulting damage to the spacecraft, or damage to electronics and communication devices.

Spacecraft charging simulations and estimations can be performed using one or more freely available (NASA) software programs,[1] such as The Charge Collector, SEE Integrated Spacecraft Charging Handbook, and NASCAP-2K, a sophisticated spacecraft-level charging analysis program developed by SAIC and NASA. NASCAP-2K can be used to explore integrated spacecraft effects such as geometry, component interactions, and others.

The Solar Probe mission is a NASA mission, part of the Living with a Star program, which will fly by the sun, observing the solar plasma, probing the solar wind and acquiring scientific data.[2] Its trajectory to the sun includes a pass around Jupiter and long periods of time in deep space. The overall mission will first travel to Jupiter for a gravity assist before entering solar orbit at perihelion 4 $R_s$. The mission is planned to orbit the sun twice, returning to deep space after the first pass. During the mission, the spacecraft will be subjected to opposite extremes of both radiation and temperature. For example, the Jovian environment is very cold (~77 K) with a primarily electron environment. The temperature near the sun, at 4 $R_s$ is very hot (> 2100 K) and has an environment consisting of protons, intense UV light, and other components that will be further studied by the mission.

The current spacecraft design consists of a large, conical heat shield, with an instrument bus attached via struts (see Figure 1). At closest approach to the sun, the heat shield is designed to keep the science instruments operating close to 300 K. This feat is currently slated to be accomplished using specific coatings on the heat shield which radiate heat and reflect solar light, thus lowering the equilibrium temperature of the spacecraft.[3,4] The coatings currently being considered are the ceramics alumina ($Al_2O_3$), pyrolytic boron nitride (PBN), and barium zirconium phosphate (BaZP), which are all visibly white in color.

Many spacecraft and mission survivability issues important to other missions are also relevant for Solar Probe, namely outgassing, structural stability during launch and to space debris impact, spacecraft charging, and radiation damage. Research is currently being conducted into all of these spacecraft issues for the Solar Probe mission. Preliminary spacecraft charging predictions are presented in this paper.

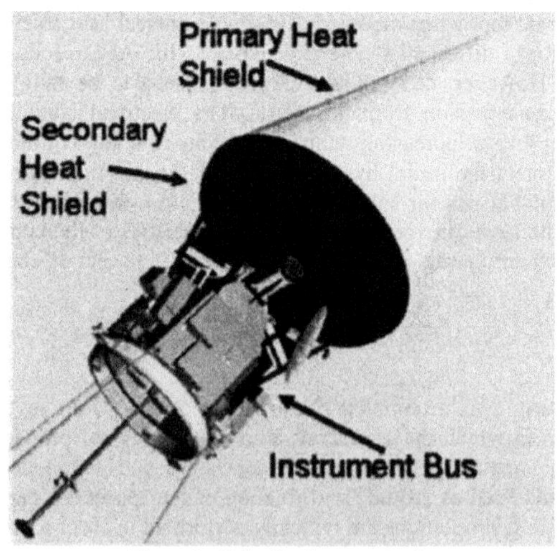


**Figure 1 - Solar Probe spacecraft.**

## EXPERIMENTAL PROCEDURE

Charging simulations and data acquisition were performed using three NASA software programs: The Charge Collector, SEE Integrated Spacecraft Charging Handbook, and NASCAP-2K. NASA's SEE Interactive Charging Handbook was used for simple charging modeling. Specifically, SEE was used to model the materials with their complex material properties as a sheet with a monoenergetic beam of electrons of defined energy striking the sheet at normal incidence. This analysis afforded an understanding of how alumina charges relative to conductive carbon in various electron environments, and served as a reference and "sanity check" for NASCAP-2K results.

Simulations were performed for two materials: conductive carbon-carbon composite (approximated as graphite), and conductive carbon-carbon coated with 100 microns of $Al_2O_3$. The ready availability of material property data dictated the choice of materials. Materials properties used in NASCAP-2K simulations were obtained from NASA references and from the literature.[5]

Radiation environments used for charging simulations were determined from literature data and from discussions with space radiation experts at NASA/GSFC and NASA Solar Probe STDT.[6,7] A finite-element model of the spacecraft was generated for compatibility with the Java-based NASCAP-2K. NASCAP-2K charging simulations were run for 4 specific mission trajectory points: near-solar (0.3 AU), deep space at 2.5 AU, deep space at 5 AU, and the Jovian environment, extrapolated to 12 $R_j$. The radiation environment 4 $R_s$ at present is not well understood, however the temperature of the ceramic at 4 $R_s$ will be high enough for the coating to be electrically conductive. For these reasons, 0.3 AU, the start of critical science along the Solar Probe trajectory was chosen for preliminary investigations. Radiation energies and densities used to represent various simulated radiation environments are shown in Table 1. Solar intensity approximations used in NASCAP-2K, governing charging due to photoemission are also shown in Table 1.

**Table 1. Radiation and plasma environment parameters used in NASCAP-2K simulations.**

| | Near Sun | Deep Space | | Jupiter 12 $R_j$ | |
|---|---|---|---|---|---|
| | 0.3 AU | 2.5 AU | 5.0 AU | 12 $R_j$ (Double Maxwellian) | |
| p+ (m-3) | $3.5x10^7$ | $7.0x10^6$ | $7.0x10^6$ | $6.0x10^5$ | $1.3x10^6$ |
| e- (m-3) | $3.5x10^7$ | | | $2.0x10^5$ | $1.2x10^6$ |
| $T_{e-}$ (eV) | 35 | 15 | 15 | 200 | $2.8x10^4$ |
| $V_{p+}$ (m/s) | n/a | $4.3x10^5$ | $4.3x10^5$ | n/a | n/a |
| $T_{p+}$ (eV) | 44 | 965.3 | 965.3 | 400 | $2.75x10^4$ |
| $T_{PHS}$,K | 460 | <100 | <100 | <100 | <100 |
| Heat flux ($W/cm^2$) | 11.1 | 0.16 | 0.04 | 0.04 | 0.04 |

The NASCAP-2K software was not designed to accommodate an unusually shaped spacecraft such as the Solar Probe, and thus it was necessary to model the conical heat shield of the spacecraft as a tapered cylinder. This impact of this approximation can be seen when viewing the charging results, and appears as marked gradations in the voltage potential along the cone.

RESULTS

Radiation-induced charging of spacecraft coatings depends on material properties such as resistivity, dielectric constant, and photoelectron and secondary electron emission. Surface charging is a complex function of these and other parameters, including spacecraft geometry, orientation, sunlight intensity and distribution, and specific radiation environment.

Initial analysis based on evaluation of material properties as a function of temperature indicated that insulators naturally become more conductive as temperature increases. Insulator resistivity is typically inversely proportional to temperature; thus, the conductivity of insulators is proportional to temperature. Near the Sun, specifically within ~0.2 AU, the $Al_2O_3$ is relatively conductive,[8] such that its resistivity thickness product falls below $2 \times 10^9$ for a 4- to 5-mil coating. Ergo it falls within NASA specifications for partially conductive coatings applied over conductive substrates.[9] Temperature-dependent resistivity data for the specific proposed coatings (plasma sprayed, with or without dopants, etc.) is required for precise analysis. Alumina was chosen for more detailed analysis because a complete set of literature data material properties was available. This information was unavailable for the other materials being considered (PBN and BaZP).

The response of $Al_2O_3$ in near-solar (0.3 AU), Jovian, and deep space (2.5 AU) radiation environments was investigated using NASCAP-2K charging analysis to determine the differences in voltage potential between a bare carbon–carbon heat shield and one coated with $Al_2O_3$ using NASCAP-2K. Analytical predictions indicate relatively low differential charging (relative to conductive substrates) < 30 V for the coated heat shield, as shown in Figure 2.

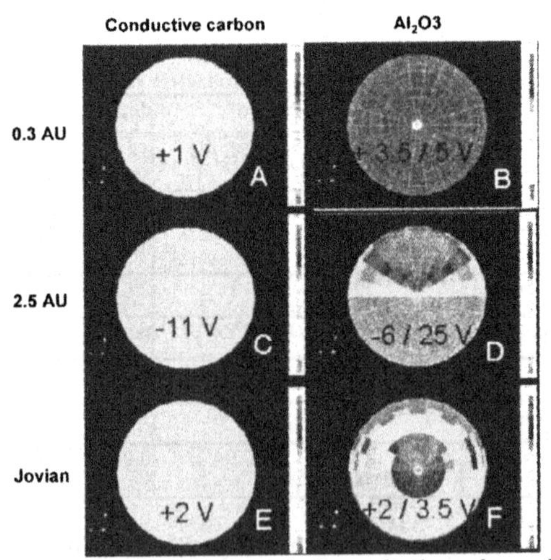


**Figure 2. Cone-end view NASCAP results for conductive carbon and $Al_2O_3$.**

Figure 2 shows charging results for the cone-shaped Solar Probe spacecraft's heat shield only, from an end-on view (tip looking towards base of cone). When the spacecraft is facing the sun at closest approach, it will face the sun head-on. However, at most other points along the mission trajectory, the spacecraft antennas will face the earth, causing sunlight to be incident on one side of the cone-shaped heat shield only, in the absence of rotation about the spacecraft's centerline. The direction of incident sunlight relative to the spacecraft was accounted for in the NASCAP-2K model. The first column in Figure 2 shows charging simulation results for a carbon-carbon composite cone, and the second column shows results for an $Al_2O_3$-coated carbon-carbon composite cone. The top row in Figure 2 shows results for relatively near-solar conditions (0.3 AU), the middle row shows results for a deep space condition, and the bottom row shows results for an estimated Jovian environment at ~12 Rj.

In general, Figure 2 shows that the conductive carbon substrate charges homogeneously, whereas the insulating $Al_2O_3$ substrate charges differentially with the more positive potential in the direction facing the Sun. In particular, as can be seen by comparing Figure 2A and Figure 2B, the conductive carbon charges homogeneously to a very low potential (+1V) whereas in the same radiation environment, the $Al_2O_3$-coated heat shield develops a low differential potential of +1.5 V. Similar results can be seen by comparing Figure 2C and Figure 2D, which represent charging of the heat shields in a deep space environment at 2.5 AU. Here both heat shields develop a slightly higher negative potential, -11 V for carbon and a differential potential of 31 V with a maximum potential of 25 V for the $Al_2O_3$-coated heat shield. Comparing Figure 2E with Figure 2F shows that the conductive carbon charges homogeneously to a low potential (+2V) whereas in the same radiation environment, the $Al_2O_3$-coated heat shield develops a low differential potential of +1.5 V in the Jovian environment.

The most obvious feature of the results shown in Figure 2 is that the conductive carbon cone does not develop a differential charge, evident in its uniform color in each case, while the alumina coated cone does develop a differential charge. In the near-solar case, the highest charge on the coated cone is at the tip, closest to the sun, and in the deep space and Jovian cases, the highest charge is accumulated on the sunlit side of the cone. Alumina is known to be a very good emitter of secondary electrons, or electrons emitted in response to electrons impacting the spacecraft. For this reason, the alumina does not charge to the large negative potentials previously expected for the electron-rich environment around Jupiter. In fact, it charges slightly positive, particularly in the sunlight part of the heat shield because of electrons lost to the photoemission process. Carbon is not a remarkable secondary electron emitter, and thus charges to more negative potentials in the presence of some plasmas. The specific voltage potentials developed depend on the exact radiation parameters used in the NASCAP-2K model.

The NASCAP-2K software was not designed to accommodate an unusually shaped spacecraft such as the Solar Probe, did not include ceramics as material options, nor was it designed to include temperature dependence of material properties. Thus it was necessary to adapt the code as well as possible. The conical heat shield of the spacecraft was modeled as a tapered cylinder and $Al_2O_3$ material properties were obtained from the literature. Temperature dependent material properties will be included in future studies.

Figure 3 shows results for the charging of the entire spacecraft in the relatively near-solar (0.3 AU) environment. A finite element model composed of more than 4000 elements was built for this NASCAP-2K charging analysis. The results of NASCAP runs for the entire spacecraft having an alumina coating on the heat shield are shown in Figure 3. Note that the sunlight is incident on the cone from the Y-axis as shown, such that the instrument bus is in shadow. The results show that the sunlit cone charges to a positive potential, mostly due to photoemission, while the portion of the spacecraft in shadow charges to negative potential. The overall potential differential is 40 V. Radiation-induced destructive arc-discharges typically occur with potentials < 100 V, so, potentials shown in Figure 3 may not represent an arc-discharge-type charging hazard.[9]

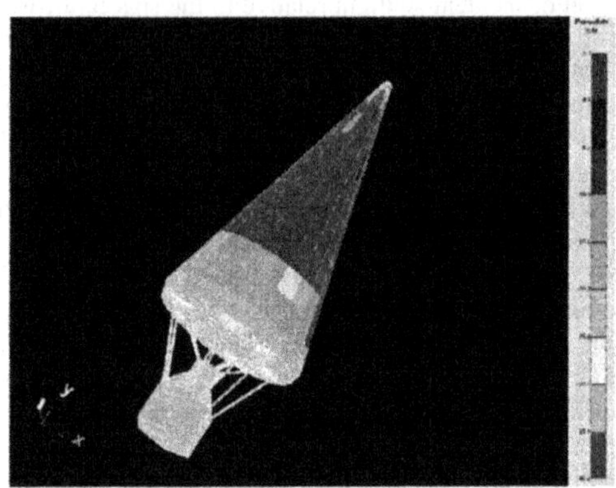

**Figure 3. Entire spacecraft NASCAP results for $Al_2O_3$-coated cone at 0.3 AU radiation environment. The potential difference ranges from +5.0 V (cone, pink) to -35.0 V (bus, orange).**

A limitation in this first version of the finite element model likely affects the absolute potentials obtained from NASCAP-2K, specifically, the lack of multi-component definition and definition of resistivites between grounded components, which will be discussed in the next section. In addition, the data shown in Figures 2 and 3 have been autoscaled. Potential differences that appear to be stark or abrupt may correspond to differences of 0.01 to 0.1 V. However, asymmetric potentials observed in Figures 2 and 3 likely result from the finite element griding details and are currently under investigation and improvement.

## DISCUSSION

As can be seen in the above charging predictions, differences exist in material response at the near-solar and Jovian/deep-space environments. Most notably, the conductive carbon substrate does not develop a differential charge, even though only one side is in sunlight in the 2.5 AU and Jovian cases. This homogeneous charging is due to the electrical conductivity of the conductive carbon, which distributes accumulated charge very quickly and uniformly. The ceramic does, however, develop a differential potential across the cone when the sunlight illuminates one side (as it does in the 2.5 AU and Jovian cases). Near the Sun, the insulative characteristics of the ceramic decrease due to temperature effects, to the point of being effectively electrically conductive. Thus this temperature-dependent behavior allows equilibration of the charge, as in the bare carbon heat shield. In addition, at this point in the trajectory, the spacecraft will be pointed such that sunlight is incident evenly across the cone i.e., the cone tip is pointed at the Sun.

Further analysis is required to determine the charging effects for the integrated spacecraft considering material properties as a function of temperature, resistivity between grounded components, radiation environments as a function of trajectory, etc. NASCAP-2K software will be used as an analytical tool to predict differential charging between components for critical radiation environments, such as start of science, the Jovian encounter, and deep space. Analog experiments will be performed to more accurately determine material properties, validate analytical models, and determine material responses. These analyses will enable system engineers to identify and address issues associated with spacecraft electronics, communications, and sensitive science functions. Overall, this charging analysis indicates that both carbon–carbon and ceramic coated carbon-carbon present manageable charging situations.

The limitations of the current work will be addressed during the Solar Probe risk assessment study. The level of effort devoted to charging studies will be expanded upon as well. The primary limitation is in the accuracy of the finite element model of the spacecraft used in NASCAP-2K. For the preliminary studies presented in this paper, the spacecraft was modeled as one conductor, meaning according to the finite element model, the entire spacecraft is molded out of one large piece of carbon-carbon composite, with no junctions, joints, etc, having all parts completely electrically shorted to each other with zero resistance. While the assumption can be made that joints, junctions, etc will be in good electrical contact with each other, i.e. $< 10^9$ ohms/square, the specific resistances will affect and must be incorporated into the model. Those changes are anticipated to enable us to make a more accurate prediction of entire spacecraft charging. However, the charging simulations for heat shield cone-only structures remain accurate for comparison purposes, as no other components are necessary. During the risk assessment program, we will improve the NASCAP-2K charging simulations by adding in separate conductors, by better approximating material properties and radiation environments used in the model, and by simulating charging behavior of PBN and BaZP. Additionally, we will determine temperature dependence of material resistivities and incorporate them into the model. Total mission charging simulations and simulations at closest approach (4 $R_s$) are also planned.

Several mitigation strategies are being considered for charging on the Solar Probe mission. The most obvious charging mitigation strategy in spacecraft design is to make all surfaces conductive whenever possible. To this end, electrically conductive white ceramics such as ZnO could be incorporated at some weight percent into the alumina coating; however concerns about thermal stability at near solar temperatures persist. The search for a suitable conductive dopant must balance several factors in material choice: required optical properties (high UV reflectivity and high IR emissivity), structural integrity, stability with respect to chemical composition, low outgassing, and low susceptibility to radiation damage are all required characteristics. The white ceramics outlined in this paper were selected for this purpose, and any additional electrically conductive dopants must fulfill these requirements as well.

Conductive nanostructures could be used as dopants within the plasma sprayed white ceramic optical coatings. Nanostructures have the unique property of being small enough not to significantly impact optical properties in small concentrations (<1%). Thus, a percolative network of conductive nanostructures such as carbon nanotubes or conductive oxide nanowires can be incorporated into the coatings to improve conductivity as a charging mitigation strategy.

## CONCLUSIONS

We have investigated charging characteristics of the candidate white ceramic coatings and investigated their potential impact on the function of the Solar Probe spacecraft. These insulative ceramic coatings were evaluated relative to highly conductive carbon. While ceramics such as $Al_2O_3$, PBN, and BaZP function to reduce the overall equilibrium temperature of the primary and secondary heat shields, an identified risk has been that they would enable voltage potentials to develop leading to unmanageable differential spacecraft charging. Charging simulations using NASCAP-2K were performed for both conductive carbon and alumina-coated conductive carbon heat shields in various radiation environments representing points along the solar probe mission trajectory. The results indicate that charging is differential on the alumina-coated heat shield, and homogeneous on the carbon heat shield, however, total voltage potentials are similar. From an initial assessment of the results, we believe that the predicted charging for either material is manageable and can be tailored through selective design of component interfaces and material additives.

## ACKNOWLEDGMENTS

We are grateful to the Solar Probe Science and Technology Definition team (led by Dr. Dave McComas of Southwest Research Institute) for their helpful discussions and support. We appreciate the NASA support provided by Haydee Maldonado of the NASA-Goddard Space Flight Center under NASA/GSFC contract #NAS5-01072 for the engineering trade studies. We also acknowledge Dr. Dennis Nagle (JHU), Dr. Ed Sittler (NASA/GSFC), Dr. John Cooper, and Dr. Ralph McNutt (JHU/APL) for their continuing guidance and support.

## REFERENCES

[1] The Charge Collector, Version 2.0, Interactive Spacecraft Charging Handbook, Version 3.01, NASCAP-2K, Version 3.0, all courtesy of NASA Space Environments and Effects (SEE) Program.

[2], http://solarprobe.gsfc.nasa.gov/, *Solar Probe STDT Final Report,* prepared for NASA Headquarters, September 2005.

[3] D. Drewry, et. al., "Spacecraft thermal management via control of optical properties in the near solar environment," Proceedings of the 30th International Conference on Advanced Ceramics and Composites, Cocoa Beach, FL, 2006.

[4] D. King, et. al., "AL2O3 Optical Surface Heat Shield for Use in Near Solar Environment," Proceedings of the 30th International Conference on Advanced Ceramics and Composites, Cocoa Beach, FL, 2006.

[5] The Charge Collector, Version 2.0, NASA Space Environments and Effects (SEE) Program; X. Meyza, et. al., "Secondary electron emission and self-consistent charge transport and storage in bulk insulators: Application to alumina," *J. Appl. Phys.*, 94, 5384, 2003.

[6] Dr. John Cooper, NASA GSFC unpublished data, Dr. Ed Sittler, Astrophysicist, NASA GSFC.

[7] Bagenal et al., "Jupiter, The Planets, Satellites, and Magnetosphere," Cambridge Planetary Science, 2005.

[8] W. H. Gitzen, "Alumina as a Ceramic Material," The American Ceramic Society, Special Publication No. 4, 1970.

[9] "Design Guidelines for Assessing and Controlling Spacecraft Charging Effects," C.K. Purvis, H.B. Garrett, A.C. Whittlesey, N.J. Stevens, NASA Technical Paper 2361, 1984.

# SPACECRAFT THERMAL MANAGEMENT VIA CONTROL OF OPTICAL PROPERTIES IN THE NEAR SOLAR ENVIRONMENT

David Drewry, Don King, Jennifer Sample, Dale Clemons, Keith Caruso, Ken Potocki, Doug Eng, Doug Mehoke, Michael Mattix, and Michael Thomas
The Johns Hopkins University Applied Physics Laboratory
11100 Johns Hopkins Road
Laurel, MD 20723

Dennis Nagle
The Johns Hopkins University Advanced Technology Laboratory
810 Wyman Park Drive, Suite G010
Baltimore, MD, 21211

## ABSTRACT

The Johns Hopkins University Applied Physics Laboratory (JHU/APL) is currently evaluating multi-functional ceramic coatings for spacecraft operating in the solar environment. Ceramics were selected based on chemical stability; and inertness to radiation damage and hydrogen degradation. Investigations have focused on "white" ceramics such as aluminum oxide, pyrolytic boron nitride, and barium zirconium phosphate which also possess desirable optical characteristics.

Engineered ceramic coatings have been found through testing to possess both high visual reflectivity and infrared emissivity. These optical properties make this a prime passive thermal management solution for NASA's Solar Probe spacecraft. This future mission is designed to send a spacecraft within $4R_s$ (4 solar radii from the center of the Sun) to collect and analyze solar wind and dust. Notionally, the uncooled carbon-carbon spacecraft ($\alpha_S/\varepsilon_{IR} = 1$) is predicted to reach equilibrium temperatures of 2100 K. This temperature results in three major problems: carbon-carbon mass loss leading to a compromise of science data; increased thermal protection system mass leading to higher launch costs; and incorporation of immature insulation materials leading to increased mission risks.

These ceramics with $\alpha_S/\varepsilon_{IR} < 0.6$ have been shown to notionally reduce equilibrium temperatures to between 1300 and 1850 K. Operational requirements for this mission imply that the ceramics must structurally survive launch, thermal cycling, and particulate impact; be inert to radiation damage; have manageable spacecraft charging; and be chemically stable. Preliminary testing of these ceramics in simulated space environments (including structural and radiation conditions) has yielded promising results. Higher fidelity studies will be continuing focused on reducing the risks associated with implementing these candidate materials.

## INTRODUCTION

JHU/APL is the mission lead and system integrator for NASA's Solar Probe engineering studies. Solar Probe is being designed to provide a near-Sun flyby at $4R_s$, as depicted in Figure 1, in order to collect fundamental science data relative to the formation of solar wind. The science objectives include: determine the structure and dynamics of the magnetic fields and sources of the fast and slow solar wind; trace the flow of the energy that heats the solar corona and accelerates the solar wind; and determine what mechanisms accelerate, store, transport energetic particles [1]. Science instruments will require shielding from the intense radiation while still enabling solar wind particle and dust measurements, wave and field measurements, and coronal and surface imaging (as illustrated in Figure 2).

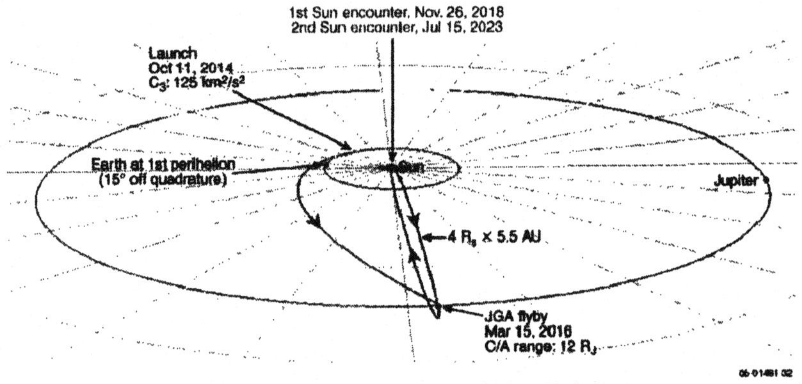


Figure 1: Solar Probe Near-Solar Flyby

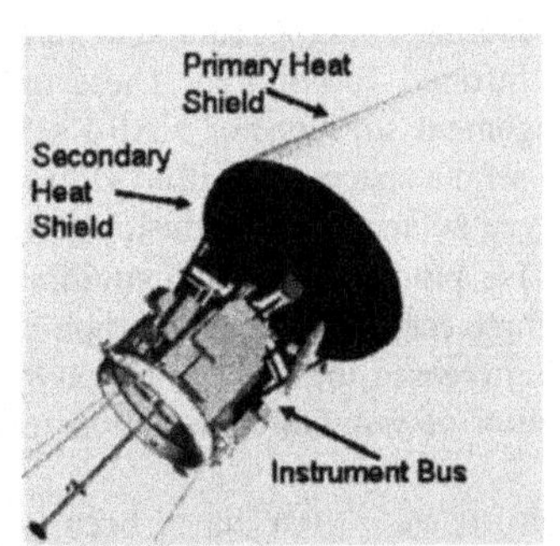


Figure 2: Solar Probe Spacecraft Configuration

Fundamental research supporting materials research for near-solar spacecraft has been funded by The Johns Hopkins University Applied Physics Laboratory IR&D since 2002 and integrated with engineering studies funded by NASA-Goddard Space Flight Center (NASA/GSFC). The research has specifically addressed the applicability of high temperature optical surfaces as a means of providing spacecraft thermal management in the near solar environment. Structures for solar applications are required to function in the temperature regime ranging from deep space at 77 K to 2100 K at $4R_s$. Mass loss is a critical issue for the Solar Probe mission as spacecraft contamination could potentially compromise critical science measurements. Mass loss is composed of thermal-vacuum (out-gassing), radiation induced sublimation, chemical erosion (H+ attack), and physical sputtering components. Initial out-gassing experiments performed by the Jet Propulsion

Laboratory (JPL) indicate that total spacecraft mass loss allocations will be exceeded by the carbon-carbon structures alone if equilibrium temperatures >2200 K are reached.

When reasonable design, environment description, and testing uncertainties are applied to the initial out-gassing data, the carbon-carbon structures (mainly the primary heat shield) notionally exceed mass loss limits at temperatures as low as 1900 K. Consequently, a lower risk design solution was sought by the APL research team to not only eliminate potential out-gassing issues (by reducing spacecraft temperatures to below 1900 K), but also to provide protection from other temperature-driven environmental effects such as chemical or radiation interactions.

While the reduction in contamination sources was a primary driver early in the research, the ability to significantly reduce spacecraft equilibrium temperature was quickly realized to have a substantial impact on the overall spacecraft design and mission. The APL approach incorporates a white ceramic coating (which is highly reflective in the visible and highly emissive in the IR) applied to the carbon-carbon spacecraft primary heat shield structure. Studies predict that this passive thermal management solution enables a significant reduction in the equilibrium temperatures of critical thermal protection system (TPS) components to <1900 K.

The reduction in the primary heat shield equilibrium temperature directly impacts high risk components such as the secondary heat shield (that blocks primary heat shield thermal radiation to the bus) and struts (that connect the primary heat shield to the bus). In early design studies, the secondary heat shield was proposed to incorporate emerging materials (such as aerogels) in order to meet thermal conductivity and mass/volume budgets. Lower temperatures enable use of mature insulation systems (such as carbon foams). Also, the struts were originally configured to be pyrotechnically severed in order to create a thermal short after launch. In addition to increasing risks due to the severance operation, this approach resulted in significant mass penalties. A new strut design was enabled by the cooler heat shield that does not require severance. The result to the mission is a lower risk spacecraft with significant launch mass (and cost) savings.

Our basic material research has been highly integrated with Solar Probe engineering studies in an attempt to make the findings timely and value-added as a means of risk reduction. Material, mission, and science requirements were obtained from the Solar Probe Science and Technology Definition Team (STDT headed by Dr. Dave McComas of Southwest Research Institute). The research has specifically addressed the following technology issues related to the functionality of an optical surface in space:

- Optical properties ($\alpha_S/\varepsilon_{IR}$) and relationship with predicted primary heat shield equilibrium temperature
- Radiation damage to optical surfaces
- Particulate damage and impact on equilibrium temperature
- Structural integrity of the coating and substrate under launch and flight conditions
- Mass loss from thermal-vacuum and radiation effects
- Spacecraft charging from irradiation ($p^+$, $e^-$, ions, and extreme ultraviolet - EUV)

Three candidate "white" ceramics; aluminum oxide or alumina ($Al_2O_3$), pyrolytic boron nitride (PBN), and barium zirconium phosphate ($BaZr_4P_6O_{24}$, herein identified as BaZP) have been investigated as thermal-optical surfaces (see Table I). Alumina was selected as the baseline coating due to its overall technical maturity level and supporting database. PBN and BaZP were retained as backup solutions.

| Candidate Material | Optical | | | | | | | Structural | | | Mass Loss | | | | Charging | | |
|---|---|---|---|---|---|---|---|---|---|---|---|---|---|---|---|---|---|
| | Baseline | Temperature Effects | Radiation Effects | | | Particulate Impact Damage | PHS Temperature Predictions | Thermal Cycling | Vibration | Cumulatove Damage | Temperature Effects | Thermal-Vacuum Effects | Chemical Interactions | Radiation Effects | Temperature Effects | Radiation Effects | Spacecraft Effects |
| | | | EUV | Electron | Proton | | | | | | | | | | | | |
| $Al_2O_3$ | E,A | E,A | E | E | E | A | A | E | E | | E | A | A | A | A | A | A |
| PBN | E,A | E,A | E | E | E | A | A | | | | E | A | A | A | A | | |
| BaZP | E,A | E,A | | E | E | A | A | | | | E | A | A | A | | | |

E = Experimental; A = Analysis

Table I: Summary of Research Investigations

# DISCUSSION OF INVESTIGATIONS

## Optical Characteristics

The conical primary heat shield is designed to present a small view factor to the Sun (and reduce the amount of solar irradiance absorbed) while providing a significant surface to radiate to deep space. Combining this design approach with optical surfaces that reflect a large portion of the visible solar irradiance (> 400 W/cm$^2$) and emit the remaining energy (in the IR band) results in a significant reduction to the heat shield equilibrium temperature. The maximum predicted temperature of the spacecraft's primary carbon–carbon heat shield (with no reflective coating, i.e., a blackbody) is on the order of 2100 K (assuming a ratio of solar absorptivity to IR emissivity, $\alpha_S/\varepsilon_{IR} = 1$) at 4 solar radii of the Sun. The optical property of $\alpha/\varepsilon$ needs to be <<1 to provide significant temperature reduction.

To assess the thermal management approach, optical data as a function of temperature and wavelength were collected by APL. Reflectance was measured at specific wavelengths over a range of temperatures (from room temperature to the predicted equilibrium temperature) and the solar absorptance or IR emittance calculated using Kirchoff's Law. Optical property testing (up to 1773 K and using an Argon cover gas) was performed on C–C coupons coated with $Al_2O_3$ and PBN and on BaZP. The $\alpha_S/\varepsilon_{IR}$ results for these coupons are shown in Figure 5. In addition, Figure 5 shows the resulting primary heat shield temperature (at a distance of 4 solar radii from the Sun) as a function of $\alpha_S/\varepsilon_{IR}$. The use of optical coatings (all with $\alpha_S/\varepsilon_{IR} < 0.6$) is seen to reduce the heat shield temperature to < 1850 K using alumina (to as low as 1300 K in the case of BaZP [2].

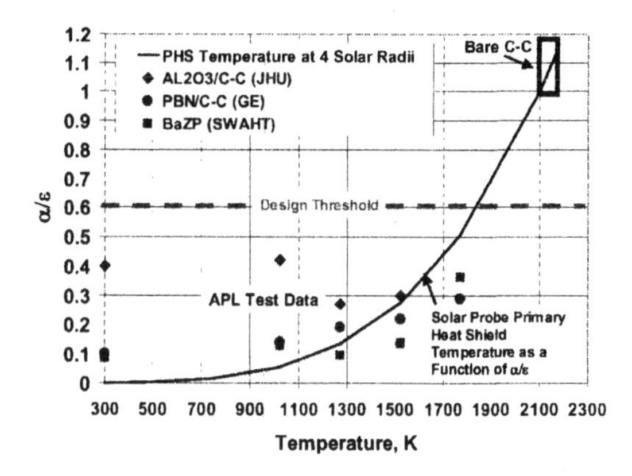


Figure 3: Optical Property Data for $Al_2O_3$, PBN, and BaZP

Radiation Damage

Radiation tests were performed on both ceramic only and ceramic coated carbon-carbon coupons. Preliminary tests were structured to evaluate material performance for the "worst case" radiation conditions of deep space, Jupiter, and solar environments along a notional spacecraft trajectory. Radiation sources include; EUV, electrons ($e^-$), protons ($p^+$), and neutrons. It is critical that these tests be performed at representative temperatures (i.e., cold for Jupiter and hot for near solar) to properly capture dynamic material responses as summarized in Table II.

| | **Near Sun** | | **Deep Space** | | **Jupiter 12 Rj** | |
|---|---|---|---|---|---|---|
| | 4 Rs | 0.3 AU | 2.5 AU | 5.0 AU | 12 Rj (Double Maxwellian) | |
| p+ per $m^3$ | TBD | $3.5 \times 10^7$ | $7.0 \times 10^6$ | $7.0 \times 10^6$ | $6.0 \times 10^5$ | $1.3 \times 10^6$ |
| e- per $m^3$ | TBD | $3.5 \times 10^7$ | | | $2.0 \times 10^5$ | $1.2 \times 10^6$ |
| $T_{e-}$ (eV) | TBD | 35 | 15 | 15 | 200 | $2.8 \times 10^4$ |
| $T_{p+}$ (eV) | TBD | 44 | 965.3 | 965.3 | 400 | $2.75 \times 10^4$ |
| Photons | $7.34 \times 10^{21}$ | EUV does not apply past 4Rs for our design purposes | | | | |
| N (nm) | 2.5 - 200 | | | | | |
| $T_{PHS}$, (K) | 2100 | 460 | <100 | <100 | <100 | <100 |

Table II: Conditions Used to Simulate Anticipated Spacecraft Environments

Of primary interest from the radiation testing, is microstructural damage that affects optical properties, structural integrity, or leads to the generation of excessive mass loss. Initial screening tests at NIST (EUV), Indiana University (protons), and NASA/GSFC (protons and electrons) were structured to distinguish between radiation damage to the fundamental ceramic materials and other potentially complex effects such as material impurities or test chamber contamination.

Space materials such as organic based coatings (for thermal control) experience solar UV degradation in near-Earth orbit. This degradation is commonly referred to as "color center" formation because these materials often darken with UV exposure. "Color centers" in the case of polymer materials are caused by energetic UV radiation breaking organic bonds resulting in a loss of oxygen atoms. Materials can be permanently degraded by UV, but coloration can also be reversed when oxygen is reintroduced upon re-entry into the terrestrial atmosphere.

Color centers can also be formed in ceramics in the presence of ionizing or displacing radiation. Anions or cations are removed from the crystal lattice and electrons or holes are trapped at the resulting vacancies (a different mechanism than UV degradation in near-earth orbit). Enough trapped electrons or holes can change the fundamental UV/VIS absorption characteristics of the crystal lattice. There are many types of color centers that can be formed (i.e., F centers, $F^+$ centers, and V centers), each with its unique absorption and emission spectra. Figure 4 shows coloration of plasma sprayed and fully dense alumina due to electron irradiation at 150 keV energy and $2.8x10^{14}$ $p/cm^2$ fluence.

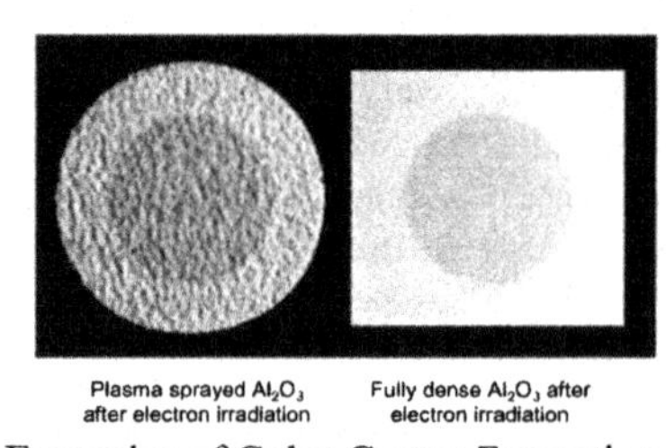


Figure 4: Examples of Color Center Formation in Alumina

However, note that the rate and extent of color center formation is expected to be significantly less in the near-solar environment since irradiation will occur at elevated temperatures. As mentioned previously, the rate and extent of color center formation is dependent on temperature; released anions or cations "floating" in the band gap recombine with vacancies more readily at increased temperatures. Some color centers are relatively unstable and will readily recombine at ambient conditions as soon as irradiation is stopped. Other point defects are more stable and require thermal bleaching to move the displaced atoms back into their lattice positions, returning ceramics to their original white color.

In our research, the alumina, PBN and BaZP were exposed to large electron and proton fluences (2.8E14 $p/cm^2$) to simulate "worst case" near-solar and Jovian environments. Proton irradiation (the predominant near-solar component) did not affect optical properties (as shown in Figure 5) as no visible color change was observed (see Figure 6). Electron irradiation (the predominant Jovian component) resulted in coloration of all three ceramics, most noticeably in alumina. Optical properties of PBN and BaZP were only marginally affected, while alumina showed an increase in UV and visible absorption (at room temperature) due to the formed color centers [3]. Heat treatment at 200°C for 1hr removed the electron coloration, bleaching the coatings back to their original white color (Figure 6). Our hypothesis is that as the heat shield equilibrium temperature increases, $\alpha_S/\varepsilon_{IR}$ (in particular the reflective component) will not be measurably affected by radiation sources.

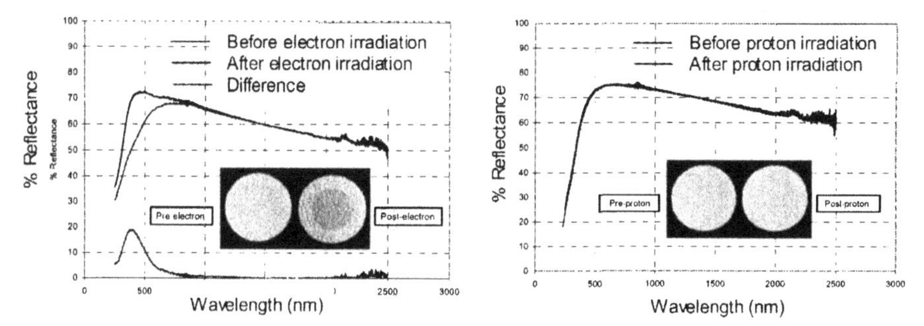


Figure 5: Optical Properties Pre and Post Proton and Electron Exposure

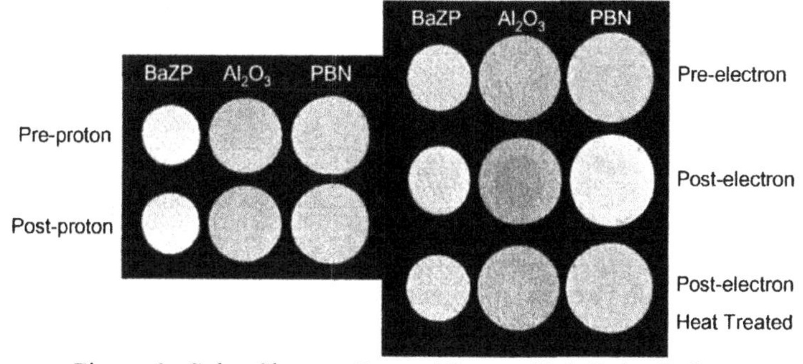


Figure 6: Color Changes Due to Electron Hole Point Defects

Particulate Impact Damage

System-level trade studies were performed to evaluate the effects of particulate impact (and subsequent degradation of the optical surface) on the primary heat shield equilibrium temperature. A finite element thermal model of a single impact location was developed and parametric studies conducted to rank the effects of hole diameter, radius of spall zone, thermal conductivity of the substrate, and $\alpha_S/\varepsilon_{IR}$ ratio of the primary heat shield. The results revealed that any significant temperature increase would be limited to the immediate region of the damage (where the optical surface has been removed and $\alpha/\varepsilon$ is that of the underlying carbon–carbon substrate, that is, $\alpha_S/\varepsilon_{IR} \sim 1$). Since the (uncoated) carbon–carbon interior surfaces have high IR emissivity and as the thermal radiation "view factor" between the interior surfaces is unity, radiation heat transfer between the damaged and undamaged sides of the heat shield serves to restrict any temperature increase to a small region around the damaged area.

After characterizing the local temperature change due to damage, the effect of damage on system thermal performance was parametrically evaluated by varying the total damage to the optical surface and calculating the resulting average equilibrium temperature of the primary heat shield. Damaged areas were treated as local hot spots (with elevated temperature), while undamaged areas remained cooler. An area-weighted average temperature was then calculated for the heat shield as a function of damaged area. The results, shown in Figure 7, reveal the robustness of the design. Separate studies performed by Dr. Cesar Carrasco (of The University of Texas El Paso) predicted primary heat shield damage levels much less than 0.1%, thus indicating that no significant temperature increase is expected. This is based on an estimated solar dust fluence which is primarily comprised of (molten and solid) silica particulate ranging in size from 300 micron to submicron.

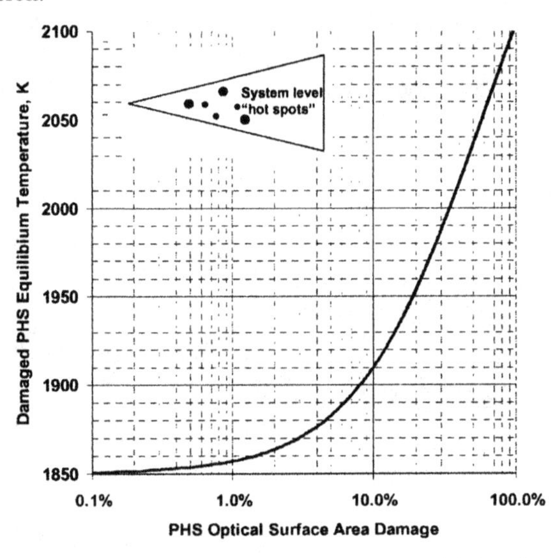


Figure 7: Predicted Heat Shield Temperature Rise with Coating Damage

Structural Integrity

The candidate coating materials represent larger classes of ceramics; namely those which are applied via plasma spray ($Al_2O_3$ and BaZP) and those applied by chemical vapor deposition (PBN and BaZP) processes. These two methods develop quite different coating microstructures and substrate interfaces. Plasma spray produces a toughened, porous coating with graded, mechanical interlocking features at the substrate interface. Conversely, CVD produces a brittle, dense coating with a well defined, chemically bonded interface to the substrate. For the proposed application, the interface must retain structural integrity during launch where random vibrations (and resulting cyclic displacements) will be imposed at room temperature; followed by thermal cycling from 4 to 1800 K where coefficient of thermal expansion effects will become pronounced; coupled with the potential for radiation degradation during prolonged mission times (>15 years). The structural interface (a prescribed distance into the substrate and coating) can be tailored for the predicted structural loads by changes to: substrate surface density;

substrate carbon phase; substrate densification method; substrate ply angle; barrier/transition coatings, and variations to the coating microstructure.

Initial structural screening tests focused on rudimentary thermal cycling and vibration experiments for plasma sprayed $Al_2O_3$. Thermal cycling tests (to simulate flight loads) were performed using $Al_2O_3$ coated carbon-carbon (1-in. diameter by 0.125 in. thick) coupons. These coupons which had significant free edge stress effects by design were cycled three times from 77 to 1600 K (in subsequent optical property testing, the coupons were exposed in multiple cycles to 1800 K). No coating or coating-to-substrate interface damage was observed, as shown in Figure 8.

Vibration tests were designed (and conducted at room temperature) to produce strain fields representative of the primary heat shield response to the launch vehicle environment. As shown in Figure 8, test specimens ($Al_2O_3$-coated carbon-carbon, nominally 1 in. wide x 9 in. long x 0.05 in. thick) were affixed with simply supported ends to a vibration table. Laser velocimetry and strain gages were used to measure displacements and strain. In addition to multiple sweeps at varying frequencies and acceleration levels, the coupon was excited by 6 g's acceleration at the measured natural frequency of 164.3 Hz to achieve a desired strain level of 330 $\mu$in/in at the center of the specimen (equivalent to a +/- .080" primary heat shield displacement). This level was maintained for 15 minutes with no coating or coating-to-substrate interface damage observed.

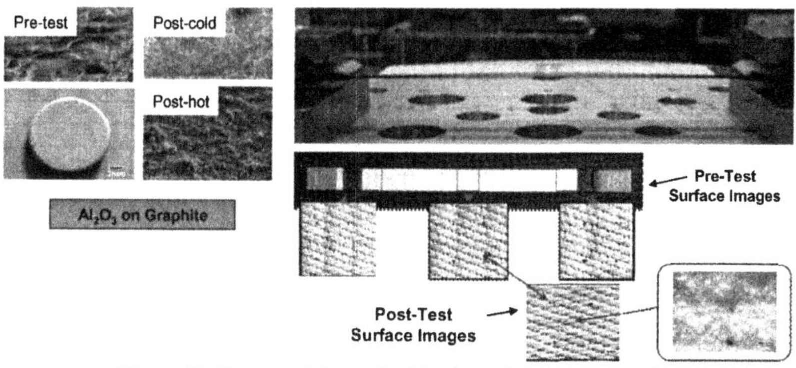


Figure 8: Structural Integrity Testing of Ceramic Coating

## Mass Loss and Contamination

Mass loss in the near solar environment can be attributed to both thermal vacuum and radiation effects. Thermal vacuum effects include outgassing (removal of moisture and organic volatiles), thermal vaporization and chemical interactions; radiation effects include physical sputtering, radiation enhanced sublimation, and chemical erosion [3,4]. The radiation effects of chemical erosion (methane formation) and radiation enhanced sublimation are specific to H+ ions interacting with carbon, and are not a consideration for ceramic coatings or other types of radiation [4]. A literature review suggests that carbon yields (mass loss) due to radiation effects are relatively small compared to mission mass loss requirements. Also, outgassed products such as moisture and organics will be depleted at relatively low temperatures during solar approach. Consequently, initial screening tests for the ceramic materials were focused on thermal vaporization.

Small samples of $Al_2O_3$, PBN, and BaZP were tested using a thermal gravimetric analysis (TGA) system. These materials (surrounded by a nitrogen cover gas) were taken to elevated temperatures in an attempt to characterize total out-gassing and out-gassing rate. Measurable mass loss occurred for all samples up to 773 K, but this mass loss was attributed to contamination due to collection of organics and water during storage. Repeated cycles to 773 K showed no additional measurable loss, which would have occurred if the material had been out-gassing as a result of a degradation process. All the materials were subsequently exposed to higher temperatures (up to 1400 K for BaZP and 1773 K for alumina and PBN) and no additional mass loss was detected (that is, any changes in the measured masses were within the TGA measurement resolution, ±0.1% of the sample weight) as shown in Figure 9 for $Al_2O_3$. These results agreed with predictions (based on literature review and analysis) that these ceramics were stable within the proposed operational envelope. Higher fidelity testing of these ceramics and carbon-carbon is planned using NASA/GRC's Thermogravimetric Analysis (TGA) and Knudsen Cell Mass Spectrometry (KCMS) equipment, both capable of "hard" vacuum (10E-6 – 10E-8 Torr) and temperatures up to 3000°C. Additional testing of carbon-carbon materials will also be performed at NASA/GSFC to determine the mass loss associated with radiation degradation.

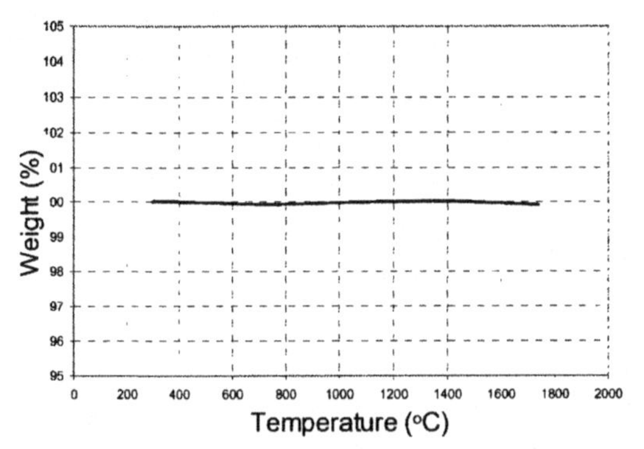


Figure 9: Mass Loss Profile of $Al_2O_3$ at Elevated Temperature

## Charging Due to Irradiation

The proposed ceramic coatings have unique charging characteristics which can dramatically impact function of spacecraft electronics and communication devices as well as sensitive science instruments. Charging depends on fundamental material characteristics including; material resistivity, dielectric constant, photoelectron emission, and secondary electron emission as well as spacecraft geometry, orientation, sunlight intensity and distribution, and specific radiation environment. It is commonly perceived that these ceramics which are typically electrically insulating at room temperature can notionally be problematic for spacecraft as they can collect charge and/or develop unmanageable voltage potentials (both magnitude and gradients) leading to arcing.

Charging for near solar spacecraft is further complicated by the requirements for specific spacecraft functions and when they occur during the trajectory. Spacecraft electronics and communication devices (which inherently can survive voltage potentials on the order of 100's of volts) need to function during the Jupiter encounter and in deep space (2.5AU) where the spacecraft is cold. During this time period, no science data is collected. However, at approximately 0.3 AU (where the spacecraft is hot) science data collection initiates and lower voltage potentials are necessary (10's of volts) [5].

A finite element model composed of more than 4000 elements was built for the charging analysis using the NASCAP-2K code. The entire Solar Probe spacecraft was modeled as having an alumina coating on top of the conductive carbon heat shield as shown in Figure 10 and was exercised over selected points of the trajectory. As an example, at 0.3 AU shows the complexity of the interaction between fundamental material response, temperature, radiation environment, and spacecraft orientation. At this point in the trajectory, the spacecraft will be pointed such that sunlight is incident evenly across the cone (the tip is pointed at the Sun) and the instrument bus is in shadow. The results indicate that the sunlit cone charges to a positive potential (due to photoemission), while the portion of the spacecraft in shadow (the bus) charges to a negative potential.

From a materials perspective, ceramic insulators become more conductive as temperature increases. Near the Sun, the insulative characteristics dramatically decrease to the point of being effectively electrically conductive. This effect creates an equilibration of the charge. The voltage potential for this location in the trajectory is < 40 volts which is within the range of instrument functional requirements. Additional studies are planned to characterize material charging responses in radiation environments with temperature effects included. Analog test articles are being designed and will be tested in both electron plasma (at the University of Puerto Rico) and electron and proton beams (at NASA/GSFC).

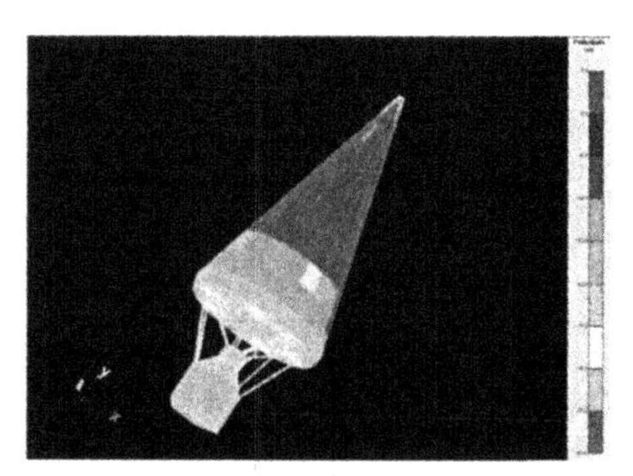

Figure 10: Entire spacecraft NASCAP results for $Al_2O_3$-coated cone at 0.3 AU radiation environment; potential difference ranges from +5.0 V (pink) to -35.0 V (orange).

## CONCLUSIONS

The Johns Hopkins University Applied Physics Laboratory has investigated ceramic coatings as a means of providing thermal management for spacecraft used in near solar exploration. These engineered coatings (such as alumina, pyrolytic boron nitride, and barium zirconium phosphate) have the unique characteristics of being strongly reflective in the visible wavelengths while being strong emitters in the near to mid IR; both of these conditions are ideal for the proposed application. Additionally, these materials are inherently stable in the space environment as they suffer minimal radiation damage, are structurally robust, and are thermodynamically stable. Initial feasibility experiments and analyses performed by JHU/APL consisting of optical property testing as a function of temperature, structural integrity testing (including thermal cycling and vibration), mass loss (including thermal vacuum and radiation effects), radiation damage testing (including electron, proton, and EUV), and spacecraft charging analyses have not identified any major "show-stoppers" associated with this technical approach. A prime mission for the application of these coatings is NASA's Solar Probe where the spacecraft will be required to survive the Jupiter environment, deep space for extended durations, and multiple trips into the extreme solar environment at $4R_s$[6].

ACKNOWLEDGEMENTS

We are grateful to the Solar Probe Science and Technology Definition team (led by Dr. Dave McComas of Southwest Research Institute) for their helpful discussions and support. We appreciate the NASA support provided by Haydee Maldonado of the NASA-Goddard Space Flight Center under NASA/GSFC contract #NAS5-01072 for the engineering trade studies. We also acknowledge Dr. John Sommerer, Dr. Joe Suter, Dr. Vic McCrary, and Dr. Ralph McNutt (all JHU/APL) for their continuing support through JHU/APL IR&D funding for the development of the fundamental material technology.

REFERENCES

[1]K. Potocki, et al, "Solar Probe Engineering Concept," Solar Wind 11 – SOHO Conference, June 2005.
[2]D. King, et al, "Alumina Optical Surface Heat Shield for Use in Near Solar Environment, " *International Journal of Applied Ceramic Technology*, in press.
[3]L. Hobbs, et al, "Radiation Effects in Ceramics," *Journal of Nuclear Materials*, volume 216, pages 291-321, 1994.
[4]Eckstein, W. and V. Philipps, "Physical Sputtering and Radiation Enhanced Sublimation," Physical Processes of the Interaction of Fusion Plasmas with Solids, p. 93-133, 1996.
[5]J. Sample, et al, "Charging of Ceramic Materials Due to Space-based Radiation Environment," *Proceedings of the 30th International Conference on Advanced Ceramics and Composites*, Cocoa Beach, FL, 2006.
[6]http://solarprobe.gsfc.nasa.gov/, *Solar Probe STDT Final Report*, prepared for NASA Headquarters, September 2005.

# Multifunctional Coatings and Interfaces

# PREPARATION OF CARBON FIBER REINFORCED SILICON OXYCARBIDE COMPOSITE BY POLYPHENYLSILSESQUIOXANE IMPREGNATION AND THEIR FRACTURE BEHAVIOR

Manabu FUKUSHIMA, Satoshi KOBAYASHI* and Hideki KITA

National Institute of Advanced Industrial Science and Technology (AIST), 2266-98 Shimo-Shidami, Moriyama-ku, Nagoya, Aichi 463-8560
*Faculty of Urban Liberal Arts, Tokyo Metropolitan University, 1-1 Minami-Osawa, Hachioji-shi Tokyo, 192-0397

ABSTRACT
Silicon oxycarbide ceramics coating on carbon single fiber and matrix composite were prepared. The silicon oxycarbide ceramic was derived from hydrolysis and condensation of phenyl group substituted alkoxysilane and the subsequent pyrolysis at 1000-1700°C. In the fractural behavior of the obtained coating or composite, the relationship between heat treatment temperature and the interface of fiber/coating was investigated in terms of fiber efficiency on the tensile strength. In single fiber, the strength and efficiency of fiber heated at 1000°C showed maximum value, and by heating above 1300°C, the reaction between fiber/coating was occurred. On the other hand, the fiber efficiency in fiber bundle heated at 1500°C showed the maximum value. This is due to the effect of fiber pull out through porous matrix, since the reaction of silica and carbon, and the following gas evolution occurred above 1300°C.

## 1. INTRODUCTION

Silicon oxycarbide ceramics have gathered the great attention because of thermal stability and their simple preparations as well as high mechanical properties. This kind of polymer derived ceramic has been generally prepared from sol gel route with the precursor of organically modified silane [1-3]. Especially, the polymerized silane is expected as the precursor of the ceramic matrix composite, because of simple process and high ceramic yields [4-5]. In addition, since the structure of polymer derived ceramic can be controlled by the pyrolysis, the heat treatment temperature is found to be sensitive to achieve the desirable properties.

In the fabricating composite materials, the interface between fiber and matrix has an important role for the strength. In the case of carbon/carbon composite, the tensile strength of single fiber composite showed the strength of 90% for the original fiber. This is called as "fiber efficiency" for the fiber itself strength. In contrast, the efficiency in fiber bundle was found to decrease to 30-40% [6]. Thus, in composite material, the improvement of the interfacial strength between fiber and matrix is an important issue to be still overcome.

In this study, the carbon fiber reinforced silicon oxycarbide matrix composite was developed. In addition, the relationship between the heat treatment temperature and the interface of fiber/matrix was investigated.

## 2. EXPERIMENTAL PROCEDURE

Phenyltriethoxysilane (PTES) from Shin-Etsu Chemical Co.,Ltd. was used as precursor silicon alkoxide. Distilled water and PTES were stirred under an ethanol solvent; with molar ratio of PTES/ethanol=1/3, PTES/$H_2O$=1/2 and HCl/PTES=0.01/1. The solution after mixing for 1hr was used as coating precursor.

The carbon fiber (PAN based fiber; Toray M46JB-12K) with high modulus was used. Before coating, fiber was heated to 600°C for removing the sizing agent. The diameter of carbon fiber after removing sizing agent was 5μm. The coating was carried out by dip-coating method of sol solution. Carbon fiber was immersed into sol and pulled up at the rate of approximately 10 cm/min and dried at 60°C for 24h. After drying, polyphenylsilsesquioxane (PPSQ) hybrid coated fiber was obtained. For coating, single fiber and bundle fibers were used. Hereafter, for the part of PPSQ derived Si-C-O ceramics in the obtained products, "coating" for single fiber and "matrix in composite" for fiber bundle are called in this work.

Then, the coated fiber and the composite were heated to 1000, 1300, 1500 and 1700°C at a heating rate of 3 °C /min and kept at the respective temperature for 2hr in an Ar atmosphere. Additionally, bulk specimens were also prepared with same process. The tensile strength of the fiber bundle composite was measured with the cross head speed of 0.1mm/min and that of single fiber composite was obtained with the cross head speed of 0.5mm/min.

The fiber efficiency was calculated by according the following equation,

$$E = 100\frac{P}{\sigma_{sf} S_f N} \qquad (1)$$

where P, $\sigma_{sf}$, $S_f$ and N indicate fracture load (N), tensile strength (MPa), cross-sectional area ($m^2$) and the number of fibers. The single fiber after desizing at 600°C was used for the above calculation, where 25 tests were carried out and the average values from Weibull plot were applied.

## 3. RESULTS AND DISCUSSION

### 3-1 Pyrolysis characterization of polyphenylsilsesquioxane

Figure 1 shows the X-ray diffraction patterns of the bulk specimens heated at 1000-1700°C. The broad pattern at around 2θ=23° is due to amorphous silica, and the peaks at 2θ=36° 60° and 72° are attributed to β-SiC. The broad pattern assigned to silica decreased with the increasing of heating temperature. At 1700°C, the peaks due to β-SiC showed less broadening and more intense. These results may suggest the structural conversion of PPSQ hybrid during heating.

Firstly, the molecular structure of PPSQ hybrid after gelation is composed from $C_6H_5SiO_{1.5}$, partially accompanying with hydroxyl and ethoxy groups. Then, hydrocarbon and hydrogen by the heating of PPSQ are evolved around 600-1000°C, resulting in the formation of amorphous structure such as silicon oxycarbide, silica and free carbon [1-2, 7-8]. Finally, around 1300-1700°C, silica and oxycarbide react with free carbon to form silicon carbide [1, 9]. In the present study, the obtained results on XRD were well consistent with above mentioned structural conversions in previous reports.

### 3-2 Properties of the coated single fiber

Figure 2 shows the load-elongation curves and the fiber efficiency of the coated single fiber heated at various temperatures. The coated fiber heated at 1000°C showed the largest elongation, and the maximum load reached about 0.055 N. In contrast, the other specimens were fractured under lower load. The specimens heated at 1300 and 1500°C was failed under 0.03 and 0.02 N, respectively.

As seen from the result of the efficiency, the fiber efficiency decreased with the increasing of heating temperature. The product heated at 1000°C showed the highest fiber efficiency of 73%. However, the values on the efficiency of the samples at 1300 and 1500°C were around 30 and 20%, respectively. This fiber efficiency suggests how much the strength of product is affected by the original strength of fiber itself. The decreasing of fiber efficiency of the products by the heat treatment above 1300°C may suggest the reaction of fiber with coated layer and the degradation of fiber. Thereby, in order to clarify the morphology of fiber during heating, SEM observations were carried out.

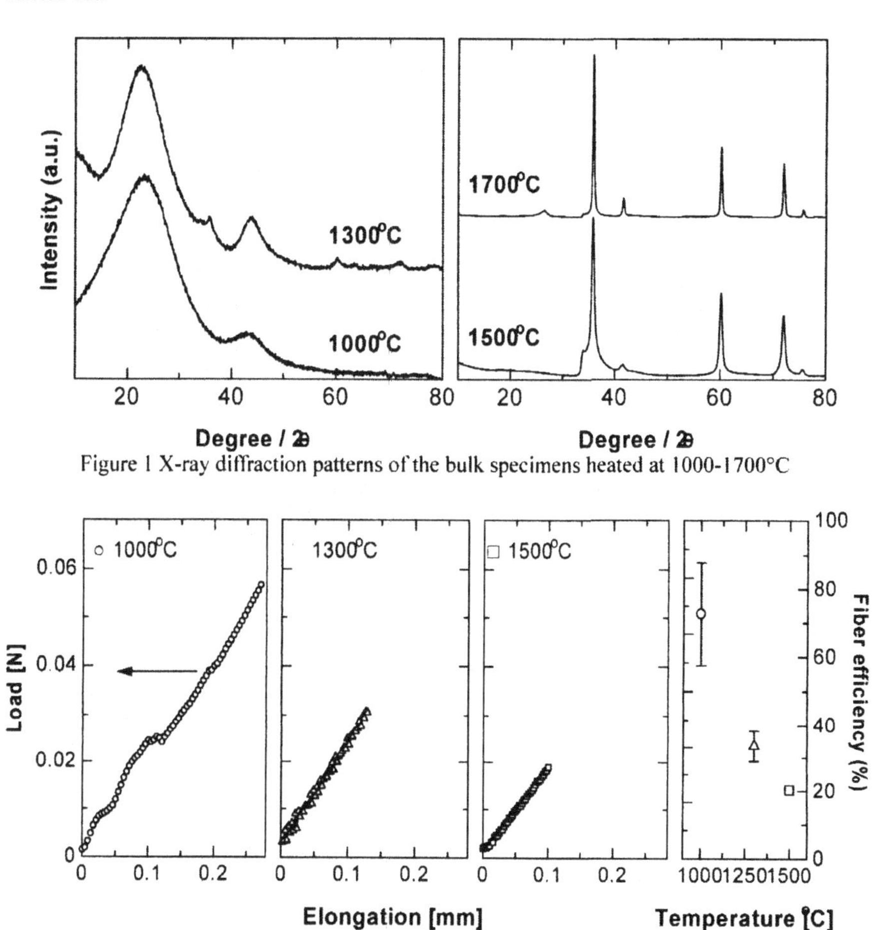


Figure 1 X-ray diffraction patterns of the bulk specimens heated at 1000-1700°C

Figure 2 Load-elongation curve (left) of the coated single fibers and their fiber efficiencies (right) as a function of heating temperatures.

Figure 3 shows the SEM observation of the fibers heated at (A) 1000, (B) 1300, (C) 1500 and (D)

1700°C. In these observations, the morphological changing of fiber by heat treatment could be clearly monitored. The smooth surface was observed for fiber at 1000°C, in contrast, the rough surface on heated fiber above 1300°C was observed. In especial, the particles on the surface of fiber heated at 1300 and 1500°C were formed. These phenomena may strongly suggest that fiber reacted with the coated layer composed of silica and oxycarbide. In general, the carbothermal reduction reaction between silica and carbon to form silicon carbide can thermodynamically occur above 1300°C, according to the equations 2-4. At 1000°C, it is considered that this reaction is difficult to be occurred. Thus, high fiber efficiency and smooth surface may be observed. However, above 1300°C, these reactions are found to progress with the consumption of carbon. In fact, the weight loss by heating above 1300°C was significantly observed.

$$SiO_2 + 3C \rightarrow 2SiC + 2CO(g) \quad (2)$$
$$SiO_2 + C \rightarrow SiO(g) + CO(g) \quad (3)$$
$$SiO(g) + 2C \rightarrow SiC + CO(g) \quad (4)$$

The evolved SiO gas (in the equation 4) has the possibilities of the recondensation to the fiber surface and the following further reaction with carbon to form SiC. In addition, the used carbon crucible and Ar flowing environment can accelerate these reactions.

The crystallization of SiC above 1300°C was monitored by XRD. As the heating temperature increases, the gradual formation of nucleation of SiC and the growth may be occurred. Therefore, the observed particles on the surface of fiber might be corresponding to the formation of SiC, though the further investigation should be carried out.

In short summary, the single carbon fiber coated by silicon oxycarbide could be fabricated through the pyrolysis of organically modified siloxane, and the coated fiber heated at 1000°C showed 70% on the fiber efficiency for strength.

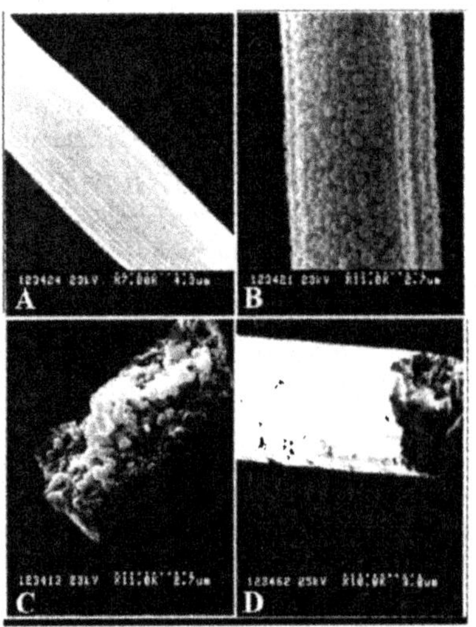

Figure 3 SEM observations of the coated single fibers heated at (A) 1000, (B) 1300, (C) 1500 and (D) 1700°C.

### 3-3 The properties of coated fiber bundle

Figure 4 shows the load-elongation curves and the fiber efficiency of composites heated at various temperatures. In contrast to the result of coated single fiber in Fig.2, the composite heated at 1500°C showed the largest elongation of 0.33 mm under 242 N. Then, the composite heated at 1000°C was fractured under 85 N. Though the strength and their efficiencies of coated single fibers were directly related with the reaction between the fiber and coated layer, for composite, the fiber efficiencies are not simply due to the reaction between fiber and oxycarbide matrix. In general, the fracture of fiber reinforced composite depends on the bonding between matrix and fiber, which can be confirmed by the microstructure such as pull out of fiber. Figure 5 shows SEM observations of the fracture surfaces of the composite heated at (A) 1000, (B) 1300, (C) 1500 and (D) 1700°C.

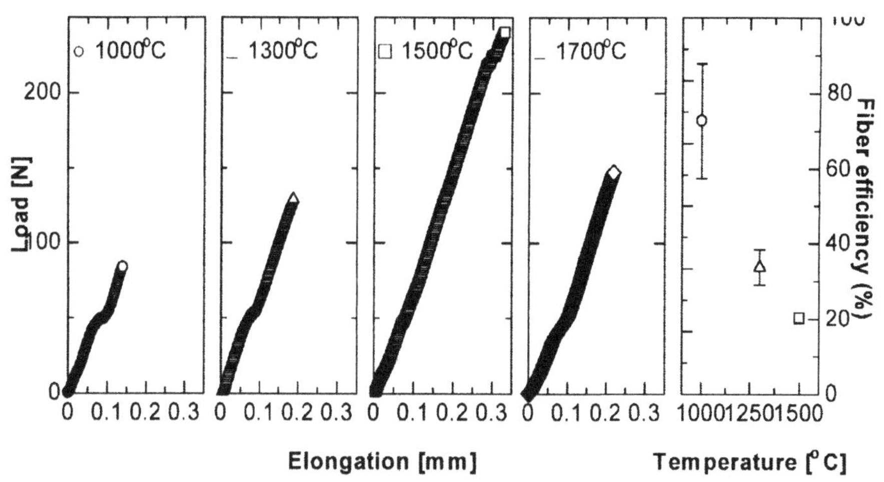


Figure 4 Load-elongation curve (left) of the coated fiber bundles and their fiber efficiencies (right) as a function of heating temperatures.

The fiber pull-out from composite heated at 1000 or 1300°C was not significantly observed and the adhesion of matrix to fiber was also confirmed. This may indicate that the bonding of matrix to fiber was strong, in other words, the fracture of matrix and fiber was simultaneously occurred by strong bonding between fiber and matrix. On the other hand, in the composites heated above 1500°C, fiber pull-out was clearly observed. In addition, these composites showed the higher fiber efficiency than those of composites heated at 1000-1300°C. Thus, the bonding between fiber and matrix was found to be quite lower than that of composites heated at lower temperatures.

In the XRD results, the crystallization rapidly progressed above 1500°C, which was thought to be related to the carbothermal reduction reaction. In addition, this reaction in coated single fiber provided the damaged fiber. Also in composite, it is reasonable to consider that the reaction between fiber and matrix occurs. In the fundamental theory of the fracture of composite, the occurrence of the reaction between matrix and fiber resulted in the strong bonding between fiber and matrix. However, by heating above 1500°C where the reaction of silica and carbon can occur, the bonding between fiber and matrix was found to be lower. In addition, the

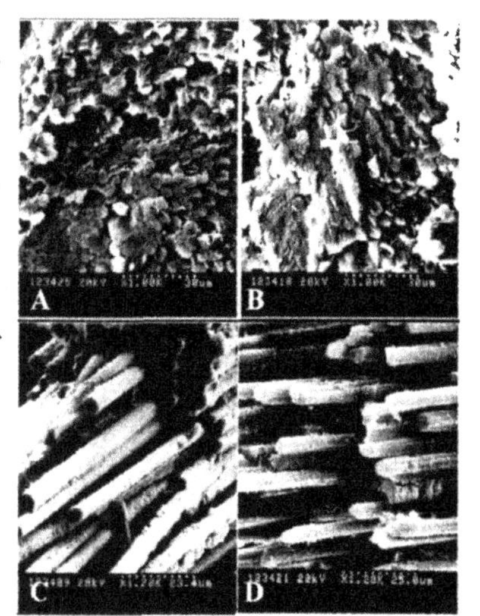


Figure 5 SEM observations of composites heated at (A) 1000, (B) 1300, (C) 1500 and (D) 1700°C.

crack produced inside matrix was grown along the fiber, since the matrix might be porous and lower strength by the gas evolution of the reaction of silica and carbon. Thus, the resultant strength is found to be higher because of the effect of fiber pull-out.

In contrast, the heat treatment around 1000-1300°C led to the simultaneous fracture of fiber and matrix due to the strong bonding between fiber and matrix.

## 4. CONCLUSIONS

In this study, silicon oxycarbide coating on carbon single fiber and fiber reinforced ceramic composite were fabricated by the heating at 1000-1700°C and their fractural behavior was examined. As the precursor of oxycarbide, phenylsilsesquioxane derived from phenylgroup substituted silicon alkoxide was used. In the fractural behavior of the obtained coating on single fiber, the relationship between heat treatment temperature and the interface of fiber/coating was investigated. The single fiber heated at 1000°C showed the smooth surface of coating and the maximum fiber efficiency for tensile strength. In contrast, the damage fiber heated above 1300°C by the reaction between fiber/coating was observed. The fiber efficiency in fiber bundle heated at 1500°C showed the maximum value. This is due to the effect of fiber pull out through porous matrix, since the reaction of silica and carbon, and the following gas evolution occurred above 1300°C.

## ACKNOWLEDGMENT

The authors would like to appreciate Mr. H.WATANABE in Tokyo Metropolitan University for his experimental supports.

## REFERENCES

(1) Fukushima,M., Yasuda,E., Manocha,L.M., Manocha,S.M., Nakamura,Y., Akatsu,T. and Tanabe,Y., J.Ceram.Soc.Jpn., 110, 12, 1044-1047 (2002).

(2) Fukushima,M., Yasuda,E., Nakamura,Y. and Tanabe,Y., J.Ceram.Soc.Jpn., 111, 11, 857-859 (2003).

(3) Fukushima,M., Yasuda,E., Nakamura,Y., Teranishi,Y., Nakamura,K. and Tanabe,Y., J.Ceram.Soc.Jpn., 112, 5, s1531-s1534 (2004).

(4) Manocha,L.M., Yasuda,E., Tanabe,Y. and Manocha,S, Adv.Compo.Mat., 9, 4, 309-318 (2000).

(5) Manocha,S, Vashistha,D. and Manocha,L.M., Key Eng.Mat., 164, 81-84 (1999).

(6) Kogo,Y., Kikkawa,A. and Saito,W., Proc.8th Japan International SAMPE Symposium, p825-828.

(7) Radovanovic,E., Gozzi,M.F., Goncalves,M.C. and Yoshida,I.V.P., J.Non-Cryst.Solids, 248, 37-48 (2000).

(8) Takamura,N., Taguchi,K., Gunji,T. and Abe,Y., J.Sol-Gel Sci.Tech., 16, 227–234 (1999).

(9) Soraru,G.D. and Sutto,D., J.Sol-Gel Sci. Tech., 14, 69-74 (1999).

# INTERFACIAL PROCESSING VIA CVD FOR NICALON BASED CERAMIC MATRIX COMPOSITES

Christopher L. Hill, Justin W. Reutenauer, Kevin A. Arpin, Steven L. Suib, and Michael A. Kmetz

University of Connecticut
Department of Chemistry
55 North Eagleville Rd
Storrs, CT, 06250

## ABSTRACT

The design of a coating interface for high temperature Ceramic Matrix Composites (CMCs) is vital to achieving the high level of operating requirements for CMCs in aerospace applications. Increasing toughness and oxidative resistance, especially for SiC based fiber reinforced composites, require a coating system with strict control over thickness and processing conditions to prevent damage during deposition and promote creep resistance. A boron nitride coating was deposited on Nicalon™ fabric using a Low Pressure Chemical Vapor Deposition (LP-CVD) reactor. Following previous investigations, an environmental protective layer consisting of $Si_3N_4$ was also deposited. SiC/SiC single tow unidirectional mini-composites manufactured by Chemical Vapor Infiltration (CVI), with and without the coating system, were processed. Correlations between thickness and tensile strengths of both the fiber and mini-composites are presented. Variations in BN and $Si_3N_4$ coating thicknesses represent a strong influence on overall tensile strengths of the processed fiber and mini-composites. Field-Emission Scanning Electron Microscope (FE-SEM) and Scanning Auger Spectroscopy were used in characterizing the interfacial composition and morphology. Thermo-Gravimetric Analysis was used to determine if a thin layer of $Si_3N_4$ could decrease the moisture problems that turbostratic BN exhibits. The overall coating process on Nicalon fiber did not significantly decrease the preprocessed fiber tensile strengths while successfully promoting high temperature usability in typical environments met by CMC components in aerospace applications.

## INTRODUCTION

Extending the temperature range for structural components involves careful selection of materials based on the thermo-mechanical properties in the appropriate environment. CMCs have been heavily investigated during the last twenty years in order to improve the thermal stability and resistance to oxidation for aerospace applications. A variety of SiC based fibers are used for reinforcement in SiC/SiC CMCs. Interfacial processing represents a critical aspect of controlling CMC properties, determining composite strength and toughness[1,2]. BN has been shown to be an acceptable material based on morphology and its ability to facilitate a weak fiber matrix bond improving composite toughness along with good oxidative resistance[3]. Extending the oxidative resistance of CVD turbostratic BN, t-BN, while maintaining desirable composite properties, has received much investigation[4-8]. Premature composite failure attributed to interface degradation occurs at intermediate temperatures, around 800 °C in dry conditions, which is exasperated by the addition of moisture[4]. t-BN deposited around 800-1000 °C at low pressure, exhibits significant instability even at room temperature in the presence of moisture[3]. These issues represent significant drawbacks for the widespread use of BN and necessitate further optimization of the interface for use in SiC/SiC composites. Increasing the degree of crystallinity of BN through post deposition heat treatment or by highly ordered deposition processing provided attractive methodology to improve oxidative properties. Although, processing of this nature on Nicalon™ may prove severe due to temperature limitations inherent to the fiber[8-10]. The doping of BN, or the use of novel precursors also warranted investigation and may lead to additional processing optimization of the standard t-BN[6, 8, 11]. Previous work utilizing the combination of BN and $Si_3N_4$ has been examined and appears to be an effective method for interfacial processing of Nicalon™ fabric[5, 6].

Selected for this investigation, a traditional t-BN coating was applied to Nicalon™ fabric with the addition of a thin layer of silicon nitride to improve wet/dry oxidative stability. Research on the interface system is complimented with the examination of unidirectional single strand composites processed from tows taken directly from the coated fabric. CVI SiC was chosen as the method to produce the SiC/SiC composites. The impact of such a coating system on thermo-mechanical properties of unidirectional composites as a test for interface compliance will be discussed.

## EXPERIMENTAL

### Fiber Coating

Interfacial processing of 8H-S CG-Nicalon™ fabric occurred in a hot walled LP-CVD reactor utilizing $BCl_3$ and $NH_3$ precursors for t-BN deposition. The hot zone length was 122 cm. Nitrogen was also used as a dilution gas. MKS™ mass flow controllers delivered the reactant and dilution gases. Pressure was maintained at 0.1 kPa with the use of an MKS™ throttling valve. Deposition temperature was approximately 850 °C monitored with thermocouples as close to fabric locations as possible, generally 5 cm away. A thin coating of silicon nitride was deposited from the reaction of $SiCl_4$ and $NH_3$. $N_2$ was again used as a dilution gas. A MKS™ electronic butterfly valve throttled pressure to maintain 0.1 kPa. Coating thickness for BN ranged from 150-800 nm while the $Si_3N_4$ was varied from 20-150 nm. Typically 1-2 meters of CG-Nicalon™ cloth was coated per run. Fabric with $Si_3N_4$ coatings thicker than 200 nm showed significant room temperature tensile strength reduction and were not used in this study.

Fiber tensile testing was carried out on tows taken from the woven processed fabric and compared with the preprocessed material. Tows were adhered with epoxy to cardboard tabs. An acrylic binder was applied to the gauge length of the tow aiding in even load transfer between the approximately 500 filaments[1]. This method has been acceptable at monitoring strength degradation strictly from a comparative standpoint. Gauge lengths and crosshead speeds were 2.54 cm and 0.508 mm/min respectively measured on an Instron model 1101. For measuring the fabric prior to interfacial processing, tows were desized with acetone. Coating thickness was measured using a FE-SEM on a Zeiss DSM 982 Gemini FE-SEM with a Schottky Emitter at an accelerating voltage range from 2 to 4 kV and a beam current of about 1 mA. Tows were sampled from multiple locations throughout the fabric. Thermo-Gravimetric Analysis, on a 951 TGA DuPont Thermal Analyzer, of the both BN and $BN/Si_3N_4$ coated fabric were examined with roughly 20 mm tows. A platinum sample holder was used. An isothermal treatment at 600°C was used, 40 °C/min ramp rate, for 250 minutes under argon, air, and air bubbled through water at 70 sccm. Scanning Auger Microscopy (SAM) was used to analyze overall concentrations of components along with depth profiles. SAM data was collected on a Physical Electronics PHI 610 scanning Auger spectrometer with a 5 kV beam voltage. SAM depth profiles corroborated with coating thickness measurements obtained with FE-SEM. Tantalum oxide standard was used to calibrate the sputtering rate for depth profiles.

Mini-composites

Single tow unidirectional composites were made by CVI in a 1" diameter atmospheric pressure reactor. Hydrogen was bubbled through methyltrichlorosilane, 100 sccm, for 5 hours at 1050 °C to produce the SiC/i/SiC mini-composite. Composites of both SiC/BN/SiC and SiC/BN/$Si_3N_4$/SiC were produced using fiber tows taken from woven fabric. Tow specimens were 5 cm in length. Matrix thickness and morphology was investigated with both FE-SEM and optical microscopy. Fiber volume fraction was measured to be around 40%. Composite tensile testing was carried out using an Instron, model 1101, with a 2.54 cm gauge length and crosshead speed of 0.254 mm/min, about 10 samples for each type of composite were tested. Fractured cross sectional areas were measured using SEM and optical microscopy. In an effort to correlate interfacial coating thickness and thermo-mechanical stability, mini-composites were subjected to wet oxidizing environments at elevated temperature of 600 °C and 1000 °C for 24 hours. Both ends of the mini composites were cut exposing fibers before heat treatment to allow the wet oxidizing conditions, 5% $H_2O$ in air, to have access to the interface. Post oxidation tensile testing was then carried out to investigate interfacial degradation leading to any change in composite properties.

RESULTS

Chemical analysis of BN/$Si_3N_4$ coating on Nicalon™ fabric utilizing SAM clearly showed two distinct interphases between the outer $Si_3N_4$/BN coating and BN/Fiber region. Figure I depicts the SAM depth profile for typical coating processes described.

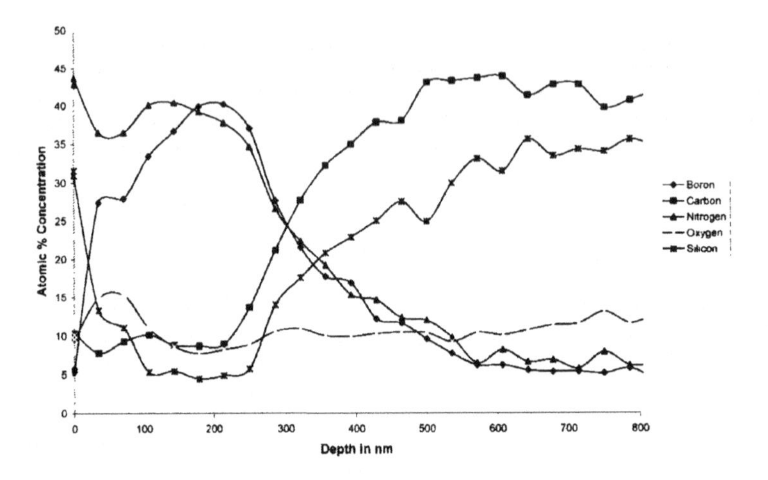


Figure I. Scanning Auger Depth Profile for $BN/Si_3N_4$ coated on a Nicalon$^{TM}$ Fiber

The slight increase in oxygen concentration at the $Si_3N_4/BN$ interphases may be due to the removal of the fabric from the reactor for analysis prior to the final $Si_3N_4$ deposition. The observed oxygen increase may not be indicative of a continuous coating process. Two pre-sputter surveys were taken prior to ion milling of the sample.

After chemical analysis, coated fibers were examined for thickness and morphology. FE-SEM micrograph of coating morphology is shown in Figure II.

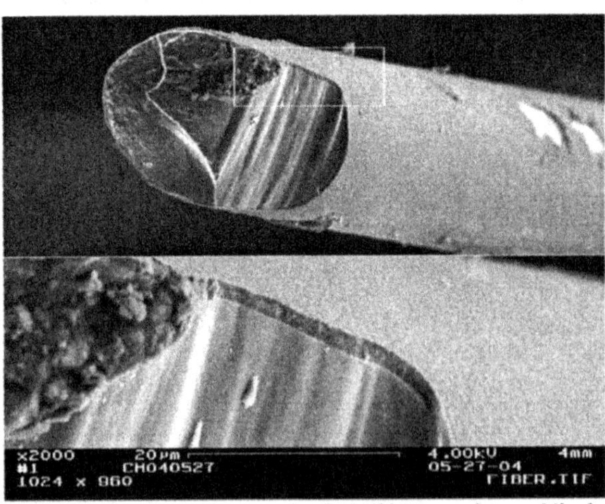


Figure II. FE-SEM micrograph of $BN/Si_3N_4$ coating on Nicalon$^{TM}$ Fiber

Quality of the interfacial deposition represents an extremely important part of the fabrication of composites with desirable properties. Deposition conditions including processing temperature and precursor selection can have adverse effects on fiber surfaces unless strict control of interfacial processing is demonstrated. Previous investigations on composite failure led to the conclusion that nearly all fiber failure occurs due to surface flaws or defects[2]. Fabric selected for tensile testing and composite processing exhibited uniformity in thickness and morphology, shown in Figure II, while deposition times were varied to produce material with different interfacial thicknesses.

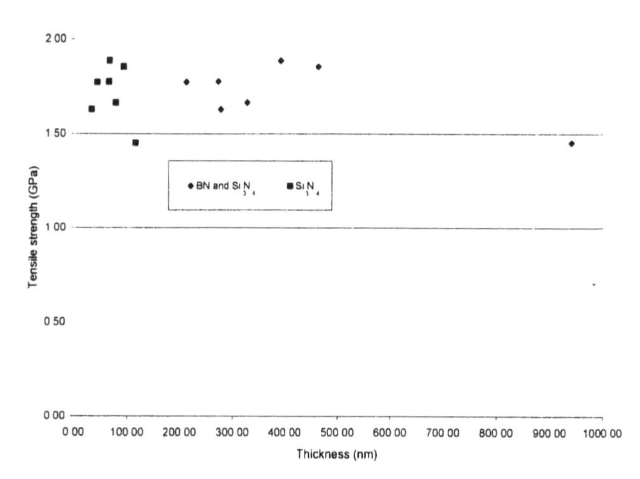


Figure III. Tow tensile strength variation with coating thickness

The variability of coating thickness with tensile strength taken from over thirty coated fabric samples is shown in Figure III. Using the tow test method, the as-received fiber strengths were 2.02 GPa. Published results for CG-Nicalon are in the 3.0 GPa range[12]. Woven fabric may exhibit a lower tensile strength caused during fabric processing. Scattering of the tensile-thickness data (Figure III) suggests a limited or no correlation between total coating thickness and tensile strength for the coating system and deposition parameters in this investigation. Fabric coated with only BN showed similar results. The $Si_3N_4$ overcoat for fibers of equal BN coating thickness continued to display significant scatter within the range of 50-100 nm. Average tow tensile strength was 1.72 GPa, a 15% decrease from the as-received fabric. This could be attributed to the high modulus silicon nitride coating acting as a crack initiator. Other investigations reported a much higher reduction in coated fiber tensile strength using interfacial systems with composite or duplex coatings differing from the deposition described here[5,6,8,11].

Regarding thermal analysis involving the sampling of material with different coating thicknesses, only processed fibers treated in the TGA with the air/water mixture displayed a difference in thermal stability.

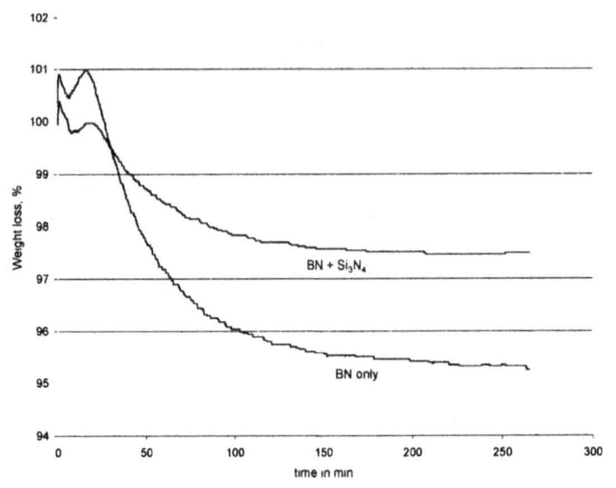


Figure IV. TGA of both BN and BN/$Si_3N_4$ coated Nicalon™ tows

TGA plots for both BN and BN/$Si_3N_4$ coated tows showed a nearly 2 % mass loss difference between the t-BN coating and the t-BN with $Si_3N_4$. This was observed for multiple runs. Since the maximum temperature and analysis time were limited to 600 °C and 250 minutes, the lack of a weight gain may not be indicative of the thermal stability of the coated fibers. Thermal studies need to be carried out for longer times to investigate the eventual oxidation of the fiber. The loss in mass, presented in Figure IV, may be attributed to the evolution of $NH_3$ and $H_2O$ from the coating. The BN/$Si_3N_4$ appeared to improve the thermal stability and may improve the wet oxidative capabilities of the BN interface.

Microscopy of the post oxidized fibers presented in Figures V and VI shows the possible effect of wet oxidation on BN and the lack there of on BN/$Si_3N_4$ coated tows.

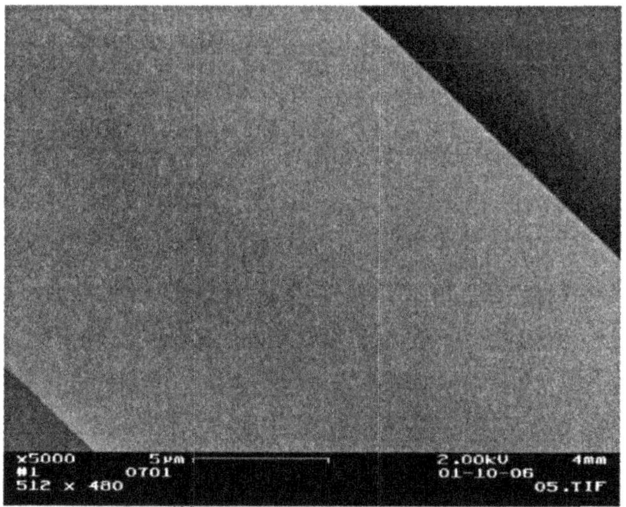


Figure V. Nicalon™ tow coated with BN and $Si_3N_4$ after wet oxidation at 600 °C

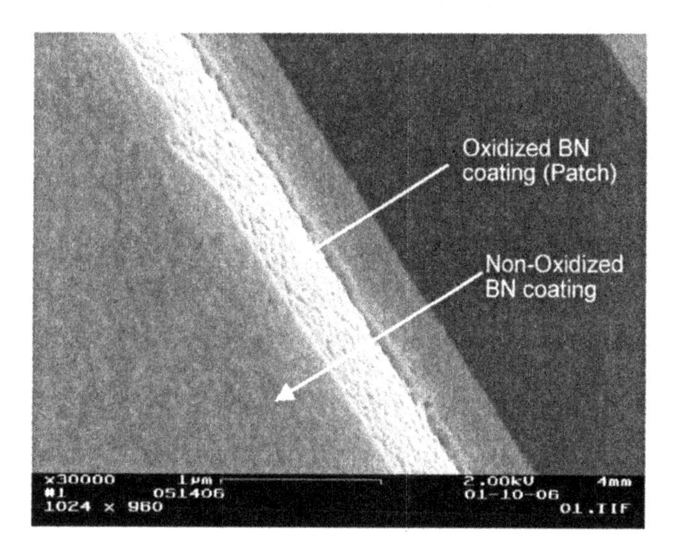


Figure VI. Nicalon™ tow coated with t-BN after wet oxidation at 600 °C

Multiple fibers in the t-BN coated tow exhibited the distinct patches, Figure V, several microns in length. No features shown in Figure VI were observed in the BN/$Si_3N_4$ coated fibers (shown in Figure V). These patch features were never observed in other as-possessed fibers and have been attributed to the oxidation of BN. It is also interesting to note, there were no signs of a debris field indicating that debonding of the interface did not occur. Therefore it is postulated that the wet oxidation is responsible for the patch features found on BN coated tows

Unidirectional single strand mini composites produced with a CVI SiC matrix experienced differing thermo-mechanical properties. Much investigation into the damage mechanisms involved in SiC/BN/SiC CMCs have potentially been explained where composite toughness was reduced in the intermediate temperature range of 600-1000 °C[11, 13-15]. Composites with the duplex coating exhibited lower fracture strength. The reason for this is inconclusive at this time. From the composite survey investigating varying thicknesses of the $Si_3N_4$ outer layer (<200 nm), composite tensile strength exhibited similar scatter. Figures VII and VIII are shown depicting fractured sections of mini composites for BN & BN/$Si_3N_4$ coated tows.

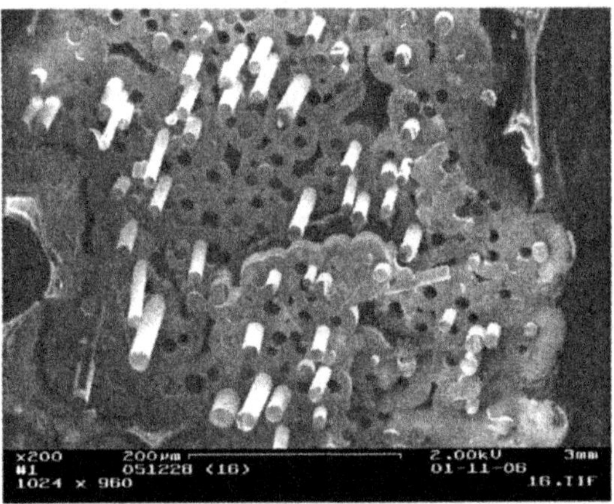


Figure VII. Room temperature fracture of SiC/BN/SiC composite (non-oxidized)

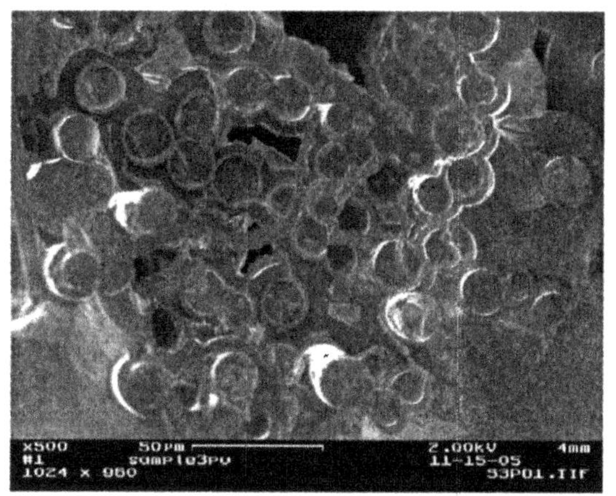

Figure VIII. Fractured $SiC/BN/Si_3N_4/SiC$ (non oxidized)

From the two micrographs of the fractured mini-composites shown in Figures VII and VIII, the composite with the BN only interface (Figure VII) demonstrates significantly more fiber pullout compared with the $BN/Si_3N_4$ interface composite (Figure VIII). The coating composition played a role in reducing the overall strength of the $BN/Si_3N_4$ coated fiber reinforced mini-composites. Figure IX shows the percent contribution of the $Si_3N_4$ top layer to the duplex coating as a function of composite ultimate tensile strength (UTS) within the coating thickness parameters. In addition, the duplex coating did not overly degrade the mechanical properties of the Nicalon$^{TM}$ fabric shown by tow testing (15% reduction). This may indicate that the reduced toughness of the $BN/Si_3N_4$ composites is a result of a less compliant interface and not due to strength degradation caused by interfacial processing.

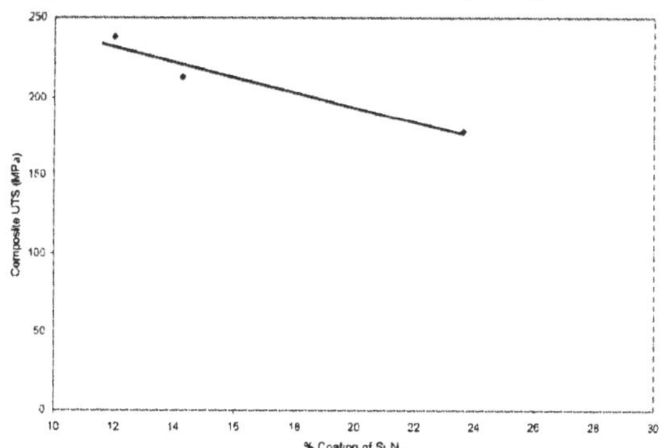


Figure IX. Percent composition of $Si_3N_4$ in $BN/Si_3N_4$ coating with decrease in strength of mini-composites

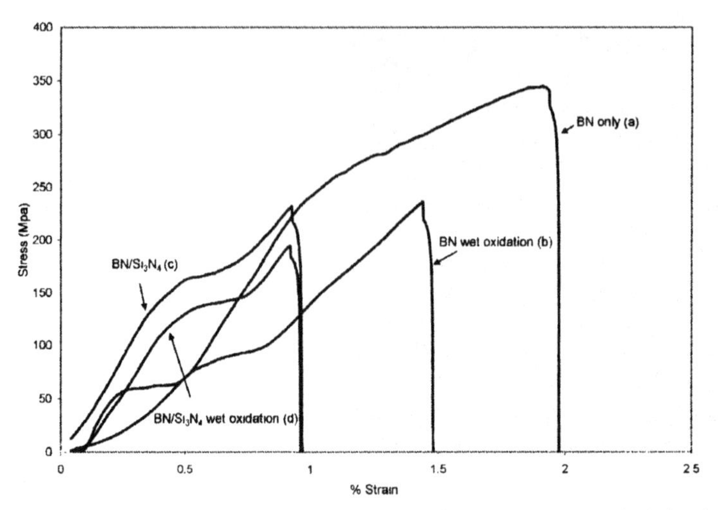


Figure X. Room temperature and wet oxidation (24 hr) stress-strain behavior for $SiC/BN/Si_3N_4/SiC$ and SiC/BN/SiC composites

There is difficulty in predicting composite behavior based on coating thickness for the dual $BN/Si_3N_4$ interface, although there may be a trend in the reduction in ultimate tensile strength. (UTS), with increasing silicon nitride coating thickness (Figure IX). Fiber tensile strength was severely degraded when silicon nitride thickness approached 200 nm. This is consistent with the reduced UTS for the mini-composites fabricated.

Room temperature stress-strain plots for a BN and $BN/Si_3N_4$ interface unidirectional single strand mini-composite are shown in Figure X. The composite fabricated with a BN only interface, **a**, had a proportional limit, $\sigma_{PL}$, around 250 MPa (36 KSI) and a UTS of 350 MPa (51 KSI) strained to failure at 2%. Comparing the stress-strain behaviors of **a** and **b**. both the ultimate tensile strength and toughness of the treated composite were decreased. Plots **c** and **d** are for the $BN/Si_3N_4$ coated reinforced fiber composites which exhibit typical stress-strain behavior for this type of CMC[16]. UTS for the non oxidized and wet oxidized composites are 240 MPa (34 KSI) and 200 MPa (29 KSI), respectively. A difference in $\sigma_{PL}$ for **c** and **d** occurred following wet oxidation; the $\sigma_{PL}$ of 145 MPa (21 KSI) and 119 MPa (17.2KSI) was determined for untreated and wet oxidized, respectively. Although **a-d** exhibit composite behavior, there appears to be a less significant reduction in composite properties exhibited for $BN/Si_3N_4$ coated fibers of composites **c,d** than for BN coated CMCs represented in plots **a** and **b**. This may be attributed to the higher thermal stability of the $BN/Si_3N_4$ interface demonstrated by the thermal analysis of the two coatings.

## CONCLUSION

Tow testing was used to show the effects of two different interfaces on the strength of Nicalon™ cloth. The deposition of t-BN with and without $Si_3N_4$ on Nicalon™ fabric displayed roughly a 15% tensile strength decrease from preprocessed fabric, 2.02 GPa down to 1.72 GPa. Tow testing of the wide survey indicated little to no correlation between coating thickness and tensile properties for the processed material based on the variation of thicknesses. With the dual BN/$Si_3N_4$ interfacial system, uniform coatings were produced with no apparent fiber bridging and the $Si_3N_4$ layer was deposited evenly throughout the fabric examined by SAM.

Wet oxidative studies potentially indicated a protection offered by the outer $Si_3N_4$ layer during the initial oxidation of BN during thermal analysis. Micrographs of BN only coated fibers post TGA treatment displayed unusual patch formations not present on dual coated BN/$Si_3N_4$ fibers of any thickness.

Single strand unidirectional mini-composites fabricated from tows of woven Nicalon$^{TM}$ fabric were processed and exhibited fracture strengths around 34 KSI (240 MPa) for BN/$Si_3N_4$ and 51 KSI (350 MPa) for BN coated reinforced composites. From the wet oxidation studies on the composites, a significant reduction in fracture toughness was exhibited by the BN coated fiber reinforced composite where a less prominent decrease was exhibited by the BN/$Si_3N_4$ coated material. The combination of $Si_3N_4$ and BN as an interfacial system for SiC/SiC composites may be an attractive CVD processing methodology for CMC fabrication.

## ACKNOWLEDGEMENTS

The authors would like to thank Kathleen Sinnamon for mechanical testing and Pratt & Whitney for supporting this work.

## REFERENCES

[1]M. Kmetz, J. Laliberte, W. Willis, and S. Suib, "Synthesis, Characterization, and Tensile Strength of CVI SiC/BN/SiC Composites," *Ceram. Eng. Sci. Proc.*, **12**, 2161-2174 (1991).

[2]J. Singh, D. Singh, and M. Sutaria, "Ceramic composites: roles of fiber and interface," *Composites; Part A*, **30**, 445-450 (1999).

[3]S. Gallet, G. Chollon, F. Rebillat, A. Guette, X. Bourrat, R. Naslain, M. Couzi, and J. Bruneel, "Microstructural and microtextural investigations of boron nitride deposited from $BCl_3$-$NH_3$-$H_2$ gas mixtures," *J. European Ceram. Soc.*, **24**, 33-44 (2004).

[4]T. Matsuda, "Stability to moisture and chemically vapour-deposited boron nitride," *J. of Mater. Sci.*, **24**, 2353-2358.

[5]W. Kowbel, and L. Tsou, "A Chemical Vapor Deposition (CVD) BN-$Si_3N_4$ Interfacial Coating for Improved Oxidation Resistance of SiC-SiC Composites," *J. Mater. Synt.*, **3**, 121-131 (1995).

[6]N. Bansal and R. Dickerson, "Tensile strength and microstructural characterization of HPZ ceramic fibers," *Mater. Sci. Eng.*, **A222**, 149-157 (1997).

[7]F. Rebillat, A. Guette, R. Naslain, and C. Brosse, "Highly Ordered Pyrolytic BN Obtained by LPCVD," *J. European Ceram. Soc.*, **17**, 1403-1414 (1997).

[8]F. Hurwitz and R. Shinavski, "Alternative interphase coatings for improved durability of SiC/SiC composites," *Ceram. Eng. Sci. Proc.*, **24**, 231-237 (2003)

[9]S. Jacques, A. Lopez-Marure, C. Vincent, H. Vincent, and J. Bouix, "SiC/SiC minicomposites with structure-graded BN interphases," *J. European Ceram. Soc.*, **20**, 1929-1938 (2000).

[10]F. Rebillat, A. Guette, L. Espitalier, C. Debieuvre, and R. Naslain, "Oxidation Resistance of SiC/SiC Micro and Minicomposites with a Highly Crystallised BN Interphase," *J. European Ceram. Soc.*, **18**, 1809-1819 (1998).

[11]F. Hurwitz, J. Scott, and P. Chayka, "Progress of BN and doped-BN coatings on woven fabrics," *Ceram. Eng. Sci. Proc.*, **22**, 389-397 (2001).

[12]G. Kister and B. Harris, "Tensile properties of heat-treated Nicalon and Hi-Nicalon fibres," *Composites: Part A*, **33**, 435-438 (2002).

[13]R. Bhatt, Y. Chen, and G. Morscher, "Microstructure and tensile properties of BN/SiC coated Hi-Nicalon, and Sylramic SiC fiber performs," *J. of Mater. Sci.*, **37**, 3991-3998 (2002).

[14]G. Morscher, D. Bryant, and R. Tressler, "Environmental durability of BN-based interphases (for $SiC_f/SiC_m$ composites) in $H_2O$ containing atmospheres at intermediate temperatures," *Ceram. Eng. Sci. Proc.*, **18**, 525-534 (1997).

[15]M. Lebaroux, L. Vandenbulcke, V. Serin, J. Sevely, S. Goujard, and C. Robin-Brosse, "Oxidizing Environment Influence on the Mechanical Properties and Microstructure of 2D-SiC/BN/SiC Composites Processed by ICVI," *J. European Ceram Soc.*, **18**, 715-723 (1998).

[16]J.W. Holmes, B.F. Sørensen, "Fatigue Behavior of Continuous Fiber-Reinforced Ceramic Matrix Composites"; pp 261-326 in *High Temperature Mechanical Properties of Ceramic Matrix Composites*. Edited by S.V. Nair, K. Jakus Butterworth-Heinemann, 1995.

# COATINGS OF Fe/FeAlN THIN FILMS

Yuandan Liu, R. E. Miller, Tao Zhang, Qiquan Feng, W. Votava, Dingqiang Li, L. N. Dunkleberger, X. W. Wang*
Alfred University
Alfred, NY 14802

R. Gray, T. Bibens, J. Helfer
BTI
Rochester, NY 14586

K. Mooney, R. Nowak
SUNY Buffalo
Buffalo, NY 14260

P. Lubitz
Naval Research Laboratory
Washington, DC 20375

Yanwen Zhang
Pacific Northwest National Laboratory
Richland, WA 99352

## ABSTRACT

We report new results on materials properties of Fe/FeAlN thin films. Films were fabricated via a pulsed DC sputtering technique. Sputtering target materials were $Fe_xAl_{1-x}$, where x varies from 0.025 to 1. Film thickness varied from 0.01 to 4 micrometers, and particle size varied from 5 to 100 nanometers, depending on the fabrication conditions and target materials. A film with 22 layers was also fabricated via the sputtering technique. Materials properties of the films were analyzed by SEM/EDS, XRD, RBS and ESCA (XPS).

## INTRODUCTION

Since 1990, nano-sized materials have generated substantial interests in many areas for applications.[1,2] In our previous work, nano-magnetic FeAl, FeAlO and FeAlN films with different Fe/Al ratios were fabricated via a sputtering technique.[3-6] It was observed that magnetic properties of the films are related to chemical composition, film thickness and fabrication conditions. When the iron molar ratio in bulk FeAl alloy is less than 70%, the bulk materials exhibit very weak magnetism.[7] In contrast to the bulk materials, a thin film material and/or a nano-sized material may exhibit different magnetic properties.[8-9] For example, a nano-magnetic material may exhibit larger coercive force than the corresponding bulk material.[9] A sputtering fabrication technique is utilized in this study to fabricate Fe/FeAlN thin films. In this study, chemical compositions and sublayer structure of FeAlN thin films with different Fe/Al ratio are analyzed via Rutherford backscattering spectroscopy (RBS). The films are also analyzed by SEM/EDS, XPS and XRD.

## EXPERIMENTAL PROCEDURE

A Kurt J. Lesker Super System III deposition system outfitted with Lesker Torus magnetrons was utilized for the process[10] as shown in Figure 1. The vacuum chamber of the system is cylindrical, with a diameter of approximately one meter and a height of approximately 0.6 m. The base pressure is 2 μTorr. In these experiments, each target is a disk with a diameter of approximately 0.07-0.1 m. The sputtering gas is argon. A pulsed DC power source is utilized at a power level of 500-2,000 W. The magnetron polarity switches from negative to positive at a frequency of 100 KHz, while the pulse width for the positive or negative duration can be adjusted to yield suitable sputtering results.

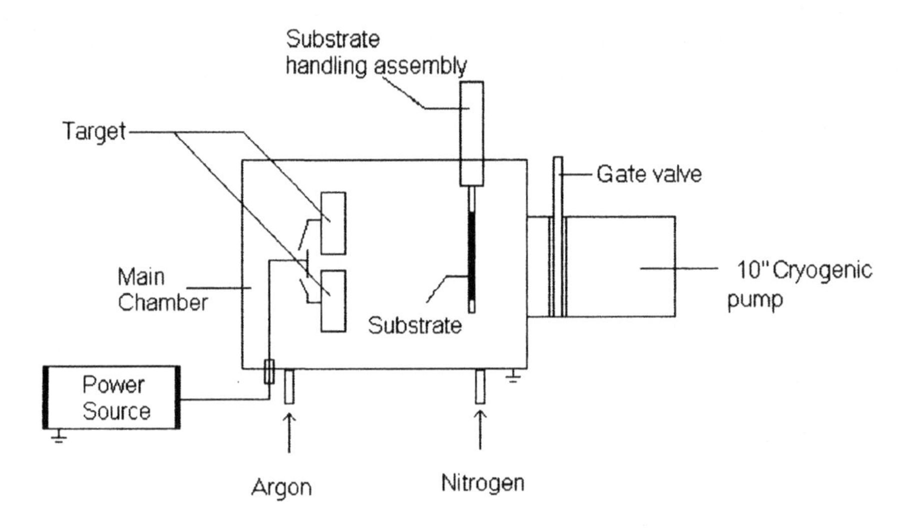


Figure 1. Schematics of the sputtering system (not to scale)

Weight ratios of Fe/Al in the targets are respectively: 5/95(L), 11/89(W), 17/83(W), 24/76(W), 67/33(W), 82.5/17.5(W) and 100/0(L).[11] Besides argon flowing at a rate of 15-45 sccm, nitrogen is supplied as a reactive gas with a flow rate of 15-50 sccm. During fabrication, the pressure is maintained at 2-10 mTorr. This pressure range is found to be suitable for nano-magnetic material fabrication. The substrates are silicon wafers and fused silica slides. The flat silicon wafers are bare without a thermally grown silicon dioxide layer, and have a diameter of 0.1-0.15 m or less. The shape of the flat fused silica slides is square with a dimension of 1 inch ×1 inch. The distance between the substrate and the target is 0.05-0.26 m. During the deposition, the wafer or the slide is fixed onto a substrate holder. The substrate holder is rotated at a rotational speed of 0.01-0.1 rps, and is moved slowly up and down along its symmetrical axis at a maximum speed of 0.01 $m.s^{-1}$. To achieve a film deposition rate on the flat wafer of 0.5 $nm.s^{-1}$, the power required for the AlN films and the FeAlN films is 500 W. Film thickness is between 100 nm and 4 μm with a typical deposition time between 200 and 2,000 s. In some film structure designs multilayers are fabricated. The thickness and compositions of the films are measured by a scanning electron microscope (FEI Quanta 200F ESEM) equipped with an energy dispersive X-ray system (Oxford Inca EDX Premium Si). To estimate the Fe/Al ratios, K-lines obtained in EDS are utilized. The phase formation and grain size estimation are analyzed using X-ray diffraction (Siemens D500 X-ray diffractometer) and a transmission electron microscope (JEOL JEM-2000FX). The chemical composition and depth profile of the films are analyzed by Rutherford backscattering spectroscopy (RBS) and an X-ray photoelectron spectrometry (XPS, PHI Quantera SXM).

## RESULTS

### SEM/EDS

Figure 2 shows a cross-sectional view of a two-layer film (FeAlN/AlN) with a film thickness of approximately 4 μm. For the FeAlN layer fabricated with 82.5wt%Fe-17.5wt%Al target, the thickness is approximately 2.4 μm. For the AlN layer, the thickness is approximately 1.6 μm. Figure 3 shows a cross-sectional view of another FeAlN film fabricated with 82.5wt%Fe-17.5wt%Al target, which has a film thickness of approximately 1.75 μm. These two images reveal that the FeAlN/AlN and/or FeAlN growth is somewhat columnized. The chemical compositions of thin films are analyzed by EDS with a spot size of approximately 1 μm. An EDS spectrum acquired from the middle of cross-sectional area in Figure 3 is shown in Figure 4. Peaks of nitrogen, iron and aluminum are from the FeAlN film. The peak of carbon is presumably due to surface contamination, and oxygen peak also is presumably due to surface contamination and film oxidation during or after deposition.

In Figure 5, a cross-sectional view of a film with 22 layers is illustrated. From the bottom to the top, the layer sequence is FeAlN, ten repeats of AlN and Fe, and AlN. Portions of the first layer on the bottom and the 22nd layer on the top are shown respectively. The FeAlN layer is fabricated with 11wt%Fe-89wt%Al target and the Fe layer is fabricated with 100wt%Fe target. Total thickness of the film is approximately 1.1 μm. The thickness of the 1st layer of FeAlN is approximately 100 nm. The thickness of the 2nd layer of AlN is approximately 100 nm. For the ten repeats of AlN and Fe, the thickness of each Fe layer (layer #3, 5, 7, 9, 11, 13, 15, 17, 19 and 21) and the thickness of each AlN layer (layer# 4, 6, 8, 10, 12, 14, 16, 18, 20 and 22) are kept nearly constant, respectively. Figure 6 shows an enlarged cross-sectional view of the film. The columnized growth is somewhat truncated. According to the EDS measurement, iron concentration varies along the depth of the film from the top to the bottom.

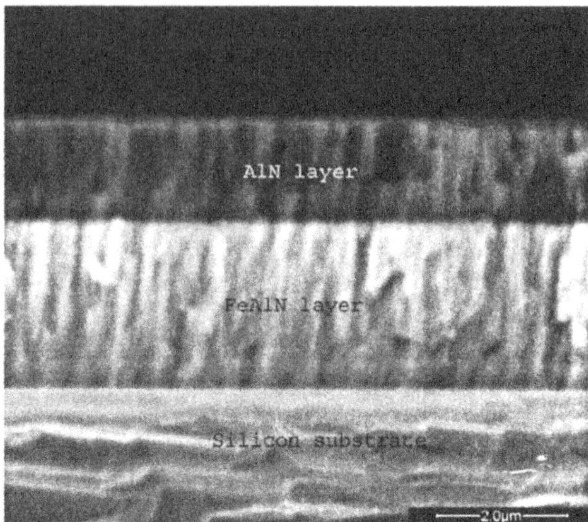


Figure 2. Cross-sectional SEM image of a two-layer film (FeAlN/AlN). The FeAlN layer is fabricated with the 82.5wt%Fe-17.5wt%Al target.

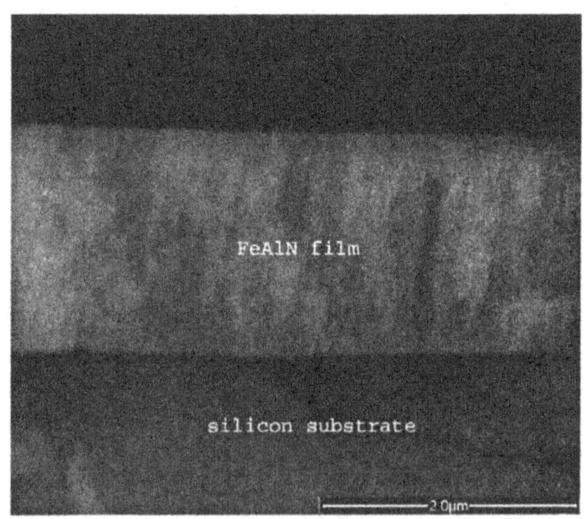


Figure 3. Cross-sectional SEM image of an FeAlN film. The FeAlN film is fabricated with the 82.5wt%Fe-17.5wt%Al target.

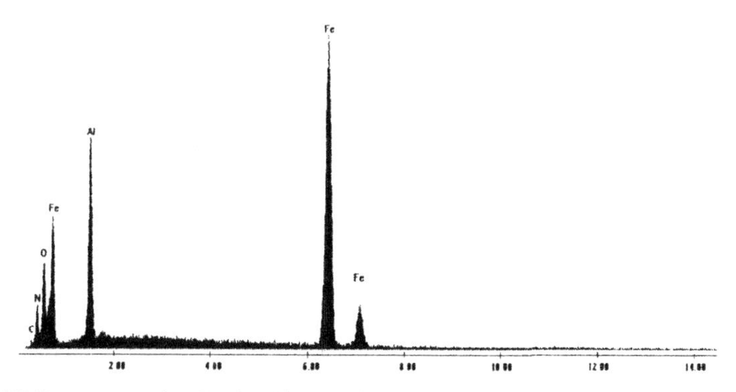


Figure 4. EDS spectrum of an FeAlN thin film fabricated with the 82.5wt%Fe-17.5wt%Al target.

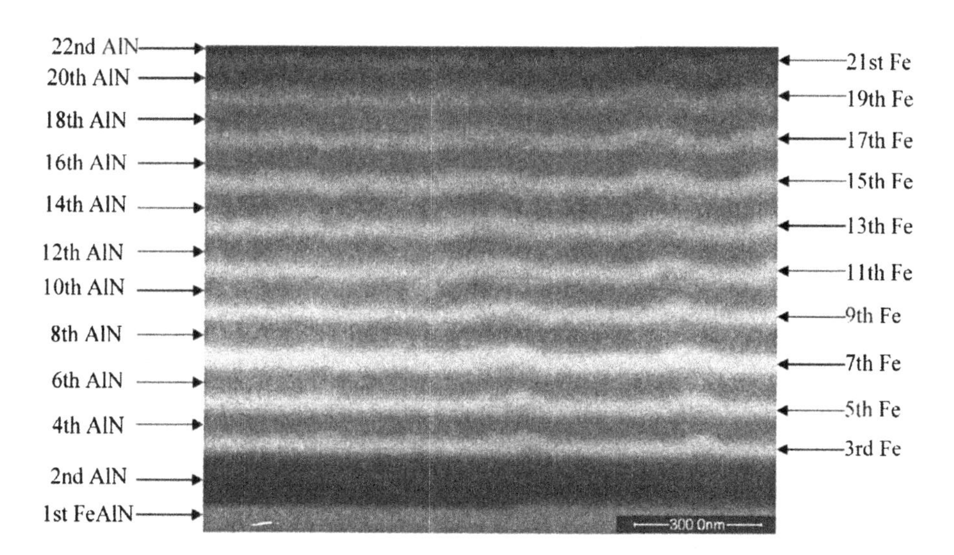


Figure 5. Cross-sectional SEM image of a film with 22 layers. For the ten repeats of AlN and Fe, the bright area is Fe layer and the dark area is AlN layer. The top AlN layer (22nd layer) is partially shown. The bottom FeAlN film (1st layer, partially shown) is fabricated with the 11wt%Fe-89wt%Al target.

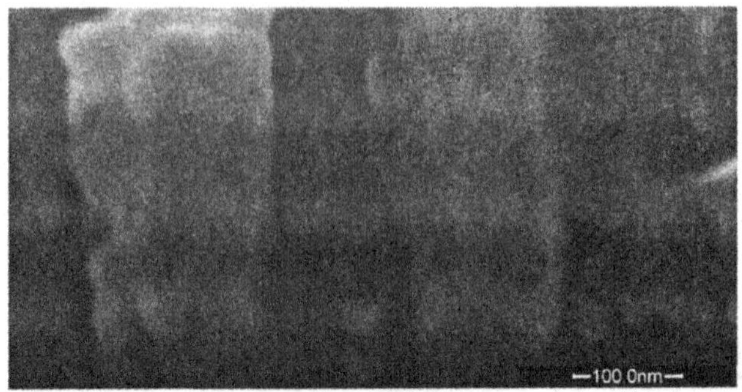


Figure 6. Enlarged and selected cross-sectional SEM image of the film in Figure 5 to show truncated columnized growth.

XPS/RBS

A typical area for XPS analysis was approximately 3mm × 3mm. An argon ion gun at 5keV was used to sputter the surface of the film to remove the contaminated surface and to reveal the depth profile. The atomic concentrations of the films were calculated using C (1s), N (1s), O (1s), Si (2p), Al (2p) and Fe (2p) photoelectrons. In Figure 7, an XPS spectrum of AlN layer acquired from the surface of the film with 22 layers is shown. Aluminum, nitrogen and/or oxygen signatures are revealed. The additional argon signature is presumably due to the sputtered argon ion. Depth profile of the film with 22 layers shows that the iron concentration varies, which is consistent with that of the EDS measurement. In Figure 8, an XPS spectrum of an FeAlN film deposited with 11wt%Fe-89wt%Al target is shown. Aluminum, iron, nitrogen and oxygen signatures are revealed. The additional argon signature is also presumably due to the sputtered argon ion.

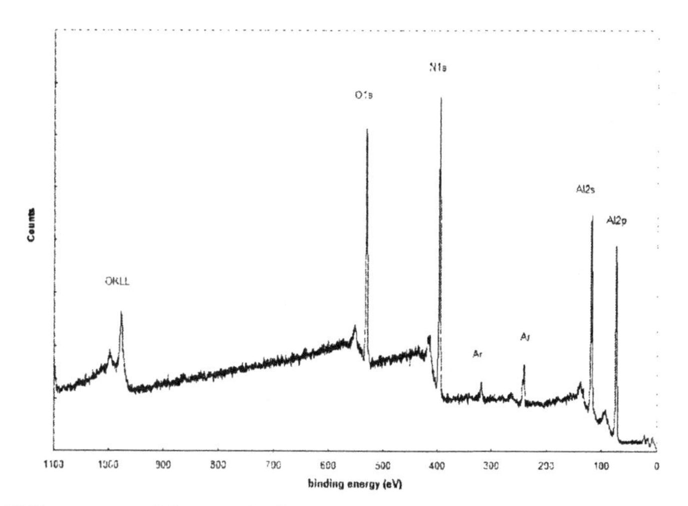


Figure 7. XPS spectrum of the top AlN layer from the film with 22 layers.

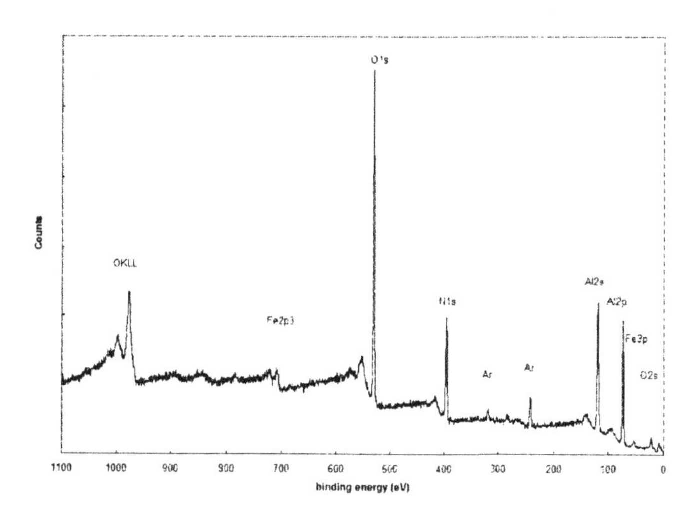


Figure 8. XPS spectrum of an FeAlN film fabricated with the 11wt%Fe-89wt%Al target.

RBS measurements with 2.0 MeV $He^+$ beam were carried out using the 3.0 MV tandem accelerator facilities within the Environmental Molecular Sciences Laboratory (EMSL) at the Pacific Northwest National Laboratory (PNNL), with a Si detector located at a scattering angle of 165° relative to the incoming beam. RBS results are shown in Tables I and II. Table I tabulates film structures in sublayers of four films determined by RBS technique. Each film has two or more sublayers. Sublayer 1 (close to silicon substrate) in each film is less than 50 nm.

For four films fabricated with the targets of 5, 11, 17 and 67wt%Fe, the average iron concentrations in the films are somewhat close to the target material concentrations.

For sublayer1, the iron concentration in the film with 11wt%Fe target is substantially the same as that of the target materials; the iron concentration in the film with the 5wt%Fe target is slightly higher than that of the target materials; the iron concentration in the film with 17wt%Fe target is lower than that of the target materials; the iron concentration in the film with the 67wt%Fe target is substantially lower than that of the target.

For the film with the 11wt%Fe target, the iron concentration of sublayer 2 is the same as that of the sublayer 1. For the film with the 5wt%Fe target, iron concentration in sublayer 2 is slightly lower than that of sublayer 1. For the film with the 17wt%Fe target, the iron concentration of sublayer 2 is larger than that of sublayer 1. For the film with the 67wt%Fe target, the iron concentration of the sublayer 2 is larger than that of the sublayer 1. The iron concentration of each of the subsequent sublayers is larger than that of the previous sublayer.

In Table II, the chemical compositions of sublayers in each film are summarized in terms of Fe, Al, N, O and Si (substrate). Based on the chemical composition tabulated in Table II, sublayer 1 is related to the interfacial layer between the film and the silicon substrate.

Table I. Film structure of four films.

| Target nominal composition Fe/Al in weight ratio | Nominal film | Film structure | Thickness ($\times 10^{15}$ atom/cm$^2$) | Thickness scaling to Angstrom | Measured Fe/Al in weight ratio | Average Fe/Al in weight ratio |
|---|---|---|---|---|---|---|
| 5/95 | FeAlN | sublayer 2 | 7400 | 8689 | 6/94 | 6.1/93.9 |
| | | sublayer 1 | 300 | 352 | 8/92 | |
| 11/89 | FeAlN | sublayer 2 | 8550 | 10105 | 12/88 | 12.0/88.0 |
| | | sublayer 1 | 224 | 265 | 12/88 | |
| 17/83 | FeAlN | sublayer 2 | 8896 | 10225 | 17/83 | 16.9/83.1 |
| | | sublayer 1 | 224 | 257 | 12/88 | |
| 67/33 | FeAlN | sublayer 4 | 11240 | 13019 | 66/34 | 64.0/36.0 |
| | | sublayer 3 | 800 | 927 | 54/46 | |
| | | sublayer 2 | 300 | 347 | 42/58 | |
| | | sublayer 1 | 200 | 232 | 24/76 | |

Table II. Chemical composition of four films.

| Target nominal composition Fe/Al in weight ratio | Nominal film | Film Structure | Fe (atm%) | Al (atm%) | N (atm%) | O (atm%) | Si (atm%) |
|---|---|---|---|---|---|---|---|
| 5/95 | FeAlN | sublayer 2 | 1.5 | 45.0 | 53.5 | 0.0 | 0.0 |
| | | sublayer 1 | 1.0 | 25.0 | 27.5 | 25.0 | 21.5 |
| 11/89 | FeAlN | sublayer 2 | 3.0 | 45.0 | 52.0 | 0.0 | 0.0 |
| | | sublayer 1 | 1.07 | 16.3 | 33.1 | 38.6 | 11.0 |
| 17/83 | FeAlN | sublayer 2 | 4.58 | 45.0 | 30.8 | 19.6 | 0.0 |
| | | sublayer 1 | 1.07 | 16.3 | 33.1 | 38.6 | 11.0 |
| 67/33 | FeAlN | sublayer 4 | 21.6 | 22.8 | 35.1 | 20.5 | 0.0 |
| | | sublayer 3 | 13.1 | 23.5 | 34.0 | 21.7 | 7.76 |
| | | sublayer 2 | 3.5 | 10.0 | 34.0 | 21.7 | 30.8 |
| | | sublayer 1 | 1.5 | 10.0 | 34.0 | 21.7 | 32.8 |

## XRD/TEM

Figure 9 shows three XRD patterns of FeAlN films deposited on fused silica substrates. The nominal target compositions are 82.5wt%Fe-17.5wt%Al, 67wt%Fe-33wt%Al and 24wt%Fe-76wt%Al respectively. There is a broad peak around 22° 2θ for each pattern. For the film fabricated with the 82.5wt%Fe-17.5wt%Al target, there is a hump near 44° 2θ, which presumably is related to phases of $Fe_aN_b$ ($Fe_3N$, $Fe_4N$ and $Fe_2N$) and/or Fe. This hump disappears in the XRD pattern for the film fabricated with the 67wt%Fe-33wt%Al target. For the film fabricated with the 24wt%Fe-76wt%Al target, peaks are shown around 33°, 38° and 58° 2θ, which are related to AlN diffraction peaks (PDF#025-1133).

The average crystallite size of $Fe_aN_b$ and/or Fe phases in the FeAlN film (82.5wt%Fe-17.5wt%Al) is estimated to be approximately 2 nm.[13] The average crystallite size of AlN phase in the FeAlN film (24wt%Fe-76wt%Al) is estimated to be approximately 12 nm. The dominant phase in the FeAlN film (67wt%Fe-33wt%Al) appears to be amorphous. TEM studies reveal the different shapes of the crystals in the films and the average size is 2-10 nm.

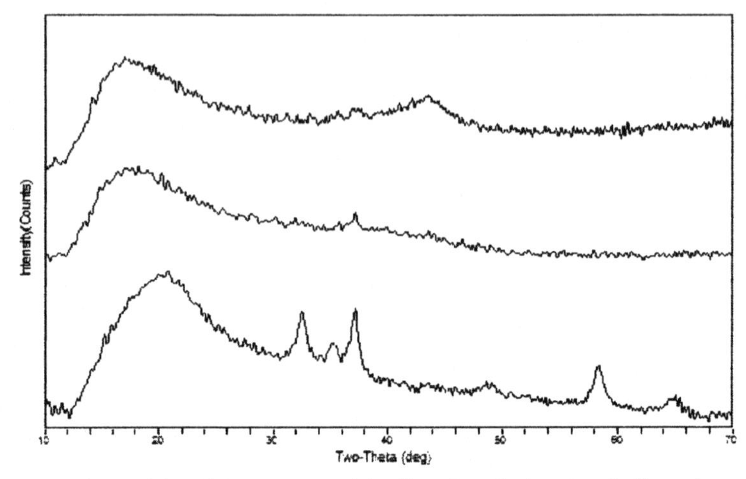


Figure 9. Comparison of three XRD patterns of the films deposited on fused silica substrates. The targets utilized from top to bottom: 82.5wt%Fe-17.5wt%Al, 67wt%Fe-33wt%Al and 24wt%Fe-76wt%Al.

## DISCUSSION AND CONCLUSION

Various Fe/FeAlN films were fabricated via a sputtering process. As revealed by SEM, the columnized growth of FeAlN/AlN films was observed. A film with 22 layers was also fabricated. EDS and XPS analyses of Fe layers and AlN layers show the variation of Fe/(Fe+Al) along the film depth from the top to the bottom of the film. The signatures of Fe, Al, N and O are revealed from the surface of an FeAlN film fabricated 11wt%Fe-89wt%Al target. The sublayer structure of FeAlN thin films is revealed via RBS. The variations of Fe/(Fe+Al) are shown in different sublayers in the most of FeAlN films. Measured iron concentration is close to nominal iron concentration for the films with 5, 11, 17 and 67wt%Fe targets. For a film fabricated with the target of 82.5wt%Fe, a hump in XRD pattern around 44° 2θ is revealed, which presumably is related to phases of $Fe_aN_b$ ($Fe_3N$, $Fe_4N$ and $Fe_2N$) and/or Fe. The crystallite size is approximately 2 nm, which is somewhat close to the crystal size measured by TEM.

## ACKNOWLEDGEMENTS

Work partially supported by BTI, NYSTAR-CACT, and NSF-CGR. The operational support for the EMSL facilities was provided by the Office of Biological and Environmental Research, U.S. Department of Energy. Pacific Northwest National Laboratory is operated by Battelle for the U.S. Department of Energy under Contract No. DE-AC06-76RLO 1830.

FOOTNOTES

*Corresponding author: Tel: +1-607-871-2130, Email address: fwangx@alfred.edu (X. W. Wang).

REFERENCES

[1]M. Solzi, M. Ghidini, and G. Asti, "Macroscopic Magnetic Properties of Nanostructured and Nanocomposite Systems," pp. 124-201 in *Magnetic Nanostructures*. Edited by S. Nalwa. American Scientific publishers. Stevenson Ranch. California, 2002.

[2]A. Carl and E. F. Wassermann, "Magnetic Nanostructures for Future Magnetic Data Storage: Fabrication and Quantitative Characterization by Magnetic Force Microscopy," pp. 59-92 in *Magnetic Nanostructures*. Edited by S. Nalwa. American Scientific publishers, Stevenson Ranch. California, 2002.

[3]Y. Liu, R. E. Miller, D. Li, Q. Feng, W. Votava, T. Zhang, L. N. Dunkleberger, X. W. Wang, R. W. Gray, T. Bibens, J. L. Helfer, K. P. Mooney, R. Nowak and P. Lubitz, " Materials Properties of Nano-sized FeAlN Particles in Thin Films," *The 6th Pacific Rim Conf. on Cera.&Tech. (PacRim-6)*, submitted (2005).

[4]Y. Liu, C. Daumont, E. Pavlina, R. E. Miller, C. McConville, X. Wang, R. W. Gray, J. L. Helfer, K. P. Mooney, and P. Lubitz, "Magnetic Properties of FeAlN Thin Films with Nano-sized Particles," *Proc. 9th Inter. Conf. on Ferrites (ICF 9)*, 119-29 (2004).

[5]C. Daumont, Y. Liu, E. Pavlina, R. E. Miller, C. McConville, X. Wang, R. W. Gray, J. L. Helfer, K. P. Mooney, and P. Lubitz, "Characterization of FeAlN Thin Films with Nano-sized Particles," *Ceramic Transactions*, **159** 157-64 (2005).

[6]X. W. Wang, R. E. Miller, P. Lubitz, F. J. Rachford, and J. H. Linn, "Nano-magnetic FeAl and FeAlN Films via Sputtering," *Ceramic Eng. & Sci. Proc.*, **24** [3] 629-36 (2003).

[7]R.S. Tebble and D.J. Craik, "Magnetic Materials", pp. 81-88, Wiley-Interscience, New York, 1969.

[8]G. G. Bush, "The Complex Permeability of a High Purity Yttrium Iron Garnet (YIG) Sputtered Thin Film," *J. Appl. Phys.*, **73** [10] 6310-1 (1993).

[9]M. DeMarco, X. W. Wang, R. L. Snyder, J. Simmins, S. Bayya, M. White, and M. J. Naughton, "Mossbauer and Magnetization Studies of Nickel Ferrites," *J. Appl. Phys.*, **73** [10] 6287-9 (1993).

[10]The Kurt J. Lesker Company, Clairton, PA. The magnetron is Torus 3 or 4.

[11]The suppliers of the targets are: W: Williams Puretek; L: Lesker Co.

[12]MDI Jade 7.0 [Computer Program]. Materials Data, Livermore, CA, 2004.

## POLYMERIC AND CERAMIC-LIKE COATINGS ON THE BASIS OF SIN(C) PRECURSORS FOR PROTECTION OF METALS AGAINST CORROSION AND OXIDATION

M. Günthner, Y. Albrecht, G. Motz
Ceramic Materials Engineering, University of Bayreuth
Ludwig-Thoma-Strasse 36 b
Bayreuth, Germany, 95447

### ABSTRACT

Since the price of steel and metals increased rapidly in the last few years, further approaches for protection metal substrates against wear, corrosion or oxidation are necessary. Thin ceramic coatings are promising candidates for this aim. Besides thermal spraying or CVD- and PVD-processes, simple varnish-techniques can be used to deposit thin films.

Two different special tailored solid and very good soluble precursors in the systems Si-C-N (ABSE) and Si-N (PHPS) have been developed for applications of polymer and ceramic-like coating materials. With corresponding precursor solutions in toluene or ether substrates with complex geometry can be coated by simple dip- or spray-coating-techniques. The degree of ceramisation and the properties of the resulting coatings are adjustable by special temperature programs. Processing and pyrolysis are possible either in inert atmospheres like nitrogen or in air.

Earlier results have shown that especially metals like aluminium, copper and magnesium can be effectively protected against acids, bases and aqueous salt solutions by thin SiN(C)-layers. Furthermore, hard gradient coatings up to 50 µm in thickness can be generate by the in-situ reaction of titanium substrates with SiN(C)-coatings through the formation of nitrides, silicides and carbonitrides near the surface.

New investigations exhibit a great potential of the ABSE- and PHPS-coatings to protect steel sheets (e.g. X5 Cr Ni 18-10) against corrosion (especially in acids) and against oxidation up to 1000 °C. The resulting SiON(C)-layers in the surface act as an excellent diffusion barrier against oxygen and reduce the weight gain by oxidation about two orders in magnitude.

### INTRODUCTION

The economic loss by wear, oxidation and corrosion of metals is estimated at several billion € per year. New developments in the automotive industry, especially the use of very corrosive Mg-alloys for lightweight structures and a new regulation in the electronic industry to forbid lead containing plumbs (corrosion problems at higher processing temperatures of printed circuit boards), demand coatings with excellent chemical and thermal resistance. Varnishes, paints and enamel layers have traditionally been used as protective layers. They are characterised by simple and well established application techniques. However, varnishes and paints are limited in temperature stability as well as resistance to wear and solvents whereas enamel layers are very sensitive to mechanical deformations. Nonoxide and oxide ceramic layers or even amorphous and diamond like carbon are now applied mainly with CVD- or PVD-processes to improve wear and hardness of the surface[1, 2, 3]. Drawbacks of these well established techniques are the high apparative effort (e.g. plasma and high vaccum technology) and the limited adhesion of the layers as well as the formation of cracks especially under thermal stress. In addition, it is still difficult to coat complex shaped components[4].

An alternative approach to these methods is the processing of polymeric and ceramic-like coatings by the well known pyrolysis of appropriate organoelemental compounds (precursors). In this work we report on coatings based on a tailored solid and soluble polycarbosilazane (ABSE) and on polysilazane (PHPS). By variation of the thermal treatment the applied precursor layer acts as a high temperature stable varnish, as a ceramic like coating or it can be used for generation of gradient coatings in the metal surface. The so called precursors ABSE and PHPS are synthesized in a high yield by a simple reaction of chlorosilanes with ammonia, which are both cheap products of the chemical industry. The poly(carbo)silazanes are very good soluble in ordinary organic solvents like toluene or ether. Magnesium, copper, titanium and steel substrates were dip-coated with suitable precursor solutions by variation of concentration and hoisting speed. After a curing step up to 400 °C in nitrogen or air the oxidation (steel sheets up to 1000 °C) as well as the corrosion resistance in acids, bases and in aqueous salt solution were investigated. A further heat treatment at 800 and 1000 °C of precursor coated titanium sheets in nitrogen or argon led to the formation of crystalline gradient layers in the substrates surface.

## EXPERIMENTAL PROCEDURE

The ABSE-polycarbosilazane is synthesized by ammonolysis of bis(dichloromethyl)silane in toluene as described elsewhere for other silazanes[5, 6] and is easily transferred to a pilot plant (batch size 40 l). The synthesis yield of the colourless, brittle and meltable solid is about 75 %. The commercial available PHPS-polysilazane (Clariant AG) is produced by ammonolysis of dichlorosilane.

For coating experiments by dip- and spray-coating techniques precursor solutions in toluene and ether with low viscosity were used. In the case of dip-coating process with a hoisting apparatus the coating thickness was adjusted by concentration and by variation of the hoisting speed. Subsequent annealing of the coatings was performed in air (Nabertherm® LH 60/14) and/or in nitrogen (F-A 100-500/13, GERO GmbH).

All coatings in the polymer as well as in the ceramic state were examined by light microscopy and scanning electron microscopy (Jeol JSM 6400). The layer thickness dependent on the pretreatment was measured with a Fischerscope® MMS (Helmut Fischer GmbH+Co) by the eddy current method (ASTM B244) or by the profile method (DIN EN ISO 4287 / MFW-250 Mahr GmbH). The resistance to corrosion was examined in 1 n acids (HCl, $H_2SO_4$), 1 n KOH and 5 mass% aqueous NaCl solution. The oxidation behavior up to 1000 °C was determined by loss of weight, by scanning electron microscopy, EDX (Jeol JSM 6400) and glow discharge spectrometry (GDOES / Spectruma 750). The gradient layers were additionally characterized by XRD (Seifert XRD3000P) and microhardness measurements (Helmut Fischer GmbH+Co, H100VP) as well as tribological investigations (pin-on-disk, Wazau SST).

## RESULTS AND DISCUSSION

### Precursor properties

As reported[7] the ABSE precursor structure consists of two structural elements namely very stable 5-membered carbosilazane rings and bridging linear carbosilazane groups (Fig. 1). This leads to high molecular weights (> $10^5$ g/mol) and a melting point higher than 150 °C. Further cross-linking takes place only at temperatures higher than 200 °C by separation of gaseous ammonia because of the absence of very reactive functional groups e.g. Si-H and Si-vinyl. This fact and the constitution of the polymer network result in a very good chemical stability against air and moisture, and makes a handling in air possible at room temperature. After storage of a fine grained ABSE-powder (Ø 50 μm) for 3 days in air at this temperature only 2 mass % oxygen were detected.

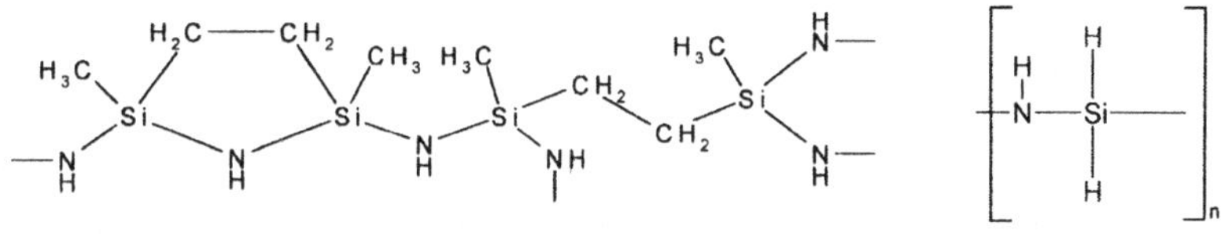

Figure 1: Basic structure units of the ABSE- (left) and PHPS- (right) precursor

The commercial available PHPS-precursor contains no organic groups (Fig. 1). Due to the reactive Si-H group the system is very sensitive to moisture and crosslinks rapidly in air. The ceramic yield is about 85 % in argon or nitrogen. The meltable and solid polysilazane can be solved in solvents like ether and toluene.

Coating properties

It is well known[8] that the coating thickness depends mainly on the concentration and viscosity of the solution, the withdrawal speed, the roughness of the substrate and the temperature and duration of the curing step. Figure 2 shows the thickness of the coatings on polished steel substrates as a function of precursor concentration and hoisting speed. The sheets were annealed in air for 1 h at a temperature of 500 °C.

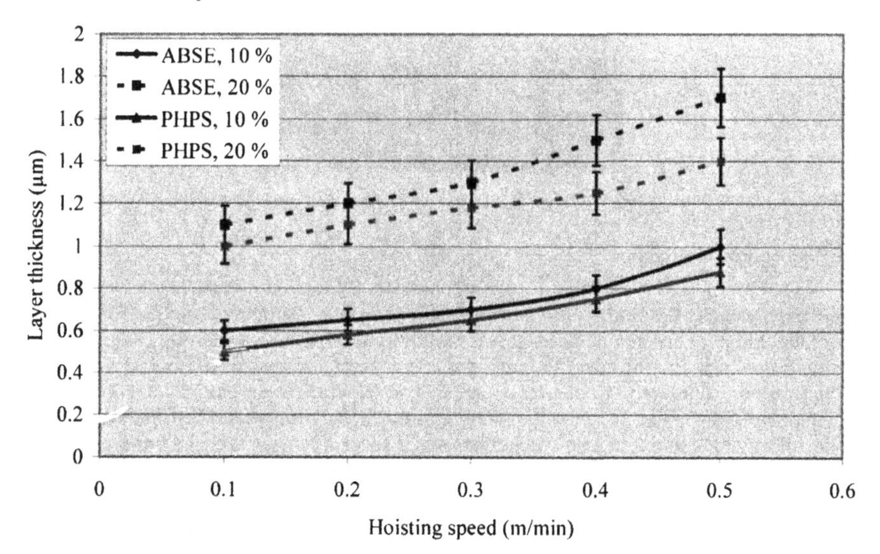


Figure 2: Thickness of the coatings as a function of precursor concentration and hoisting speed on steel samples, cured in air at 500 °C for 1 h

It is clearly seen that the layer thickness increases with hoisting speed and precursor concentration and lies between 0.5 and 2 µm. The precursor layers are dense, uniform and homogeneous. Thicker layers tend to crack, if the heating rate or temperature is too high.

As mentioned above, contrary to the PHPS-polymer the ABSE-precursor reacts very slowly at room temperature with moisture or OH-groups. But the rate of reaction increases with temperature. Therefore, polymeric coatings based on Si(C)N can react with oxidic surfaces of metals like magnesium, titanium, copper or steel due to higher affinity of silicon to oxygen in comparison to nitrogen. The direct chemical bonding between oxidic substrates and SiCN-coating leads to a pull-off strength of > 38 $N/mm^2$ (determined with adhesion pull test ASTM D 4541).

Hardness and corrosion resistance of precursor based coatings depend on pyrolysis temperature[7]. Coatings produced by using the ABSE and PHPS-polycarbosilazane reveal chemical stability against solvents, acids and bases after cross-linking at 300 °C. A typical corrosion test for magnesium in the automotive industry is the storage in a 5 mass% NaCl solution for 96 hours (DIN 50 021-SS). Figure 3 shows that Mg reacts very fast under these relatively soft conditions. In contrast, a polymeric ABSE layer (annealed in nitrogen at 300 °C) protects magnesium. Some sheets were bent after coating and stored in the NaCl solution but no corrosion was observed in the bending region.

Figure 3: Uncoated (left) and with ABSE coated Mg-sheet after storage in NaCl solution for 96 h

Gradient coatings on titanium

Titanium is a very suitable metal for implants or for light-weight structures. For bio-medical applications the limited mechanical stability and the hardness is not sufficient[9]. In the case of e.g. hip joint or knee implants it was tried to minimize the abrasion by using of diamond or ceramic coatings. But the adhesion problems and a thermal mismatch led to a separating of the coating from the substrate. An alternative method to avoid these problems is the generation of hard gradient layers in the surface of the titanium substrate by using the ABSE-polycarbosilazane and the affinity of titanium to many other elements. Especially the carbon and nitrogen content of the precursor is able to form hard phases with titanium like Ti-nitrides and -carbides. The good adhesion of the ABSE-layer enables a remarkable diffusion of Si, C and N into the titanium to form hard crystalline phases at higher temperatures in an in-situ reaction. To exclude the influence of the pyrolysis atmosphere first experiments were examined in argon. With subsequent XRD-measurements Ti-rich nitrides, silicides and carbonitrides were detected in the surface (Tab. 1). Pyrolysis in an excess of nitrogen (nitrogen atmosphere) led to an enhanced formation of nitrogen rich crystalline compounds.

Table 1: Formation of crystalline phases by reaction of the ABSE-precursor with Ti-substrates

| Metal | $T_{max}$ [°C] | Atmosphere | Time [h] | Crystalline Phases |
|---|---|---|---|---|
| cp-Ti | 550 | argon | 0.5 | $Ti_{10}N_3$ |
| cp-Ti | 800 | argon | 0.5 | $Ti_{10}N_3$, $Ti_5Si_3$, $TiC_{0.7}N_{0.3}$ |
| cp-Ti | 800 | nitrogen | 0.5 | TiN, $Ti_2N$, $TiN_{0.76}$, $Ti_5Si_3$, $TiC_{0.7}N_{0.3}$ |
| cp-Ti | 1000 | nitrogen | 0.5 | TiN, $Ti_2N$, $TiN_{0.76}$, $Ti_5Si_3$, $TiC_{0.2}N_{0.8}$ |
| TiAl6V4 | 800 | nitrogen | 0.5 | $Ti_2N$, $Ti_3N_{1.29}$, $TiSi_2$ |
| TiAl6V4 | 1000 | nitrogen | 0.5 | TiN, $Ti_2N$, $TiN_{0.76}$, $Ti_5Si_3$, $TiC_{0.2}N_{0.8}$ |

Especially the temperature of the thermal pre-treatment influences the hardness and the depth of the gradient layer. The microhardness of the pure SiCN after pyrolysis at 800 °C ($N_2$) is about 6 GPa and increases after pyrolysis at 1000 °C ($N_2$) up to 7.5 GPa. Figure 4 shows that the hardness values of the coated cp-Ti and TiAl6V4 start at the surface with 6 GPa and decrease up to hardness of the base material. Thus the TiAl6V4-alloy consists of the α- and β-phase, the hardness values of that alloy are higher than those of the cp-Ti (α-phase). The thickness of the formed gradient layer is about 30 µm. In contrast, the values of the microhardness after pyrolysis at 1000 °C ($N_2$) start also with the hardness of the pure SiCN-layer (8 GPa). The enhanced formation of crystalline phases (Tab. 1) lead to a remarkable increase in hardness up to 13 GPa. Figure 5 shows a cross section of the gradient layer on TiAl6V4. After a pyrolysis at 1000 °C the gradient layer has a thickness up to 40 µm. Only the remained very thin SiCN-coating shows cracks, whereas the gradient layer is crack free.

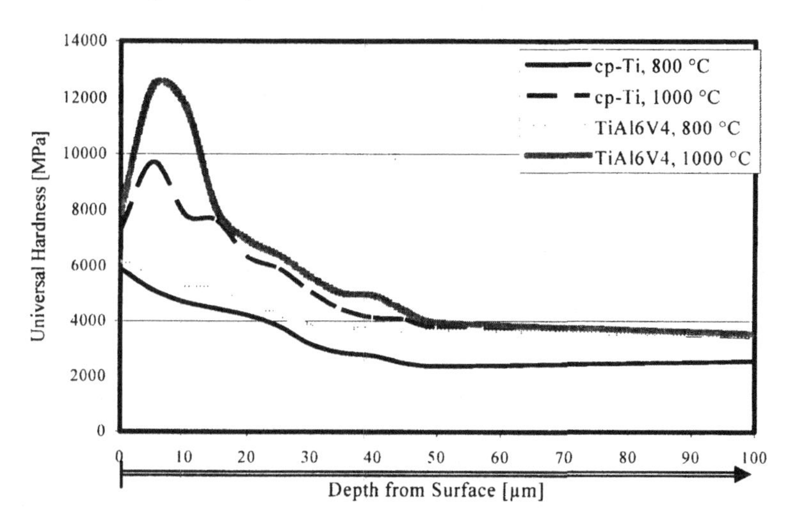


Figure 4: Universal hardness values of gradient coatings generated by pyrolysis ($N_2$-atmosphere) of different ABSE coated Ti-substrates

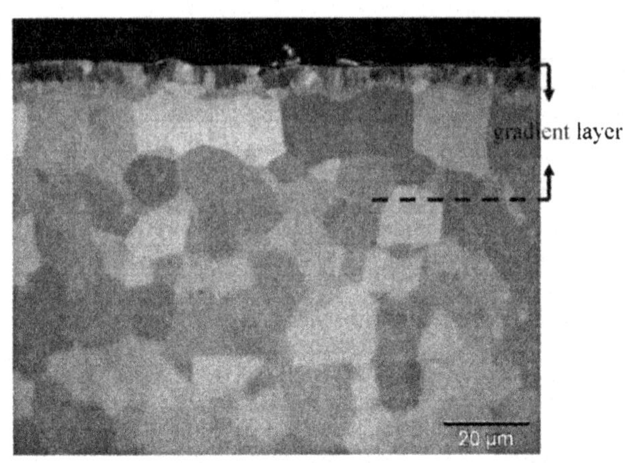


Figure 5: Cross-section micrograph of an ABSE-coated TiAl6V4 sample, pyrolysed at 1000 °C for 1 h ($N_2$-atmosphere)

The wear resistance of the gradient layer at the Ti-surface in comparison to the untreated cp-Ti was investigated with pin-on disc tests. In both cases 100 Cr 6 was used as pin. As expected a high wear coefficient $k_g$ of $4.9 \times 10^{-6}$ g/Nm of the uncoated cp-Ti disc was determined. A significant improvement of the wear behavior provides the generation of the gradient layer. After a heat treatment of the ABSE precursor coated cp-Ti disc at 1000 °C in nitrogen atmosphere for 10 h the wear coefficient is reduced to an excellent value of $k_g$: $2.0 \times 10^{-9}$ g/Nm (t: 100 min, $F_N$: 25 N, v: 5 Hz). These results and the hardness course (Fig. 4) are only caused by the in-situ formed gradient layer near the surface of the Ti-substrates. The continuous transition from metallic properties inside the metal to ceramic properties at the surface avoids adhesion problems and e.g. mismatch in the coefficient of thermal expansion.

Oxidation protection of copper and steel

In the electronic industry copper is very important as conductor path on printed circuit boards. For future applications it is necessary to protect these fine structures from oxidation and corrosive attacks. In the last years the research was focused on the improvement of conventional up to 280 °C stable circuit boards because of the Pb-free problem (e.g. higher temperature of Sn-Cu-, Sn-Sb- or Sn-Ag-solders than Pb-containing plumbs). An effective protection of copper conductor paths is necessary, because copper does not form a passivating oxide layer. To investigate the protection effect of ABSE-coatings on copper, different layer thicknesses were applied on flexible foils via dip-coating and annealed and/or oxidized at different temperatures up to 300 °C. In all experiments the best results were achieved with thinner SiCN-coatings (up to 1.5 µm). The shrinkage of the precursor layer during annealing and the thermal expansion of the copper lead to cracks and hence to the oxidation of the substrate, if the coatings are too thick. A thinner coating is more suitable to compensate thermal stresses. For oxidation up to 250 °C no thermal pretreatment of the precursor layer in $N_2$-atmosphere is necessary. However, higher oxidation temperatures require a previous annealing between 300 and 500 °C in nitrogen to cross-link the ABSE-precursor and to improve the oxidation stability (Fig. 6).

Figure 6: ABSE-coated copper sheet and copper foil after oxidation at 250 °C (left) or 300 °C (right) for 1 h

Beside of copper sheets steel can be protected very effectively from oxidation by precursor coatings. Depending on the steel grade, oxidation resistance up to 1000 °C is possible. Figure 6 shows stainless steel samples (X5 Cr Ni 18-10), which were partially coated with PHPS or ABSE and oxidized up to 900 or 1000 °C for 1 h. Stainless steel is used for instance in the food and chemical industry or in the process engineering. It can be seen in figure 7, that the uncoated region is oxidized, whereas the resulting SiON(C)-layer passivates the substrate effectively. The weight gain by oxidation can be reduced about two orders in magnitude via precursor derived coatings.

Figure 7: PHPS- and ABSE-coated stainless steel sheets after oxidation at 900 °C (left) or 1000 °C (right) for 1 h

## CONCLUSION

The special tailored ABSE precursor and the commercially available PHPS polysilazane are very suitable for processing high temperature stable varnishes via dip- and spray-coating techniques. The preceramic nature of the resulting coatings provides application temperatures up to 400 °C. The reactivity of the silazane with oxide surfaces leads to direct chemical bondings between layer and substrate. Thin ABSE-layers up to 1.5 μm on copper avoid oxidation up to 300 °C. Therefore, it is possible to protect circuit boards from corrosion and oxidation. ABSE layers on magnesium act as very effective protection from corrosion in aqueous salt solutions and can solve corrosion problems in automotive applications.

The in-situ reaction of titanium substrates with SiCN-layers at temperatures higher than 550 °C lead to gradient coatings up to 50 μm in thickness. Because of the formation of crystalline nitrides, silicides and carbonitrides especially near the surface, the universal hardness direct behind the remained SiCN layer increases up to values of about 13 GPa. The modified surface shows a very improved resistance from wear and abrasion. The combination of metallic and ceramic properties opens new fields of application for titanium components not only for implants.

The ABSE- and PHPS-coatings exhibit a great potential to protect steel sheets (e.g. X5 Cr Ni 18-10) from corrosion (especially in acids) and oxidation up to 1000 °C. The resulting SiON(C)-layers in the surface act as an excellent diffusion barrier against oxygen and reduce the weight gain by oxidation about two orders in magnitude.

REFERENCES

[1]A. Grill, "Diamond-like carbon coatings as biocompatible materials - an overview", *Diamond and Related Materials,* **12**, 166-170 (2003)

[2]K. H. Kim, B. H. Park, "Coatings and Characteristics of Ti-X-N By Plasma Enhanced Chemical Vapour Deposition", *International Workshop on Surface Engineering and Coatings NAL* (1998)

[3]C. Fernandez-Ramos, J. C. Sanchez-Lopez, M. Belin, C. Donnet, L. Ponsonnet, A. Fernandez, "Tribological behaviour and chemical characterisation of Si free and Si-containing carbon nitride coatings", *Diamond and Related Materials,* **11**, 169-175 (2002)

[4]F.-W. Bach, K. Möhwald, A. Laarmann, T. Wenz, "Moderne Beschichtungsverfahren", Weinheim (2005)

[5]G. Motz, J. Hacker, G. Ziegler, "Special Modified Silazanes for Coatings, Fibers and CMCs", *Ceramic Engineering & Science Proceeding,* **21**, 307-314 (2000)

[6]S. Traßl, D. Suttor, G. Motz, E. Rößler, G. Ziegler, "Structural Characterisation of Silicon Carbonitride Ceramics derived from Polymeric Precursors", *J. Europ. Ceram. Soc.*, **20**, 215-225 (2000)

[7]G. Motz, G. Ziegler, "Simple Processibility of Precursor-derived SiCN Coatings by optimized Precursors", *Proc. Seventh Conf. & Exhibition of the Europ. Ceram Soc.*, **1**, 475-478 (2001)

[8]C. J. Brinker, G. W. Scherer, "Sol-gel science", Boston (1990)

[9]E. Wintermantel, S.-W. Ha, "Medizientechnik mit biokompatiblen Werkstoffen und Verfahren", Berlin (2002)

ACKNOWLEDGEMENTS

The Stiftung Industrieforschung in Cologne is acknowledged for financial support and Clariant AG for supplying the precursor PHPS.

# EFFECT OF TEMPERATURE AND SPIN-COATING CYCLES ON MICROSTRUCTURE EVOLUTION FOR Tb-SUBSTITUTED $SrCeO_3$ THIN MEMBRANE FILMS

Satyajit Shukla[2], Mohamed M. Elbaccouch[1], Sudipta Seal[2], Ali T-Raissi[1*]

[1]Florida Solar Energy Center
University of Central Florida
1679 Clearlake Road
Cocoa, FL 32922-5703

[2]Advanced Materials (AMPAC) & Mechanical, Materials, and Aerospace Engineering
University of Central Florida
4000 Central Florida Blvd, Engineering 381
Orlando. FL 32816-2450

## ABSTRACT

Ceramic oxides with perovskite structures ($A^{2+}B^{4+}O_3$) have been receiving considerable attention in the solid-state electrochemical systems, such as the development of solid oxide fuel cells (SOFCs), gas sensors, and hydrogen ($H_2$) permeable membranes. The goal of this investigation is to process a terbium-doped strontium cerate ($SrCe_{0.95}Tb_{0.05}O_{3-\delta}$) (SCT) thin membrane films by spin-coating using ethylene glycol-based polymeric precursor. Continuous and dense $SrCe_{0.95}Tb_{0.05}O_{3-\delta}$ membrane thin films with neither pin-holes nor cracks are reported. The thicknesses of the membrane films are within the range of ~ 200 nm – 2 µm. For a single spin-coating cycle, the membrane film (200 nm thick) appears to be discontinues. However, the membrane films are dens for multiple spin-coating cycles. The polymeric precursor and the microstructure of the $SrCe_{0.95}Tb_{0.05}O_{3-\delta}$ membranes are characterized using scanning electron microscopy (SEM), focused ion-beam (FIB) microscopy, and x-ray diffraction (XRD). This work reveals that good film quality with uniform texture and homogeneous structure can be produced via spin-coating technique as a function of spin-coating cycles and processing temperature. Also, surface morphology and grain size strongly depend on sintering temperature with even grain size distribution for each sintering temperature. The flexibility of the present process approach demonstrates the capability of precisely controlling the thickness of the ceramic membrane films within a sub-micron range.

*Corresponding Author:
Email: ali@fsec.ucf.edu; Tel: (321) 638-1446; Fax: (321) 504 – 3438.

## INTRODUCTION

The objective of this work is to synthesize and characterize thin film membranes that separate hydrogen from gas mixtures with 100% selectivity and high hydrogen flux. The focus is on ion transport systems with a perovskite structure of mixed ionic-electronic conductivity ($A^{2+}B^{4+}O_3$-doped with a trivalent cation).[1] The membrane materials and the characterization techniques have attractive applications in producing high temperature hydrogen permeable reactors for the steam reforming of alkanes,[2] gasification of coal.[3] solid oxide fuel cells,[4] and associated sensor technologies.[5] This ceramic oxide membrane approach to hydrogen gas purification is much more selective than permeation through polymeric membranes, yet much less expensive than metallic foils such as palladium.

The targeted material is terbium (Tb)-doped strontium cerate ($SrCeO_3$). The membranes are synthesized using a polymeric-based precursor containing the metal cations and spin-coated on silicone oxide substrates. $SrCe_{0.95}Tb_{0.05}O_3$ membrane films having thicknesses within the range of 200 nm–2 μm are reported. For multiple spin-coating cycles the investigated membrane films have no pin holes, interconnected porosity, or cracks. The $SrCe_{0.95}Tb_{0.05}O_{3-\delta}$, membranes are permeable to hydrogen through the simultaneous diffusion of protons and electrons, and not permeable to $N_2$-like molecules as long as the membrane films are dense and free of pinholes. The effect of calcination temperatures and film thicknesses are characterized using scanning electron microscopy (SEM) and focused ion-beam (FIB) microscopy. X-ray diffraction (XRD) confirms the presence of perovskite phase within 400-1000°C.

## EXPERIMENTAL

### Synthesis and Spin-Coating Procedures

Details of the synthesis and spin coating techniques have been published elsewhere.[6] Briefly, the metal nitrates in the desired stoichiometric ratio were mixed with ethylene glycol, distilled water, and nitric acid. Heating polymerized the solution forming a complex mixture between the metal cations and ethylene glycol.[7] Water and other volatile materials were evaporated until a viscous gelatinous solution was formed. Silicone substrate was pre-heated at 600°C in air to form $SiO_2$ layer. All spin-coating measurements were conducted at 3000 rpm for 20 s (Cookson Electronics Equipment – Model P6204). The multiple spin-coating procedure is represented schematically in Figure 1. The films were dried at 80 and 300°C for 1 min each before spin-coating a new layer, and the final sintering was done at 400-1000°C for 4 h in air to form a high quality polycrystalline metal oxide films.

### Characterization

The surface morphology and the average particle size of the thin film membranes were analyzed using SEM (JSM-6400F, JEOL, Tokyo, Japan). To avoid any surface charging during the SEM analysis, the ceramic thin film membranes were coated with approximately 30 nm Au–Pd layer using a sputter coater (K350, Emitech, Ashford, Kent, England).

FIB (FEI FIB 200 TEM, Hillsboro, OR) milling was performed on the Tb-doped $SrCeO_3$ membrane films to estimate the film thickness as a function of temperature and number of coating-cycles. The procedure for the FIB-milling for the film thickness measurement has been already described in detail elsewhere.[6,8]

The crystal structure and purity of the ceramic membrane was determined using Rigaku XRD Technique utilizing Cu $K\alpha$ X-radiation of wavelength 1.54 Å to confirm the deposition of the perovskite ceramic membrane. The average nanocrystallite size as a function of temperature was calculated using the Scherrer's correlation,[9]

$$D = \frac{0.9\lambda}{\beta cos\theta_\beta} \quad (1)$$

where $D$ is the average nanocrystallite size in nm, $\lambda$ the radiation wavelength (0.154 nm), $\beta$ the half width at half maximum intensity (radians), and $\theta_\beta$ half the difference between the two extreme diffraction peak angles (radians) where the intensity is zero.

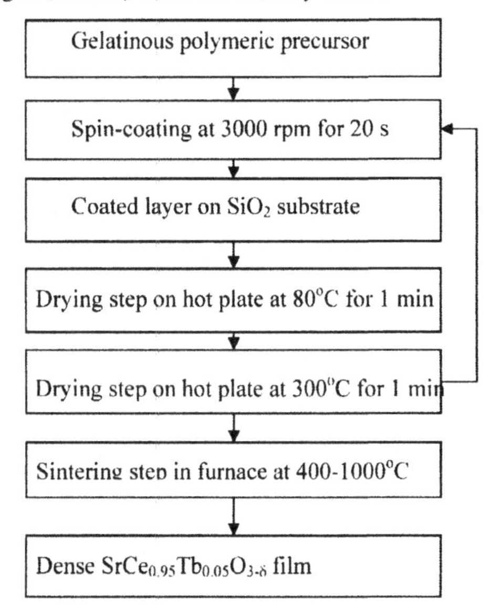


Figure 1. Schematic diagram of the multiple spin-coating procedure.

## RESULTS AND DISCUSSION

### SEM Analysis

Figure 2 shows high magnification SEM images for $SrCe_{0.95}Tb_{0.05}O_{3-\delta}$ film, as a function of the spin-coating cycles, sintered at 400°C for 4 h. The membrane thin film spin-coated for only one cycle (Figures 2a) shows a relatively dense structure at the surface with respect to those processed with more than a single cycle. By comparing Figure 2a with Figures 2b-c, it is qualitatively noted that the average particle size decreases with multiple spin-coating cycles. Typically, the average particle size of ~500 nm is observed for the thin film membrane obtained via single-cycle, while for those synthesized using ten and twenty-five spin-coated cycles, it is observed to be ~200 nm. Comparison of Figures 2b and 2c, however, reveals that, the average amount of surface porosity, surface morphology, and average particle size for the thin film membranes spin-coated for more than one-cycle remain qualitatively the same. Thus, multiple spin-coating cycles are observed to result in increase in the amount of surface porosity, but decrease in the average particle size relative to those of a single coating-cycle.

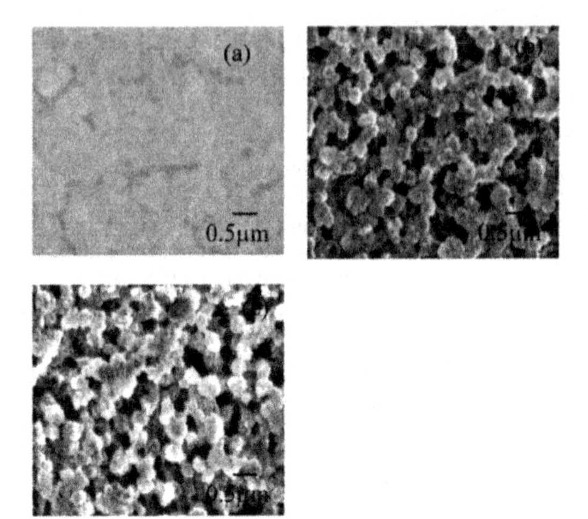


Figure 2. SEM surface morphology images at high magnifications of $SrCe_{0.95}Tb_{0.05}O_{3-\delta}$ after spin-coating at 3000 rpm for 20 s and sintering at 400°C for 4 h. (a) one spin-coating cycle, (b) 10 spin-coating cycles, and (c) 25 spin-coating cycles.

Figure 3 shows high magnification SEM images respectively for $SrCe_{0.95}Tb_{0.05}O_{3-\delta}$ as a function of the sintering temperature within the range of 400-1000°C. The figures show that temperature has a strong impact on the $SrCe_{0.95}Tb_{0.05}O_{3-\delta}$ surface morphology. Each temperature treatment has a unique effect on the texture, particle size, and amount of porosity. Figures 3d and 3a-c show that the amount of surface porosity diminishes significantly as temperature increases leading to denser surface at 1000°C. A clustered-like and diffused-like particles form on the surface matrix as the temperature increases from 400 to 1000°C.

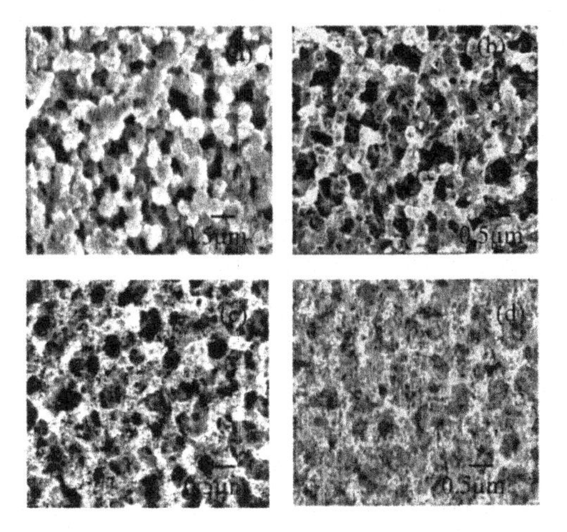

Figure 3. SEM surface morphology images at high magnifications of $SrCe_{0.95}Tb_{0.05}O_{3-\delta}$ after twenty-five spin-coating cycles at 3000 rpm for 20 s and sintering for 4 h at (a) 400°C. (b) 600°C, (c) 800°C, and (d) 1000°C.

FIB Analysis

The cross-sectional FIB images of $SrCe_{0.95}Tb_{0.05}O_{3-\delta}$ membrane thin films synthesized using multiple coating cycles with the calcinations treatment at 400 and 1000°C are respectively presented in Figures 4a-e and Figure 4f. As shown in Figure 4a, the membrane thin films with thickness of 180 nm could be synthesized using a single coating cycle. The membrane film, however, appears to be discontinuous under these processing conditions. With increasing the spin-coating cycles (Figures 4b-f) dense and continuous membrane thin films, with increased film thickness. is obtained. Figures 4b-f reveal that with increasing the number of spin-coating cycles, the porosity is progressively associated with the near surface region, while the bottom part of the thin film remains relatively denser. This suggests that the multiple spin-coating cycles fill the surface porosity associated with the previously deposited layers. Overall it appears that a single-coating cycle results in discontinuous thin film while multiple spin coating cycles result in dense, pin-hole free, and continuous membrane thin film, which appears to be the best morphology for the effective $H_2$ separation based on proton conduction.

The actual variation in the membrane film thickness as a function of the number of spin-coating cycles is quantitatively presented in Figure 5. The membrane film thickness ($t$) increases linearly in the logarithmic scale with increasing the number of spin coating cycle ($n$) following the power law relationship of the form,

$$t = 200n^{0.8} \tag{2}$$

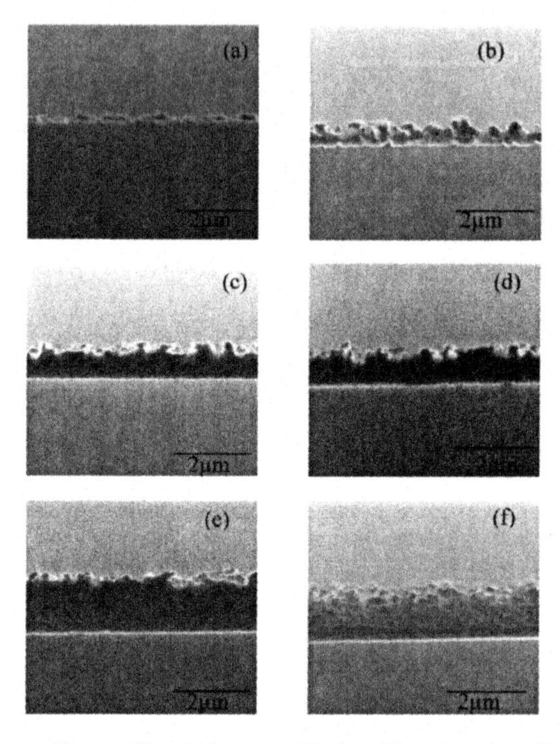


Figure 4. FIB high magnification film thicknesses of $SrCe_{0.95}Tb_{0.05}O_{3-\delta}$ after spin-coating at 3000 rpm for 20 s and sintering for 4 h: (a) 1 spin coating cycle-400°C, (b) 5 spin coating-cycles-400°C, (c) 10 spin-coating cycles-400°C, (d) 15 spin-coating cycles-400°C, (e) 25 spin-coating cycles-400°C, and (f) 25 spin-coating cycles-1000°C.

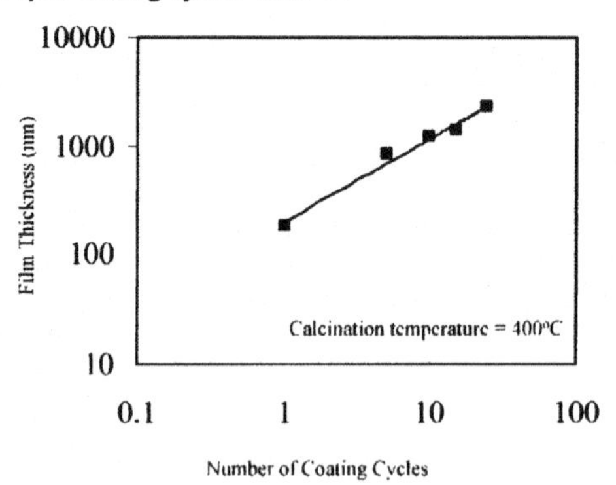


Figure 5. Variation in the $SrCe_{0.95}Tb_{0.05}O_{3-\delta}$ membrane film thickness as a function of the number of spin-coating cycles for the FIB images represented in Fig. 4.

XRD Analysis

The broad and narrow XRD line traces for the $SrCe_{0.95}Tb_{0.05}O_{3-\delta}$ films obtained with twenty-five spin-coated cycles and sintered at 400-1000°C, are presented in Figures 6 and 7 respectively.[5] In Figure 6, the presence of XRD pattern strongly suggests the crystalline nature of the membrane thin films. The diffraction peaks observed in Figure 6 are labeled according the PDF card (No. 47-1689) of JCPDS-ICDD, which confirms the pervoskite-type crystal structure of the thin film membranes.

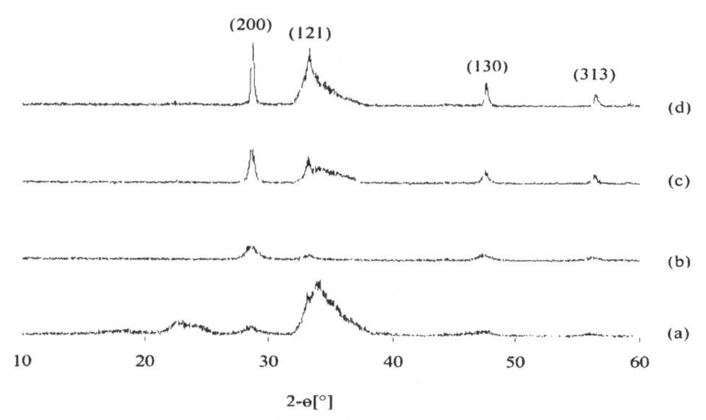


Figure 6. XRD broad scans of 25 spin-coated cycles of $SrCe_{0.95}Tb_{0.05}O_{3-\delta}$ at 3000 rpm for 20 s and sintering for 4 h at: (a) 400°C, (b) 600°C, (c) 800 °C, and (d) 1000°C.

In Figure 7, for the narrow scan analysis, the major peak (200) is chosen. The narrow scan analysis shows that, with increasing calcination temperature, the half width at half maximum intensity (FWHM) of the major peak decreases, which qualitatively suggests that, the nanocrystallite size within the thin film membrane decreases with increasing calcination temperature. The quantitative variation in the nanocrystallite size ($D$), as determined using Eq. 1, as a function of the calcination temperature ($T$) is presented in Figure 8, where the following relationship is observed to be obeyed,

$$D = 1.8\, e^{0.04T} \quad (3)$$

As observed in Figure 8, the average nanocrystallite size increases from 8 to 70 nm with increasing calcination temperature within the range of 400-1000°C. The increase in the nanocrystallite size results in less grain boundaries and less chemistry defects in the membrane matrix.

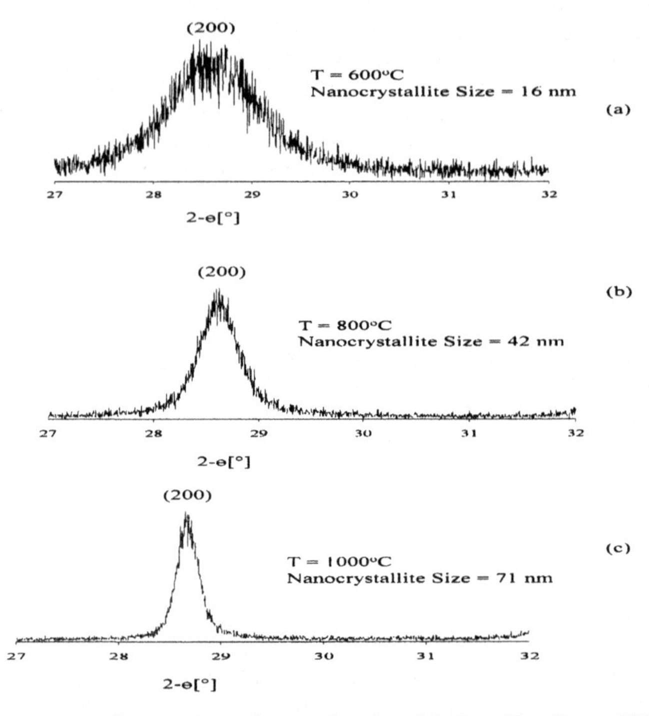


Figure 7. XRD narrow scans of twenty-five spin-coated cycles of $SrCe_{0.95}Tb_{0.05}O_{3-\delta}$ at 3000 rpm for 20 s and sintering for 4 h at (a) 600°C, (b) 800°C, and (c) 1000°C.

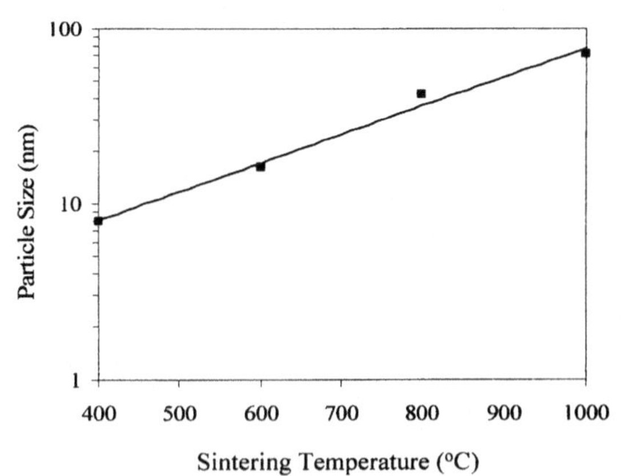


Figure 8. Variation in the particle size as a function of sintering temperature for twenty-five spin-coated cycles of $SrCe_{0.95}Tb_{0.05}O_{3-\delta}$ sintered for 4 h at 400, 600, 800, and 1000°C.

## CONCLUSIONS

The synthesis and characterization of nanocrystalline $SrCe_{0.95}Tb_{0.05}O_{3-\delta}$ membrane thin films has been demonstrated in this investigation by combining the ethylene glycol chelating process with the spin-coating technique. The polymeric precursor was deposited on silicone-based substrates and converted to dense polycrystalline metal oxide films after annealing treatments at 400-1000°C for 4 h. The ethylene glycol polymerization forms a stable single phase solution between the metal cations and the polymeric network. The present synthesis technique is a simple method to produce pervoskite-type ceramic thin films suitable for various $H_2$ separation applications.

The SEM morphological analysis shows that the films are continuous and have the same morphology across the substrates. The key point is that good film qualities with uniform texture and similar structure can be produced via spin-coating technique as a function of spin-coating cycles and calcination temperature. Also, the grain size strongly depends on the latter with even grain size distribution for each calcination temperature. The multiple spin-coating cycles covers the porosity of the previously spin-coated layer leading to a matrix that is porous only on the surface, but free of any interconnected porosity within the bulk of the membrane film. That is the multiple spin-coating technique leads eventually to a membrane film that is dense and free of pin holes

The FIB images demonstrate that by coupling the polymeric precursor synthesis with the spin-coating technique, the ceramic membrane thickness can be precisely controlled within a sub-micron (200 nm) to micron-size (2.2 μm) range, which reflects the flexibility of the present process approach. Another parameter, in addition to the number of spin coating cycles, that controls the thickness of the film is the spin-coating revolution. Increasing the spin-coating revolution (i.e. from 3000 rpm to 5000 rpm) can effectively decrease the film thickness. Since $H_2$ permeability is inversely proportional to film thickness, it is expected that the current approach of synthesizing a micron level membrane film will improve the $H_2$-flux by several order of magnitude compared to the disk geometry membranes.

## ACKNOWLEDGEMENTS

The authors acknowledge the support provided by the National Aeronautics and Space Administration through Glenn Research Center under contract No. NAG32751, and the National Science Foundation (NSF CTS 0350572). The authors also thank the Materials Characterization Facility (MCF) at the University of Central Florida for offering various analytical instruments for conducting the present research.

## REFERENCES

[1]H. Iwahara, T. Esaka, H. Uchida, and N. Maeda, "Proton Conduction in Sintered Oxides and its application to Steam Electrolysis for Hydrogen Production," *Solid State Ionics* **3/4**, 359-363 (1981).

[2]L. Garcia, R. French, S. Czernik, and E. Chornet, "Catalytic Steam Reforming of Bio-Oil for the Production of Hydrogen: Effects of Catalyst Composition," *Appl. Catal. A: Gen.* **201**, 225-239 (2000).

[3]E.D. Wachsman, and M.C. Williams, "Hydrogen Production From Fossil Fuels With High Temperature ion Conducting Ceramics," *The Electrochemical Society Interface* **Fall**, 32-36 (2004).

[4]S.J. Visco, L.–S Wang, S. Souza, and L.C. De Jonghe, "Thin-Film Electrolytes for Reduce temperature Solid Oxide Fuel Cells," Mat. Res. Soc. Symp. 369, 683-691 (1995).
[5]H. Fan, S.-H Lee, C.-B Yoon, G.-T Park, J.-J Choi, and H.-E Kim, "Perovskite Structure Development and Electrical Properties of PZN Based Thion Films," *J. Eur. Ceram. Soc.* **22**, 1699-1704 (2002).
[6]M. M. Elbaccouch, S. Shukla, N. Mohajeri, S. Seal, and A. T-Raissi, "Microstructural Analysis of Doped-Strontium Cerate Thin Film Membranes Fabricated Via Polymer Precursor Technique," *Solid State Ionics*, Submitted.
[7]C.C. Chen, M.M. Nasrallah, and H.U. Anderson, "Synthesis and Characterization of $(CeO_2)_{0.8}$ $(SmO_{1.5})_{0.2}$ Thin Films From Polymeric Precursors," *J. Electrochem. Soc.* **140 (12)**, 3555-3560 (1993).
[8]S. Shukla, S. Seal, J. Akesson, R. Oder, R. Cater, and Z. Rahman, "Study of Mechanism of Electroless Copper Coating of Fly-Ash Cenosphere Particles," *Appl. Surf. Sci.* **181**, 35-50 (2001).
[9]B.D. Cullity, S.R. Stock, "Elements of X-Ray Diffraction," Prentice-Hall, New Jersey, 2001.

# DEVELOPMENT OF BORIDIZED PASSIVATION LAYER FOR USE IN PEM FUEL CELLS BIPOLAR PLATES

K. Scott Weil, Jin Yong Kim, Gordon Xia, Jim Coleman, and Z. Gary Yang
Pacific Northwest National Laboratory
PO Box 999
Richland, WA 99352

## ABSTRACT

This paper outlines the development of a new low-cost materials concept for polymer electrolyte membrane fuel cells (PEMFCs). Employing the roll bonding process to prepare a nickel clad steel laminate, the thin outer nickel layer is then passivated using a powder pack boridization technique. Results from energy dispersive X-ray (EDX) and X-ray diffraction (XRD) analyses and from scanning electron microscopy (SEM) indicated that a relatively homogeneous $Ni_3B$ layer grows on the exposed surfaces of the nickel and that the thickness of this layer can be readily controlled through the time and temperature over which boridization takes place. At high boridization temperatures, $\geq$ 700°C, and long periods of time, a $Ni_2B$ overlayer forms on top of the $Ni_3B$. Preliminary exposure testing conducted at 80°C for 300hrs in 1M $H_2SO_4$ containing 2ppm HF demonstrates a significant increase in the corrosion resistance that is attributable to the boridization treatment.

## INTRODUCTION

The PEMFC stack is under serious consideration by nearly every major automobile manufacturer as a potential high efficiency source, low pollution power source replacement for the internal combustion engine.[1] These devices operate at low temperature (on the order of 80°C) by converting hydrogen and oxygen directly to electricity via an electrochemical reaction, with water as the only major by-product.[2] While significant progress has been made toward developing a commercially viable PEMFC system, at present the device remains only a niche product, primarily replacing batteries as a back-up source of electricity for critical stationary applications (e.g. telecommunications).[3] The key limitations to the more widespread commercialization of these power systems are the high cost of manufacture and the degradation in performance under continuous long-term operation. An additional factor that has hindered the acceptance of PEMFCs in transportation markets is their current size and weight.

The bipolar plate is the most bulky component in the stack (in both weight and volume) and one of the most expensive to manufacture. It serves not only as the electrical junction between serially connected cells, but also performs several other key functions in the device:

- Distribute the fuel and oxidant uniformly over the active areas of the cells.
- Facilitate water management of the membrane to keep it humidified, yet mitigate flooding.
- Act as an impermeable barrier between the fuel and oxidant streams (particularly $H_2$) to maintain the hydrogen gradient across the membrane necessary for high power output.
- Provide some measure of structural support for the stack.
- Remove heat from the active areas of the cells.

In the early development of PEMFCs, graphite was the material of choice for use in the bipolar plate because of its excellent corrosion resistance and surface conductivity in the low pH/hot aqueous environment of the stack.[4] However, the high cost and mechanical strength of graphite and the expense associated with machining the individual plates necessitated the development of lower cost alternative bipolar plate materials for commercial applications. To date, carbon composites, various coated metals, and uncoated stainless steel and titanium have been considered. While carbon-based composites appear to be the leading candidates, there remain several key concerns with the materials including high manufacturing cost, poor mechanical strength, and vaiable hydrogen permeability.

The use of metal-based bipolar plates in PEMFC stacks potentially offers a number advantages particularly for transportation applications including: low-cost mass-production via stamping or embossing of sheet product; fabrication in very thin form (< 200μm) to reduce weight and volume in the overall stack; impermeability to fuel and oxidant gases and to water vapor; and in general, excellent thermal conduction properties and high mechanical robustness, even as a thin stamped foil. The primary challenge with metal PEMFC interconnects is surface corrosion, and the current drive to increase the operating temperature of the stack will only exacerbate this problem. Corrosion of the bipolar plate leads to a release of metal ions that can contaminate the electrolyte membrane and poison the electrode catalysts. In addition, the formation of a passivating oxide or oxyhydroxide layer on the surface of the metal will increase the contact resistance between the bipolar plate and the adjacent graphite electrode backing layer by many orders of magnitude. Both conditions can significantly degrade stack performance. A number of research groups have investigated various schemes for protecting metallic bipolar plates, most of which rely on a thin, inert yet electrically conductive coating.[5] The greatest level of success that has been openly reported has been achieved with noble metal coatings such as gold and palladium. Unfortunately commercial use of these materials, even as thin coatings, is cost prohibitive.

Transition metal borides offer a combination of high electrical conductivity and good corrosion resistance that make them attractive for potential use as a protective surface layer in this application.[6,7] These compounds display room temperature electrical conductivities approximately one order of magnitude less than most metals, but over two orders of magnitude higher than that of graphite. Thus a thin coating of the boride will contribute negligibly to the overall resistance of the metal-based bipolar plate, which is expected to be significantly less than that of the alternative millimeter thick graphite or carbon composite plate during application. The approach being developed at Pacific Northwest National Laboratory is to fabricate a thin metal laminate sheet consisting of a middle filler layer sandwiched between two transition metal alloy layers that can be boridized to form a conductive, yet corrosion-resistant (i.e. passivating) barrier layer. Ideally the material selected for the filler, which will form the thickest of the three layers, is chosen based primarily on material cost, formability, durability, and thermal conductivity. On the other hand, the cladding material is selected based on the ease of boridization, the corrosion resistance and electrical transport properties of the resulting boride, formability, and cost. In this way, the bipolar plate can be tailored to take advantage of the merits of each material, while minimizing material and processing costs. The manufacture of clad material products is a well-established, well-understood commercial process and a number of domestic manufacturers exist that fabricate a wide variety of multilayer clad products in 50 – 300μm thick sheets.

## EXPERIMENTAL

The clad metal concept under study is shown schematically in Figure 1. The bipolar plate is essentially a composite laminate structure that consists of an inexpensive core material metallurgically bonded to an outer cladding layer via a roll bonding process (i.e. under elevated temperature and pressure). Additional layers can be formed into the laminate as necessary for improved functionality. For example, the structure illustrated in Figure 1 contains a brazing filler metal layer that would afford ready joining of two mated plates after stamping to form a bipolar plate component that contains an internal water cooling channel. As a first step in developing a clad material for PEMFC use, commercial purity nickel was selected as the clad material and 304 stainless steel (304SS) as the core. Relative to other boridizable transition metals that could be considered, such as Ti, Nb, Co, and Mo, nickel is relatively inexpensive, exhibits excellent formability (i.e. for stamping), and is readily boridized at moderate temperatures. 304SS was chosen because it is an inexpensive nickel-containing stainless steel that also displays good formability and can be readily roll bonded with a thin nickel cladding under moderate conditions. It is anticipated that once concept viability is completely demonstrated through rigorous testing, the stainless core could be replaced with an even lower cost material such as 1080 steel.

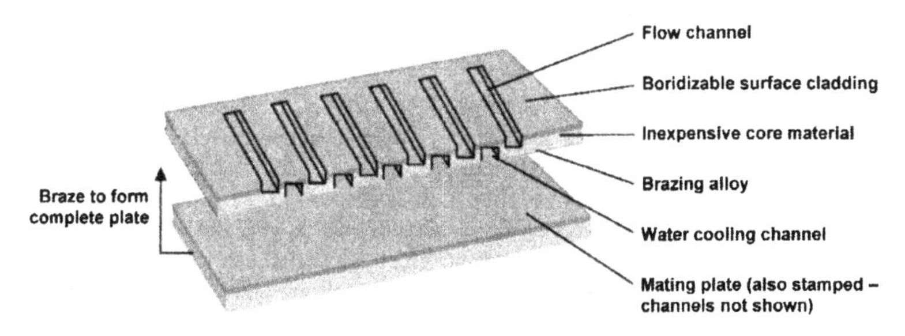


Figure 1 An illustration of the stamped clad metal bipolar plate concept. In this example, the bipolar plate component is formed from two stamped pieces that are joined via a brazing layer to form an internal water channel. Alternatively the bipolar plate can be formed without the water channel from a single laminated metal piece that is clad on both exposed surfaces with a nitridable layer.

Working with Engineered Materials Solutions Inc. (EMS; Waltham, MA), Ni/304SS/Ni clad sheet was developed with the following target dimensions: a 114.6μm (4.5mil) thick 304SS core clad with 12.7μm (0.5mil) nickel to form a sheet measuring 140μm (5.5mil) in total thickness. However, the initial boridization and subsequent corrosion studies described here were conducted using 200μm thick, 99.95% purity nickel sheet obtained from Alfa Aesar (Waltham, MA). The sheet was cut into 25cm x 25cm coupons that were lightly polished on both surfaces with 200 grit emery paper, washed with acetone and methanol, and dried at room temperature prior to boridization.

Several types of powder mixtures are commercially available for use in powder pack boridization. However, many of these contain SiC as a major phase and previous reports indicate that this compound is not inert, but participates in the coating process to form silicides.[8,9] To avoid forming this phase, the pack bed powder was prepared in-house and consisted of a mixture of 98.6% $CaB_6$ (99.9% purity; Alfa Aesar) and 1.4% $KBF_4$ (99% purity; Alfa Aesar) by weight.

The two powders were ground together in an agate mortar and pestle prior to pouring into a graphite crucible. For each boridization run, a single nickel coupon was buried into a freshly prepared powder bed. The materials were heated in a graphite resistance furnace under flowing ultra high purity helium using the following schedule: heat at 20°C/min to temperature, hold for a predetermined period of time between 2 and 8hrs, and cool at 10°C/min to room temperature.

After heat treatment, the samples were analyzed by XRD to identify the phases that had formed. The analysis was carried out with a Philips Wide-Range Vertical Goniometer and a Philips XRG3100 X-ray Generator over a scan range of 20–80° 2Θ with a 0.04° step size and 2s hold time. XRD pattern analysis was conducted using Jade 6+ (EasyQuant) software. SEM and EDX analysis were conducted to determine the microstructure and thickness of the boride coating using a JEOL JSM-5900LV equipped with an Oxford Energy Dispersive X-ray Spectrometer (EDS) system that employs a windowless detector for quantitative detection of both light and heavy elements. Accelerated exposure testing of the coupons was conducted at 80°C for 300hrs in static 1M $H_2SO_4$ containing 2ppm HF, an aqueous solution equivalent to the most aggressive pH condition in which the bipolar plate is expected to operate. To mitigate effects associated with edge corrosion of the small coupons, the cut edges were coated with a corrosion resistant epoxy such that only the flat faces were exposed to the corrodant.

## RESULTS

Shown in Figure 2 is an X-ray diffractogram for the surface of a nickel coupon boridized at 700°C for 8hrs. The primary set of peaks detected correspond to $Ni_2B$, with a smaller set attributable to $Ni_3B$. As the boridization temperature of the coupons decreased from 700°C to 500°C, subsequent XRD analysis indicated that the dominant surface phase shifted to $Ni_3B$ exclusively. At the mildest boridization condition considered in this study, 500°C/2hrs, only nickel peaks were observed in the corresponding XRD pattern. However comparison of the positions of these peaks with the standard pattern reported in the ICDD database indicated that the cubic lattice parameter is expanded by ~4.3%, likely due to diffusion and alloying of boron into the exposed nickel surface.

The depth of the boride coating is plotted as a function of boridization time and temperature in Figure 3. The thickness of the boride is found to increase with time in an approximately parabolic manner, a behavior that suggests the formation of a non-porous, uniform boride layer that acts as a physical barrier restricting subsequent boridization.[10] The Wagner expression can be used to describe the kinetics of boride growth:

$$x^2 = k_p t \qquad (1)$$

where $x$ is the thickness of the boride layer, $k_p$ is the parabolic boridization rate constant, and $t$ is the time of boridization. Fitting the data in Figure 3 to Equation (1), the parabolic rate constant for boride layer growth in nickel is approximately $9.99 \times 10^{-3}\mu m^2/s$ at 700°C, $2.85 \times 10^{-3}\mu m^2/s$ at 650°C, and $2.18 \times 10^{-5}\mu m^2/s$ at 500°C.

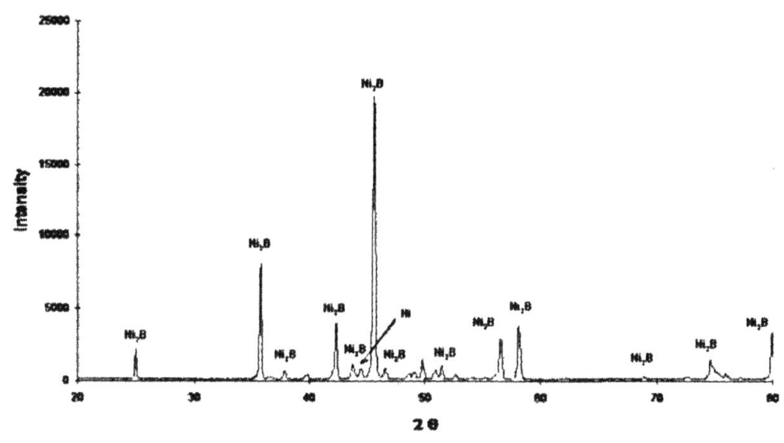


Figure 2 X-ray diffraction pattern for a sample heat treated in the boride powder pack bed for 8hrs at 700°C.

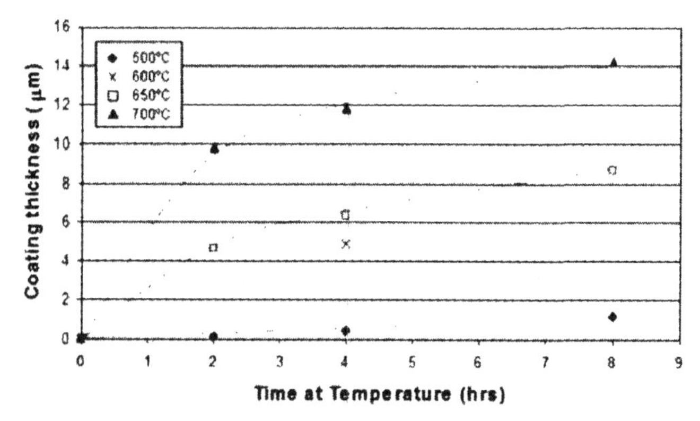


Figure 3 Boride thickness as a function of time at temperature for various soak temperatures.

The micrographs shown in Figures 4(a) – (d) display the microstructures of the nickel foil in the as-received and boridized conditions. As seen in Figure 4(a), the as-received foil exhibits an unexpected lamellar structure with a sub-dense core that contains micron-sized closed pores sandwiched on either side by a ~10μm thick dense outer layer. The origin of the pores in the Alfa-Aesar supplied foils is not known. In the sample that was boridized for 8hrs at 500°C, Figure 4(b), a uniform, ~1.5μm thick $Ni_3B$ layer (the slightly darker lamellar region) can be seen at the outer edge of the coupon. The borided regions also appears to be uniform (albeit more extensive in thickness) in the higher temperature coupons, as seen in Figures 4(c) and (d). The key differences between the two are the presence of a thin, patchy $Ni_2B$ overlayer and porosity in the underlying $Ni_3B$ in the coupon boridized for a longer period of time at 700°C. The origin of these microstructural features is currently under study, although it is suspected that their formation is interrelated.

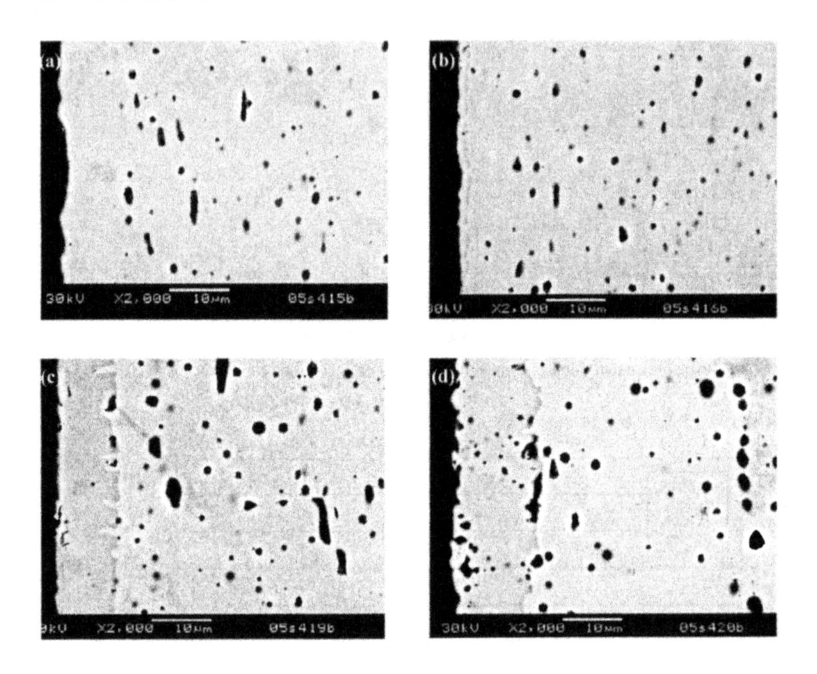

Figure 4 Cross-sectional SEM micrographs of nickel coupons in (a) the as-received state and after boridization at (b) 500°C for 8hrs, (c) 700°C for 2hrs, and (d) 700°C for 8hrs.

Shown in Figures 5(a) and (b) are cross-sectional micrographs of the EMS Ni/304SS/Ni clad material in the as-received and boridized conditions. In this case, boridization was conducted at 650°C for 4hrs. The nickel cladding (lighter regions) on either side of the as-received foil averages ~12μm in thickness, while the underlying 304SS core is approximately 115μm thick. Boridization of the clad material leads to the formation of a boride layer that measures approximately the same average thickness, ~7μm, as that observed in the pure nickel coupons under the same processing conditions. However in the clad material, this layer is not nearly as uniform. XRD analysis indicates that the boride phase consists solely of $Ni_3B$. Work is continuing to understand whether diffusion of iron and chromium from the 304SS core plays a role in modifying the morphology of the $Ni_3B$ layer.

Figure 5 Cross-sectional SEM micrograph of the clad Ni/304 SS/Ni material in the (a) as-received and (b) boridized conditions. Boridization was conducted at 650°C for 4hrs.

Preliminary exposure testing of the boridized materials was conducted in a static, 80°C acid bath of the same composition and concentration as the most aggressive pH environment anticipated in the PEMFC stack. Arguably this method of screening is an overtest, but the addition of flowing oxygen and a high applied current density that will be employed in later polarization testing suggests that these simple static corrosion conditions are a reasonable staring point in identifying promising material candidates for the PEMFC bipolar plate. As shown by the sequence of photos in Figures 6(a) – (c), boridization leads to a dramatic improvement in the corrosion resistance of nickel. The as-received nickel coupon completely dissolves during the 300hr test, forming a blue nickel sulfate solution. The coupon boridized at 700°C for 8hrs survives corrosion to a greater extent, although extensive attack occurs non-uniformly on the exposed surfaces. The 600°C/8hr boridized coupon displays far less material loss. The improvement in corrosion resistance with reduced boridization temperature is speculated to be related to a lower probability of microcracks occurring in the brittle coating. This is likely due to two factors, both of which reduce the magnitude of thermally induced tensile stresses within

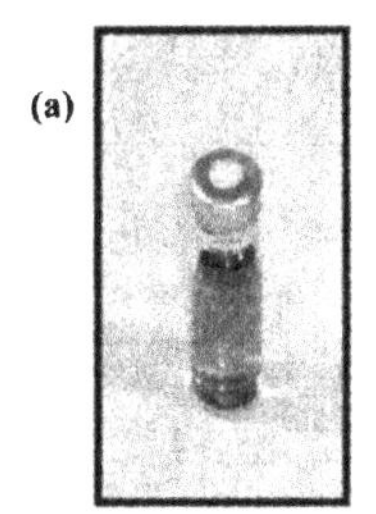

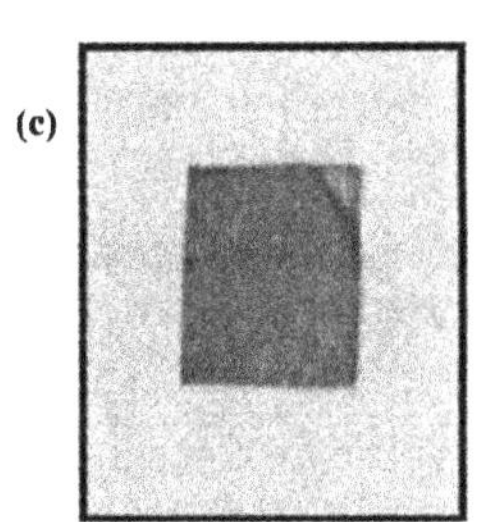


Figure 6 Photos of the corrosion effluent/coupons after exposure testing in 1M $H_2SO_4$ (with 2ppm HF) at 80°C for 300hrs. The samples were treated as follows prior to testing: (a) as-received, (b) boridized at 700°C for 8hrs, and (c) boridized at 600°C for 8hrs.

the boride layer: (1) the final process temperature is reduced by 200°C and (2) the resulting boride layer is significantly thinner. In addition, the prior XRD results suggest that the coating is more uniform in composition and therefore the potential for galvanic corrosion is mitigated. It is believed that continued reduction of the boridization temperature through modification of the coating process will lead to a completely corrosion resistant, yet electrically conductive nickel boride passivation layer.

## CONCLUSIONS

A new materials concept is being developed for application in PEMFC bipolar plates. Roll bonding can be readily employed in high production to fabricate low-cost nickel clad steel sheet product. While the nickel itself is not corrosion resistant to the PEMFC stack environment, preliminary study has demonstrated that this property can be substantially improved via boridization using the powder pack method. Based on this initial [illegible] coating process is being modified to form a corrosion impervious passivation layer [illegible] steel.

## ACKNOWLEDGEMENTS

The authors would like to thank Nat Saenz and Shelly Carlson for their assistance in sectioning and polishing the boridized in preparation for the SEM analysis work. This work was supported by the U. S. Department of Energy, Office of Energy Efficiency, and Renewable Energy. The Pacific Northwest National Laboratory is operated by Battelle Memorial Institute for the United States Department of Energy (U.S. DOE) under Contract DE-AC06-76RLO 1830.

## REFERENCES

1. S. G. Chalk, J. F. Miller, and F. W. Wagner, *J. Power Sources*, **86** (2000) 40-51.
2. V. Mehta and J. S. Cooper, *J. Power Sources*, **114** (2003) 32-53.
2. R. Borub and N. E. Vanderborgh, *Mater. Res. Soc. Symp. Proc.*, **393**, 151 (1995).
3. M. Lin, Y. Cheng, M. Lin, and S. Yen, *J. Power Sources*, **140** (2005) 346-349.
4. D. Brett and N. Brandon, *Fuel Cell Rev.*, **2** (2005) 15-23.
5. A. Hermann, Allen, T. Chaudhuri, and P. Spagnol, *Int. J. Hydr. Energy*, **30** (2005) 1297-1302.
6. W. S. Williams, *Progress in Solid State Chemistry 6*, ed. H. Reiss and J. O. McCaldin, Pergamon Press, New York, 1971, pp. 57 -118.
7. C. C. Wang, S. A. Akbar, W. Chen, and V. D. Patton, *J. Mater. Sci.*, **30** (1995) 1627 -41.
8. I. Ozbek, H. Akbulut, S. Zeytin, C. Bindal, and A. H. Ucisik, *Surf. Coat. Techn.*, **126** (2000) 166-70.
9. G. Palombarini and M. Carbucicchio, *J. Mater. Sci. Lett.*, **12** (1993) 797-8.
10. O. Kubashewskia nd B. E. Hopkins BE, *Oxidation of Metals and Alloys, 2nd ed.*, Butterworth press, London, 1962.

# Functionally Graded Materials

# CARBON-FIBRE-REINFORCED LOW THERMAL EXPANSION CERAMIC MATRIX COMPOSITES

C.M. Chan, A.J. Ruys
University of Sydney
School of Aerospace, Mechanical and Mechatronic Engineering
Building J07
Sydney, NSW 2006
Australia

## ABSTRACT

The use of carbon-carbon composites in high temperature applications has been limited by its inherent problem of susceptibility to oxidation. A lot of research has been done in developing oxidation resistant coatings, and as a result has brought about the concept of functionally graded materials (FGMs). FGMs are often utilized as thin film coatings instead of bulk-FGMs due to difficulty in controlling the gradient. A bulk carbon-fibre-reinforced carbon-carbon/fused silica FGM has been developed for the first time. EDS tests have proven that a silicon gradient does exist across the cross section of the FGM. XRD results have also revealed no devitrification in the fused silica. The SEM images taken of the fibre-matrix interface has shown the carbon matrix component to be fully dense, although the extent of the fibre-matrix adherence is weak. More tests have yet to be done to determine the mechanical properties of the FGM.

## INTRODUCTION

Carbon-carbon composites are an important class of materials where high-temperature structural applications are concerned. Some of the properties they possess include the retention of strength at elevated temperatures, low thermal expansion coefficient, high thermal shock resistance and good wear resistance in high pressure ablation environments. Their low densities ranging from 1.6 – 2.0 $gcm^{-3}$ also means they are a lightweight material, which makes them an excellent choice for space vehicles.[1] However, this is only possible provided it is not being utilized in high-temperature, oxidizing environments. Carbon oxidizes at 450°C and above, so there is a need to develop an efficient oxidation protection system.

The need for thermal barrier coatings have brought about the concept of functionally graded materials (FGMs).[2] FGMs are fabricated out of two components with a compositional gradient from one component to the other, such that the properties of both materials can be retained.[3] This method also minimizes the thermal mismatch, hence reducing the risk of failure. This study thus proposes the concept of a heat shield tile made out of a continuous carbon-fibre reinforced carbon-carbon/fused silica FGM. Fused silica has a high operating temperature of up to 1700°C, low thermal conductivity, excellent chemical inertness and a low thermal coefficient of expansion which closely matches that of carbon, hence making it an ideal substrate. The carbon fibres are needed to reinforce fused silica as well due to the low fracture strain and brittleness of monolithic fused silica.[4] One major problem is that fused silica tends to face the problem of devitrification at high temperatures, otherwise making it an ideal thermal coating. It is proposed that the fused silica act as an emergency oxidation barrier in the case of complete oxidation of the carbon facing, such that the carbon-carbon face is on the exterior. This FGM will thus serve as the base of a heat shield tile, with the exterior thermal protective coating to be developed at a later stage.

In addition, the FGM will be fabricated through the simple means of hot pressing. Current methods of fabricating carbon-carbon composites include solid pyrolysis techniques, liquid infiltration or chemical vapour deposition (CVD). Densification is achieved through chemical vapour infiltration (CVI), where hydrocarbon gases penetrate the structure and are deposited at high temperatures.[1] However the process is slow and it can take months to obtain dense carbon-carbon composites. This study also investigates the means to manufacture carbon-carbon composites (in the form of an FGM) by hot pressing carbon powder, thus cutting short and simplifying the fabrication process and as a result, produce a carbon-carbon composite with almost zero porosity.

## EXPERIMENTAL PROCEDURE

As mentioned earlier, carbon powder and fused silica were used as the matrix. The fused silica was synthesized by adding portions of tetraethyl-orthosilicate (TEOS) to a mixture of ethanol and ammonium hydroxide and mixed well under ambient conditions.[5] The mixture was then air-dried overnight in an oven at a temperature of 100°C.

The FGM carbon fibre to matrix ratio used was 1:5. Carbon fibre disks measuring 40mm in diameter were cut out and used as the reinforcements. Each carbon fibre disk was weighed and carefully laid in a graphite die. The amount of carbon powder required was then measured out and poured onto the disk of fibre and distributed manually with a screed. The matrix and fibre were further cold pressed to compact and level it. This process was repeated while varying the carbon powder to fused silica ratio for the matrix for each individual layer. The percentage of fused silica in the matrix was varied from 10 to 90 percent. To finish off, the FGM was topped with more fused silica.

The FGM was hot pressed at 10MPa; the sintering conditions consisted of a ramping rate of 400°C an hour to a temperature of 1050°C, at which the FGM was then held at that temperature for an hour before ramp-down. This was conducted in a non-oxidising environment.

## RESULTS AND DISCUSSION

The resulting FGM was cast in epoxy resin such that sections could be obtained for analysis without causing irreparable damage to the FGM. The following tests were done on the FGM sections: XRD, EDS and SEM.

### Scanning Electron Microscopy (SEM)

#### Carbon Matrix

SEM images were obtained to gain a visual insight into the microstructure of the carbon matrix, the silica matrix, and the macrostructure of the fibre-reinforced FGM. Figure 1 demonstrates that the carbon matrix was very dense, which is quite an achievement considering that it was hot pressed at 1050°C, well below the usual hot pressing temperature for carbon-based ceramics. Figure 2 demonstrates that the silica matrix was also very dense, as one would expect from hot pressing at 1050°C. It is difficult to image the regions where carbon matrix blended into silica matrix, although the EDS data given in Figure 3 demonstrates a uniform gradient. The fact that an "interface" between carbon and silica could not be imaged suggests that the two materials blended well and did indeed produce an FGM silica-carbon material. To our knowledge, this is the first time such an FGM has been reported.

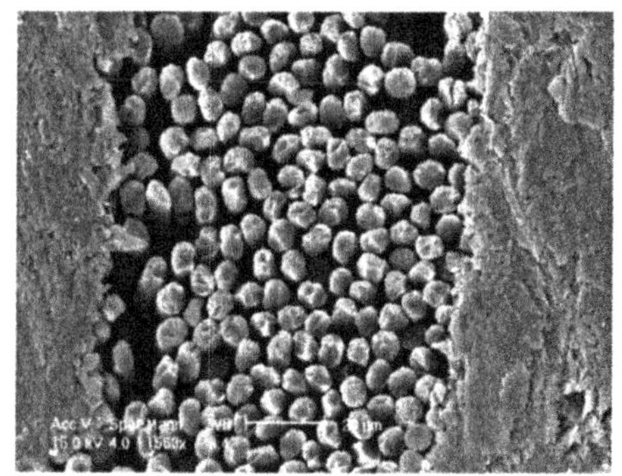

Figure 1: SEM image showing the carbon-fibre interface of the FGM

Figures 1 and 2 suggest that matrix penetration into the fibre regions was not complete. This was probably a result of the lay-up approach taken in which we laid up sheets of cloth, and layers of screeded powder in alternate fibre-powder layers (each powder layer was a different silica-carbon blend ratio), in a stepwise fashion. Perhaps future work should investigate a fibre coating approach.

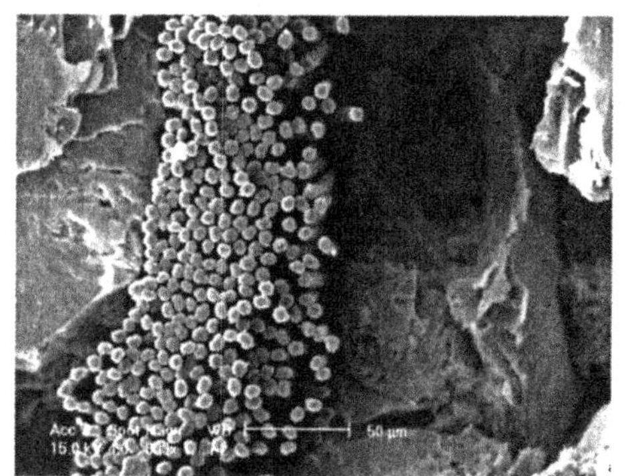

Figure 2: SEM image of carbon fibres in fused silica portion of the FGMI: large pores are observed.

EDS Results

EDS was done in sections across the FGM with only the matrix taken into consideration (i.e. excluding the carbon fibres) and the net intensities of silicon present measured. Results show that there is a variance in the net silicon intensities along the cross-section of the FGM, which proves that a gradient exists. The gradient was a very close approximation to linearity, which was a very successful outcome.

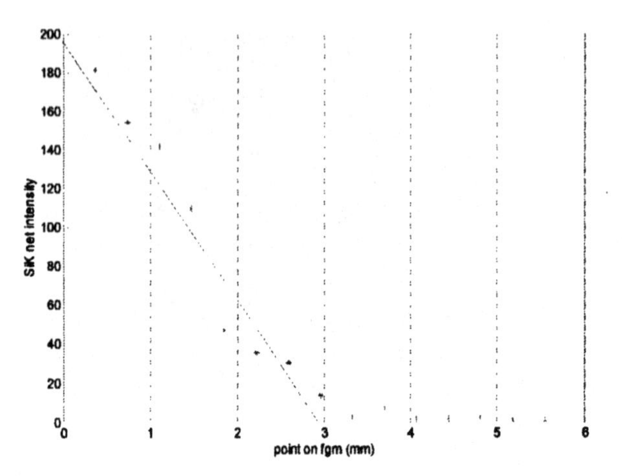


Figure 3: plot showing a decrease in the silicon (SiK) net intensities along the cross section of the FGM

Xray Diffraction (XRD)

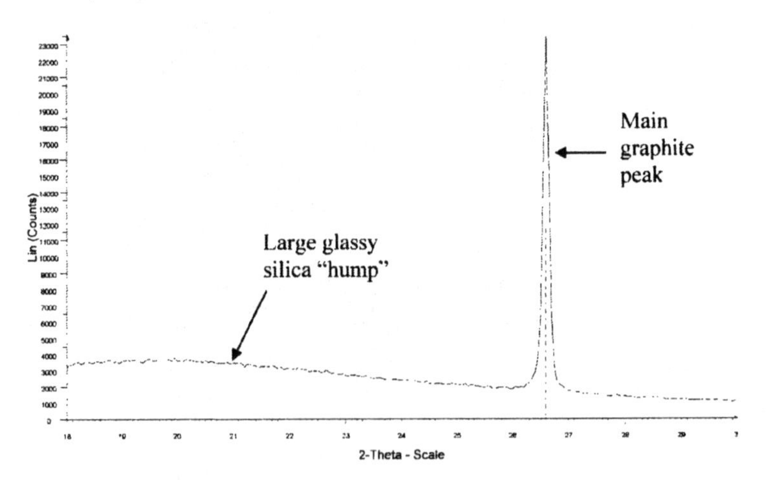


Figure 4: XRD of FGM showing the glassy peak of silica, the main graphite peak, and no evidence of crystallisation to quartz or cristobalite.

XRD was performed on sections of the FGM to determine if crystallization of the fused silica had taken place. The use of silica is sometimes limited by the fact that in the presence of nucleating agents, which can be dirt or impurities on the surface of silica refractories, it tends to devitrify to cristobalite at high temperatures. While fused silica has almost a zero thermal expansion coefficient, and thus excellent thermal shock resistance, crystalline silica has a high thermal expansion and in all forms (quartz, cristobalite, and tridymite), α-β transitions at specific temperatures that result in sudden volume changes. For example, the transformation of the higher temperature β- phase to α-cristobalite is undesirable especially when cooled to room temperature in use since it results in greatly enhanced susceptibility to thermal shock.[6] It is also possible for fused silica to devitrify into tridymite and quartz states; these possess different thermal expansion coefficients which also greatly increase thermal shock susceptibility.[7]

XRD tests were first done on pure fused silica which was pressed at 40MPa and sintered at 1200°C. A step size of 0.05 and time interval of 30 seconds was utilized over a 2theta scale of 18 to 30 degrees. The spectrum obtained showed a typical glassy phase with no crystallinity present. Furthermore, when this XRD analysis was repeated on the FGM (Figure 4), it revealed only two phases: glassy silica and graphite, demonstrating that under the processing temperature and pressure, no devitrification of the silica occurred in the presence of the carbon matrix and the carbon fibres.

## CONCLUSION

From the EDS tests, it is shown that a carbon-carbon/fused silica FGM has been successfully fabricated. The carbon matrix component has been shown to be fully dense from hot pressing at 1050°C. Although density has been determined only visually through SEM, future porosimetry trials will explore this in more depth. Furthermore, it was found that under the processing temperature and pressure, no devitrification of the silica occurred in the presence of the carbon matrix and the carbon fibres.

## ACKNOWLEDGMENTS

The assistances provided by Mr Dorji Chavara and Mr Gener Reyes in the hot-press construction and the Australian Research Council in funding this research are gratefully acknowledged.

## REFERENCES

[1]Manocha, L.M., "High performance carbon-carbon composites.", *Sadhana-Academy Proceedings in Engineering Sciences,* **28**, 349-358 (2003).

[2]Jeon, J.-H., Fang, H.-T., Lai, Z.-H., Yin, Z.-D., "Development of functionally graded anti-oxidation coatings for carbon/carbon composites.", *Key Engineering Materials,* **280-283**, 1851-1856 (2005).

[3]Ruys, A.J., Popov, E.B., Sun, D., Russell, J.J., Murray, C.C.J., "Functionally graded electrical/thermal systems.", *J. Eur. Ceram. Soc.,* **21**, 2025-2029 (2001).

[4]Jia, D.C., Zhou, Y., Lei, T.C., "Ambient and elevated temperature mechanical properties of hot-pressed fused silica matrix composite.", *J. Eur. Ceram. Soc.,* **23,** 801-808 (2003).

[5]Bogush, G.H., Tracy, M.A., Zukoski IV, C.F., "Preparation of monodisperse silica particles: control of size and mass fraction.", *J. Non-Crystalline Solids*, **104**, 95-106 (1988).

[6]Leiser, D. B., "Short-term stability of high-silica glasses.", *Ceram. Engineering & Sci. Proceedings,* **2,** 809-17 (1981).

[7]Kikuchi, Y., Hajime, S., Kuzuu, N., "Thermal expansion of vitreous silica: correspondence between dilatation curve and phase transitions in crystalline silica.", *J. Applied Physics,* **82**, 4121 (1997).

# DEVELOPMENT OF THE IMPELLER-DRY-BLENDING PROCESS FOR THE FABRICATION OF METAL-CERAMIC FUNCTIONALLY GRADED MATERIALS

D. T. Chavara, A. J. Ruys
School of Aerospace, Mechanical and Mechatronic Engineering
J07, University of Sydney,
Sydney, NSW, 2006, Australia

## ABSTRACT

Metal-Ceramic functionally graded tiles could be an ideal space plane cladding, as they combine the refractoriness of an oxide ceramic on one side, slowly graded to the ductility and strength of a metal on the other side. Fabrication methods of such continuous bulk FGMs are still proving to be a challenge. The authors have made various improvements into a low cost process known as the Impeller-Dry-Blending process, which has the ability to produce broad based gradients independent of material powder flow characteristics and density. The controlled blending appears to have worked well, as the results indicate a very even and gradual change from one material to the other. The screeding technique employed seems to have ensured that for each lateral sub-layer particles are evenly distributed across the broad area of the tile. From the results obtained in this study, the IDB process would be an ideal choice for the fabrication of continuous bulk FGMs for use as heat shield tiles.

## INTRODUCTION

Hypersonic space planes face a pressing problem of being able to withstand temperatures up to 2300°C in an oxidizing environment. Currently, space shuttle tile technologies are only limited to a maximum operating temperature of approximately 1200°C, as in the case of silica claddings, and 1500°C with borate or silicon carbide coated carbon-carbon composites. In order to counteract the problem at hand, the Japanese hypothesized the use of Functionally Graded Materials (FGMs) in 1984. A FGM basically consists of two components with a varying compositional gradient from one component to the other. A Metal-Ceramic FGM would be apt in the case of a hypersonic spacecraft. An ideal space plane cladding would be a thin tile (~ 1cm) gradually grading from a refractory oxide ceramic face to a ductile metal face, with the metal and ceramic having matching thermal expansion coefficients. Such a metal-ceramic FGM tile would be lightweight, capable of withstanding rain erosion, and yet even at extreme surface temperatures of 2300°C, be able to retain its strength and toughness.

Since 1984 various researchers across the globe have tried to manufacture FGMs for aerospace as well as other applications. FGMs fabrication methods can be classified into two main groups, Thin-Film/Interfacial FGMs as well as Bulk FGMs.

Fabrications of Thin-Film/Interfacial FGMs have been very successful using fabrication methods such as Plasma spraying[1], Vapor[2], and Electrophoretic Deposition[3].

Bulk FGMs can be further divided into another two distinct categories, known as layered bulk FGMs, and continuous bulk FGMs. Layered bulk FGMs would not be an ideal choice of a heat shield tile for hypersonic space plane skin, as immense stress concentrations would build up at each layer intersection. Layered bulk FGMs have been fabricated via the following fabrication methods: Sequential Slip casting[4], Thixotropic Casting[5], Laser Cladding[6], Slurry Dipping[7], and Layered Powder Metallurgy[8].

Thus, ideally the best option for a heat shield tile would be a continuous bulk metal-ceramic FGM. Few commercially viable fabrication methods exist for continuous bulk metal-ceramic FGM, due to the complexities involved. However various methods have been reported by researchers across the globe including centrifugal forming[9], and slip casting[10].

The authors have been working on a fabrication method for a few years, which could prove to be a viable commercial option for the fabrication of continuous bulk FGMs. This process known as the Impeller-Dry-Blending is a controlled powder blending technique similar in principal to the controlled blending techniques used to manufacture Thin-Film FGMs[11].

A major problem which any continuous bulk FGM fabrication process needs to overcome is the methodology behind densification. For example this study primarily focuses on two unique FGM combinations:

1. Magnesia-Aluminum
2. Alumina-Aluminum

Aluminum has a much lower melting point than the sintering temperature required for both ceramics. To overcome this problem we have decided to infuse the Aluminum at a later stage in both pore graded magnesia and alumina. To achieve a gradual pore graded ceramic, we will be grading both ceramics with graphite, with a burnout of the graphite at a stage when the ceramic has the strength to main the unique pore structure. With a melting point of 2800°C Magnesia would the ideal ceramic choice, as it also has excellent thermal expansion match with Aluminum.

## IMPELLER-DRY-BLENDING PROCESS

The Impeller-Dry-Blending process can be divided into three prominent sections: Feeding/Blending, Homogenization, and Deposition. Details into all the three processes are given below.

### Feeding/Blending

The feeding/blending system of the IDB process is responsible for blending the feed streams of the two powders so as to release both of them at a controlled ratio which would ensure that the FGM tile would change from being 100% Material 1 to 100% Material 2. This is achieved through the IDB process via a computer controlled blending process. The following Figures 1-4 illustrate the entire process. The computer control changes the geared disc position from Figure 1 up to Figure 3 via Figure 2 in a clockwise motion.

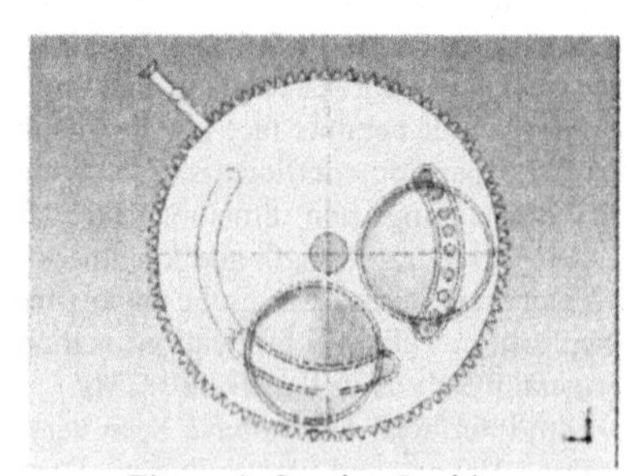

Figure 1: Starting Position

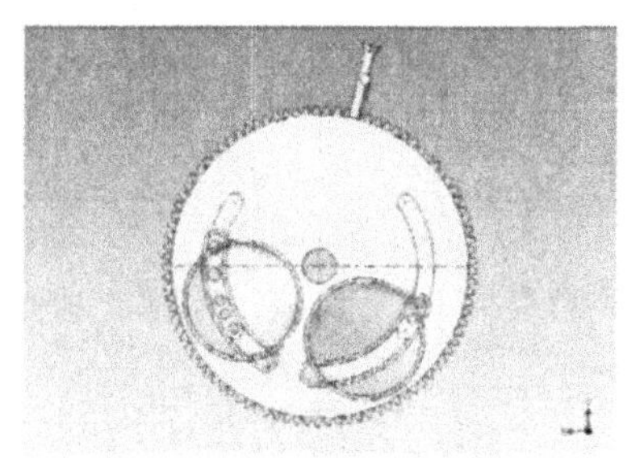

Figure 2: Transitional Position

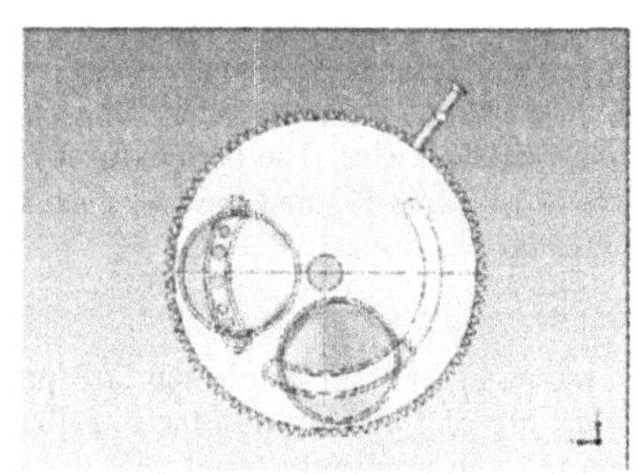

Figure 3: End Position

In Figure 1 it can be seen that only one slot (towards the right) has openings (via the aid of a set of holes), while the slot towards the left is fully closed. However when the process is finished, it can be seen that the opposite has been achieved, such that the slot towards the left is fully open, while the slot which was initially fully open is now fully closed. This simple computerized movement helps ensure that the powders are blended from 100% Material 1 to 100% Material 2.

The use of holes as exit valves for the powders into the next stage were selected by the authors to ensure as much as possible that throughout the process only the required amount of powder can go through. Thus for each particular type of powder a custom series of holes would have to be designed. Their exact position and size would also have been calculated. The selection of hole size and position along the entire arc length was achieved with the aid of the Covariance Matrix Adaptation Evolutionary Strategies (CMA-ES) algorithm developed by Nicolas Hanssen at the Berlin Technical University.

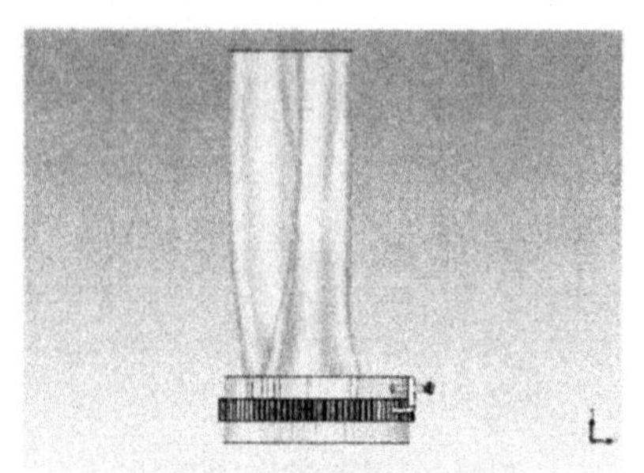

Figure 4: Feeding/Blending Section

Figure 4 shows two custom made polycarbonate tubes which act as hoppers for both powders. The custom made hoppers are positioned above the gear disc (Figure 1-3), and thus during the IDB process also move clockwise. The bottom-most disc in the above figure is the slotted disc which can be seen in Figures 1-3 and thus for each powder combination needs to changed, as each powder would have its respective flow.

Homogenization

The homogenization process is next in line in the IDB process. After the powders fall through the blending disc, they are sent directly into the impeller housing, where the blended powder streams are homogenized into a uniform blend with the aid of the impeller. Figure 5 shows a schematic of this stage of the IDB process:

Deposition

The deposition stage of the IDB process has the prime responsibility to ensure that the homogenized powders exiting from the impeller housing are deposited into the deposition mold. The ideal deposition stage would ensure that the powders are evenly deposited along each sub-lateral layer as well as longitudinally.

EXPERIMENTAL PROCEDURE:

The experimental process consists of two main steps. The first step consists of the design modification of the IDB process to ensure that the process becomes totally independent of powder densities and other powder flow characteristics. The second step involves the FGM synthesis to validate any improvements in gradients of the FGM as a result of the design modifications to the IDB process. The following two sub-sections describe in detail these two steps:

Design Modifications

The first major modification implemented into the IDB process was changes done to the control disc. Previously the control disc was a set of identical slots on either side. However this concept had its limitations on the fabrication of FGMs when both powders had different flow characteristics. To overcome this problem, the use of strategically located holes was implemented for each powder, as shown in Figures 1-3.

The deposition stage was also modified by the authors to include a screed. This was implemented to ensure that each sub-lateral layer had an even distribution of powders, and was achieved by implementing a stationary screed affixed to the deposition outlet. To ensure efficient screeding throughout the process, a rotational stage was also introduced to ensure the deposition mold could be continually spun around the screed. This rotational stage was dropped linearly at the same rate of deposition with the aid of a linear stage to ensure that the screed would not dig into already deposited material.

FGM Synthesis

As mentioned in the introduction, various powder combinations could be tried with the IDB process. The authors however have limited the study at this stage to both an Alumina-Aluminum and Magnesia-Aluminum combination as both of these would be ideal heat shields for hypersonic space plane travel. Metal infiltration is not the focus of this paper. This paper addresses the IDB blending aspects of the FGM synthesis, blending ceramic powder with combustible graphite so as to produce (after calcination) a pore-graded ceramic ready for metal infiltration. Alumina and Magnesia have been graded together with Graphite. The materials used for both powder combinations are:
Alumina: (Taimei high purity, nanoparticulate powder)
Magnesia: Causmag TGM (Super grade Magnesium Oxide)
Graphite: Grozier Flake No.2 Graphite

For both powder combinations, custom made discs had to be designed with the aid of the CMA-ES algorithm. The algorithm would select the size as well as the location of the hole along either arc. The main constraints implemented in the optimization routine revolved around the geometric/machine ability constraints, as well as flow requirement constraints. Flow requirement constraints were setup to ensure that each powder would flow from 100%-0% or vice versa. These flow requirements constraints would take into the consideration the size of the polycarbonate deposition mold (31.8mm), as well as a user defined powder blend height to ensure there is equal volume of both material deposited into the blend. For this study the blend height was set at 40mm, which would result in each material having a volume of 31.77cm3, gradually deposited in an increasing or decreasing fashion.

The respective custom-made disc was positioned under the two hoppers into which alumina/magnesisa and graphite have been placed separately. The feeding/blending process was started with the ceramic powder going through first, and gradually followed by the graphite flakes. The resulting FGM blend was collected in transparent polycarbonate molds and a pellet pressure of 200MPa was subsequently applied to the blend in the polycarbonate mold with the use of a steel die that fitted around the polycarbonate cylinder. The FGM deposit in the transparent mold was then removed from the mold and photographed so as to indicate axial uniformity, i.e., uniformity of gradient transition parallel to the deposition direction. The FGM was subsequently removed from the polycarbonate tube, and the resulting pellet immersed in epoxy and left to cure. The epoxy-encased pellet was cross-sectioned longitudinally and photographs taken to indicate the lateral uniformity.

Thus, the photographs of axial uniformity would demonstrate the effectiveness of the control disk modifications, and the photographs of lateral uniformity would demonstrate the effectiveness of the screed stage.

RESULTS AND DISCUSSION

As mentioned in the section above, the primary focus of the experiments conducted at this stage of the research revolves around improving the IDB process by making modifications and additions to the process. Continuous and gradual gradients in both the longitudinal and each sub-lateral layer need to be produced regardless of powder flow characteristics. The following three figures below show an illustration of the urgent need to modify the existing IDB process, as well as the importance of the IDB process being independent of powder flow characteristics.

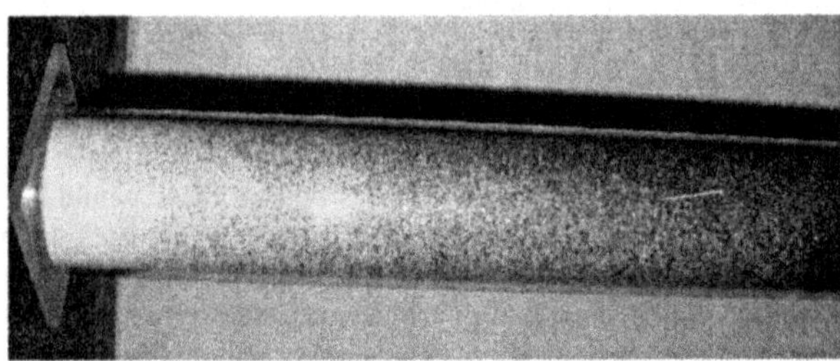

Figure 6: Brown Alumina – White Alumina blend in polycarbonate mold after IDB process

Figure 7: Silicon Carbide (Black) – Alumina (White) blend in polycarbonate mold after IDB process

From Figure 7, it can be clearly observed that the blend is very poor. The IDB process was unable to slow down the flow of the silicon carbide to match the flow of the Alumina. However when the powders have the same flow characteristics as is the case in Figure 6, the IDB process is able to produce a continuous and gradual gradient change from White Alumina to Brown Alumina. From Figures 6 and 7, it is evident that there is a need to modify the Feeding/Blending section, more specifically the control gate section.

Figure 8: Cross section view of Alumina – Stainless Steel Blend

Figure 8 shows a more promising result in the axial direction, which is a result of changes implanted in the Feeding/Blending section. However laterally, it appears that along each sub-lateral layer there is an uneven distribution of powder. This picture highlights the importance of the deposition stage of the IDB process to incorporate a screeding mechanism.

The diagrams below illustrate the results achieved for the first Alumina-Graphite powder combination attempted using the screeding process described in the experimental procedure section.

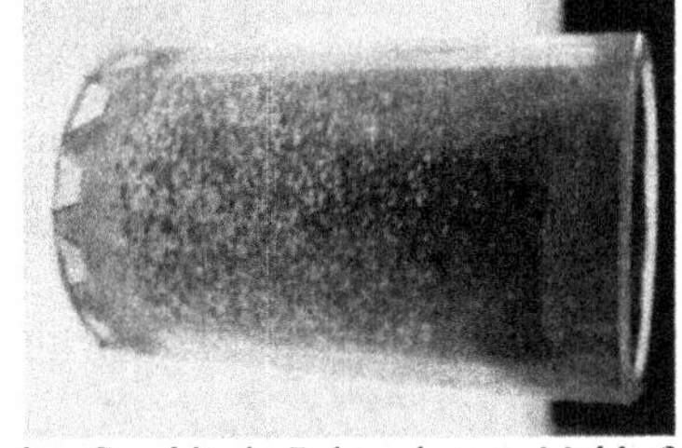

Figure 9: Alumina-Graphite in Polycarbonate Mold after IDB Process

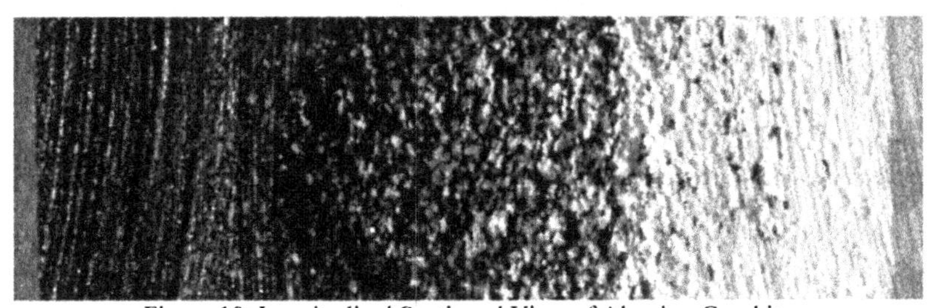

Figure 10: Longitudinal Sectioned View of Alumina-Graphite

The gradients achieved after the IDB process as seen in Figures 9 and Figure 10 (pressed to 200MPa) appear to be continuous and gradual, with the powders being distributed evenly across each sub-lateral layer in Figure 10. This is a significant contrast with the results achieved previously with two powders having different flow characteristics (Figures 7 and 8). The results seen in Figure 9 in comparison to that of Figure 7 clearly indicate that the implementation of unique control discs for each powder combination have ensured that the powders flow into the homogenization chamber at the same rate, regardless of each powders density or any other powder flow characteristics.

Likewise the comparison between Figures 10 and 8 clearly demonstrates the success of the addition of a screeding mechanisium as described in the Experimental Procedure in ensuring that each sub-lateral layer has an even distribution of powder. The curved lines which are evident in Figure 9 are a result of the diamond saw blade which was used to cut the sample. The FGM powder in the polycarbonate tube had a diameter of 31.8mm, with a height of around 56mm.

The following pictures are of the other powder combination attempted. As with the previous case, they also had similar dimensions.

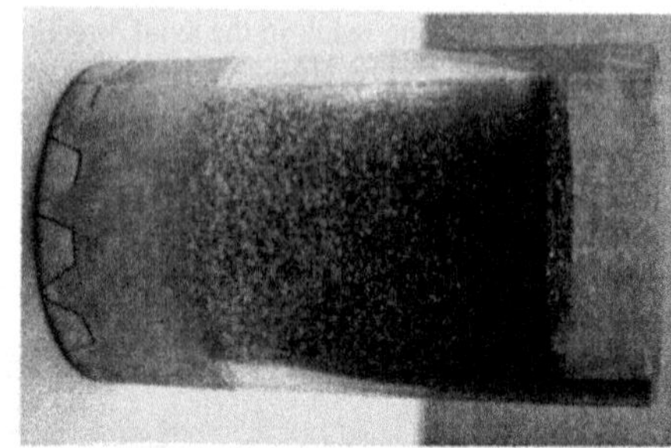

Figure11: Magnesia-Graphite in Polycarbonate mold after IDB process

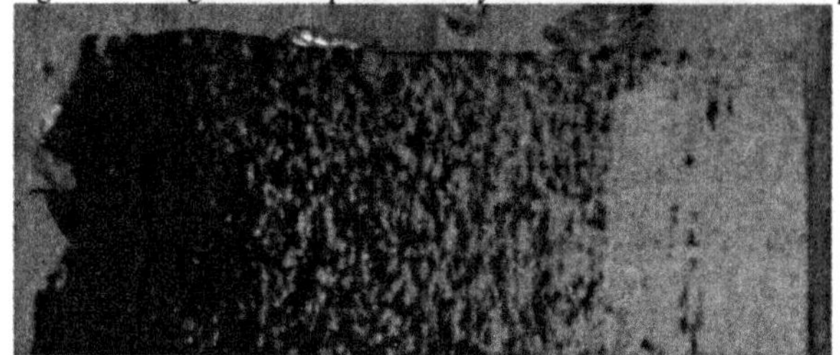

Figure 12: Longitudinal Section of Magnesia-Graphite FGM Pellet

The sectional view of this sample was cut with the aid of a small hand saw and as a result, there was less damage to the sample. The slight damage at the graphite end of Figure 12 was a result of an attempt to take the sample out of the polycarbonate mold. Likewise for the other combination this also had similar dimensions before compaction (31.8mm diameter and around 56mm in height), and the gradient achieved has been very successful. As was with the other powder combination, both changes to the Feeding/Blending section as well as additions to the deposition section have ensured a continuous and gradual change from one material to the other material in both the longitudinal and lateral directions. A continuous and gradual change is vital for a FGM material subjected to extreme heat as abrupt changes from one material to the other in either direction could lead to an unnecessary build of up thermal-stresses which could result in the failure of the FGM material. With design modifications and additions to the IDB process and from the results obtained in this study, the IDB process appears to be a commercially viable fabrication technique towards to the fabrication of continuous bulk FGMS; an ideal space-plane heat shield tile.

CONCULUSION

As the results have shown, the particular modifications made to the IDB-process appear to have achieved their design purpose. The use of a blending control disc. with a series of varying hole sizes, seem to have ensured that the required change in flow rate from the feed hoppers into the impeller housing was maintained as required. The screeding technique employed in the process seems to have distributed the powders evenly across lateral layer of the FGM tile.

Future work at this stage would immediately focus on the vacuum infusion trails into the pore graded ceramics. As these FGM tiles are being fabricated for use as heat shield tiles for hypersonic space planes, simulations of rain erosion and thermal cycling need to be done. Toughness and hardness values across the gradient also need to be studied along the gradient. In summary, based on the results produced in this study show promise, and indicate the potential of the IDB process as a cost effective continuous bulk FGM fabrication process.

ACKNOWLEDGMENTS

The authors gratefully acknowledge the technical assistance provided by Mr. Gener Reyes and Mr. Christopher Young for this study, and the assistance of the Australian Research Council in funding this research.

REFERENCES

[1]K. A. Khor, Z. L. Dong, Y. W. Gu, "Plasma sprayed functionally graded thermal barrier coatings," *Materials Letters*, **38**, 437-444 (1999)

[2]T. Hirai, M. Sasaki, "Vapor – deposited functionally gradient materials," *JSM International Journal*," **34**, 123-129 (1991)

[3]C. Zhao, L. Vandeperre, J. Vleugels, O. Van Der Biest, " Cylindrical Al2O3/TZP functionally graded materials by EPD," *British Ceramic Transactions* **99**, 284-287 (2000)

[4]S. J. Moya, A. J. Sanchez-Herencia, J. Requena, R. Moreno, "Functionally gradient ceramics by sequential slip casting," *Materials Letters*, **14**, 333-335 (1992)

[5]A. J. Ruys, S. A. Simpson, C. C. Sorrell, "Thixotropic casting of fibre-reinforced ceramic matrix composites," *Journal of Materials Science Letters*, **13**, 1323-1325 (1994)

[6]J. H. Abbound, R. D. Rawlings, D. R. F. West, "Functionally graded nickel-aluminide and iron-aluminde coatings produced via laser cladding," *Journal of Materials Science*, **30**, 5931-5938 (1995)

[7]A. Kawasaki, R. Watanabe, "Thermal fracture behavior of metal/ceramic functionally graded materials," *Engineering Fracture Mechanics*, **69**, 1713-1728 (2002)

[8]K. Atarashiya, M. Uda, "Nano-structured FGM of the system AIN-Ni," *Proceedings of the International Conference on Advanced Composite Materails*, Wollongong, 1351-1355 (1993)

[9]Y. Watanabe, T. Nakamura, "Microstructure and wear resistances of hybrid Al-($Al_3Ti$+$Al_3Ni$) FGMs fabricated by a centrifugal method," *Intermetallics*, **9**, 33-43 (2001)

[10]J. Chu, H. Ishibashi, K. Hayashi, H. Takebe, K. Morinage, "Slip casting of continuous functionally gradient material," *Journal of the Ceramic Society of Japan*, **101**, 841-844, (1993)

[11]A. J. Ruys, E. B. Popov, D. Sun, J. J. Russell, C. C. J. Murry, "Functionally graded electrical/thermal ceramic systems," *Journal of the European Ceramic Society*, **21**, 2025-2029 (2001)

# Author Index

9 780470 080535